ENVIRONMENTAL BIOLOGY

The Conditions of Life: Environmental
Selection, Extinction, Creation,
Adaptation and Overpopulation

ENVIRONMENTAL BIOLOGY

The Conditions of Life: Environmental
Selection, Extinction, Creation,
Adaptation and Overpopulation

Terry Hilleman

University of Northern Iowa
Cedar Falls, Iowa
USA

Science Publishers

Enfield (NH)   Jersey   Plymouth

Science Publishers

www.scipub.net

234 May Street
Post Office Box 699
Enfield, New Hampshire 03748
United States of America

General enquiries : *info@scipub.net*
Editorial enquiries: *editor@scipub.net*
Sales enquiries   : *sales@scipub.net*

Published by Science Publishers, Enfield, NH, USA
An imprint of Edenbridge Ltd., British Channel Islands
Printed in India

© 2009 reserved

ISBN 978-1-57808-576-7

Library of Congress Cataloging-in-Publication Data

Hilleman, Terry Bruce.
Environmental biology : the conditions of life :
environmental selection, extinction, creation, adaptation
and overpopulation / Terry Bruce Hilleman.
     p. cm.
 Includes bibliographical references and index.
 ISBN 978-1-57808-576-7 (hardcover)
1. Ecology. 2. Adaptation (Biology) 3. Evolution (Biology)
I. Title.

QH541.H534 2008
577—dc22
                                                        2008048390

For my family:

Colene: essential support.
Erika: told daddy about the tortoise types on the Galapagos.
Kurt: asked me to explain evolution, in simple terms.
Bob: serving and saving lives with Alaska's Stryker Brigade.

For my mentors:

David Fagle: introduced me to different types of life.
Arthur Trowbridge: introduced me to changes in types of life over time.
Paul Meglitsch: taught me to really look at types, and to understand what I saw.
Richard Bovbjerg: taught me most of what I know about the ecosystem, selection, adaptation and evolution.

For the authors referenced extensively in the text:

Carroll, Darwin, Diamond, Futuyma, Hesse, Kirschner/ Gerhart, Levinton, Turner, West-Eberhard and others that provided the source material

For understanding:

"Evolution is the unifying theory of all biology."
Douglas Futuyma 1997

I wrote this book in order to emphasize the importance of the environment. Environmental protection is for everyone; it is not the domain of extremists.

Contents

Prologue

The first chapter, the Overview, provides an abstract-like summary of the contents. **Portions of the overview will be repeated as bold type throughout the remaining text; alongside this will be material providing additional details and examples to the re-introduced overview.** Not everyone will want to involve themselves in all the details presented. Consider bypassing some details in the main body of text; there will only be selective interest in the more detailed areas. In other words, if reading becomes slow, move on quickly to the next chapter. Moving on quickly to areas of greater reader interest can be more enjoyable; the text is constructed for this purpose. While the quotes and details are important to some, it's a clue to others that it's time to leave this topic. I recommend doing this; it is the ideal approach. There is much to be said about keeping things simple. An effort was made to organize and clearly communicate concepts in the least-complicated manner; repetition was utilized extensively to support this effort. Terms that may be unfamiliar are defined and discussed. I have inserted comments in parentheses, sometimes even into quoted material. Different definitions for conditions of life and/or natural selection may be underlined. The epilogue poses likely questions and my answers to them. It's included for further reading on the environment and the environment's effects on life itself. It acknowledges that I may be wrong about many of the topics discussed herein. I have every confidence in hearing what is wrong. Some readers will want to go right to "the bottom line." **Skeptics and bottom-line thinkers should go directly to chapters on Genetic Perspective, Phenotype and Environment Difficulties on Theory, Environmentally-Determined Evolution, Biochemistry of Genetic Mechanisms, Do the Math, and A Choice.** After reading these chapters, consider the logical conclusions; then read the rest of the book.

Environmental biology is the study of the conditions of life (or conditions of existence) and the impact of these conditions upon and within the life it contains. This material should change the way most biologists look at life itself; nearly everything could be somewhat different. There is much more to life than DNA; that is, there are more than genes involved in the makeup of an organism. Before quickly dismissing any of the content herein as wrong, indulge the unfolding of a different perspective that actually simplifies many areas that are poorly understood. There is good reason for the poor understanding; it comes from a popular prejudice. A shift in thinking has already begun.

There is a need to better understand the impact of the environment on the life inside it, particularly that including the impact of the essential historical context. Time itself is a component of the environment. Environmental biology is strongly connected to the word, WHEN. Environments change over time; no environment is permanent. Over time, changes in the conditions of life cause changes in the types of life. Genetic change is also involved. Many believe that evolution is based on genetic change; that is a half-true belief. The role of the environment's impact upon life has been under-appreciated; a better understanding of environmental biology is essential for a better understanding of evolutionary biology. This text is a step towards fulfilling the need for better historical environmental understanding, and the environmentally-related changes in types of life over time. This is not a text on evolution; this text will reference several excellent texts on evolutionary biology. Changing environmental conditions are highly relevant; they impact the life contained by it in certain ways. If the conditions of life are not limited to the present time, environmental biology has applications to any time in the history of life on earth, past, present or future. To some degree, history tends to repeat itself.

The topic of evolution itself is a volatile one. The majority of people still reject the very idea of evolution, preferring instead the highly-conditioned religious version of creation as told by their culture. The Scopes trial in 1925 was the "trial of the century." It was featured in a History Channel Series "Ten Days That Unexpectedly Changed America". The trial represented a clash between the religious fundamentalists of the day, opposed to change, and the pragmatic progressives that embraced change as an opportunity. This unresolved clash exists even today. Religious fundamentalism is growing, worldwide. The current U.S. President (with good intentions) has advocated including the teaching of intelligent design in the school system. Many communities (with good intentions) want to ban the teaching of evolution in their schools. Only recently has intelligent design lost ground, perceived in the courts as the teaching of biased religious creationism. The ongoing clash, far from over, continues to spark controversy.

Long ago, I read Charles Darwin's *The Origin of Species by Means of Natural Selection*. Although I had strong religious convictions at the time, I believed that Darwin's theory made sense. I was confident that I understood natural selection and how it led to the origin of species. Further formal education only enhanced my understanding of natural selection and population (species) dynamics. Some religions have integrated evolution into their system of faith, some have not. I respect potential conflicts in cultural religious belief. Any religion deserves respect, but a religion deserves no more respect than it is willing to give to any other religion (or belief). Respect must be reciprocated; sadly, this means that there are more than a few religions that need to work on improving their respect for others. I do see religion as man's interpretation of God; this permits religious leaders with an agenda to use religion to empower their agenda. All religions tend to believe that their faith is the only true faith. Is there really such a thing as only one true faith (or belief)? Religion exists in the realm of philosophy; science exists in the realm of repeatable verification of theories and hypotheses by the factual findings of others. In religion, and in science, the truth can be somewhat elusive; however, science tends to self-correct; science evolves. Still, this means that at times, even scientists will have beliefs that may not be correct. And, even religions undergo changes over time; many religions seem to evolve. To their credit, most religious leaders have accepted arguments for evolution better than most biologists have accepted arguments for a design in the natural world (Richard Dawkins presented a strong argument against design in *Climbing Mount Improbable*). In the natural world, issues of evolutionary chance and religious possibilities in purpose of design have continued to cause many unresolved questions in my understanding of evolution; the odds just seem to be against continual accidental occurrence. Many people, scientists and non-scientists, are not comfortable with such a strong dependence on the sheer coincidence of a mutation alone resulting in a (genetically-determined) natural selection-based origin of species. This understandable doubt compromises the creditability of evolution itself. Nearly a half-century later, I even came to question my understanding of natural selection and the origin of species. Perhaps it is time for everyone to re-examine everything.

Environmental biology addresses environmental and other biological aspects of evolution that have been largely ignored. In her first chapter of *Developmental Plasticity and Evolution*, Mary Jane West-Eberhard notes: "One of the oldest unresolved controversies in evolutionary biology— and a source of many bitter arguments and failed revolutions—concerns the relation between nature and nurture in the evolution of adaptive design.... We have a... a poor understanding of the developmental causes of variation—and how the environment influences evolution through its effects on development." The book is a significant breakthrough in biological thinking—when it comes to environmental adaptations, the environment's effect on early development is life-changing. The biological changes upon and within a developing organism, as a result of the environment, have been well-organized and well-documented; West-Eberhard's *Developmental Plasticity and Evolution* is a thorough, well-developed, well-referenced text. Instead of referencing original research as *Developmental Plasticity and Evolution* has done (admirably), this *Environmental Biology* text will utilize some original, but more often, the most-recent references from the most relevant material on environmental studies in ecology (including biogeography and environmental science), genetics, development, anatomy, physiology, cytology, chemistry, and evolution. An attempt will be made to simplify and present related conclusions from this material. Environmental biology is particularly concerned with ENVIRONMENTAL SELECTION, EXTINCTION, CREATION, ADAPTATION and OVERPOPULATION. These topics will be discussed within the text, in detail. In that environmental adaptation is the most challenging concept, much of this text will be directed to its understanding.

"Evolution is the unifying theory of biology" (Douglas Futuyma 1997). In the evolution of life on earth, the environment played a more significant role than generally believed. One prominent authority on evolution declares that "it is unlikely that new discoveries will greatly alter the fundamental principles of evolutionary theory," while another declares "the unexpected will overwhelm our preconceived notions." *Environmental Biology* will cause everyone to evaluate a very different perspective of evolution, with a unique focus on environmental selection, environmental extinction, environmental creation, environmental adaptation, environmental overpopulation, and the environmental implications for humanity. It is naïve to think that because an adaptation is inheritable, that genetics alone creates the adaptation. Why? Not everyone will have the same areas of interest. I promise to provoke reflection, and to include something for everyone. For example, the conditions of life once gave dinosaurs an advantage over mammals. What was the advantage? The loss of habitat killed the dinosaurs; they starved to death. What were three events that caused this? Because the environment and evolution cover such a broad area, there are assuredly areas of real interest for everyone. Environmental biology is the study of the conditions of life and their impact on past, present and future organisms. In environmental biology, the focus is increasingly shifted to the changing conditions of life inside an environment, and to WHY and HOW these changes impact the organisms confined there (environmental conditions impacting the organisms' external and internal activities). Texts on environmental biology already exist, but they are little different from other texts on ecology. This is definitely not a text on ecology; there are already many excellent texts on ecology (organism roles in environmental interaction—usually for biology majors). Ecology is the broadest general topic, and it encompasses the subject of environmental biology. In the study of ecology, the focus is on the organism, and its function in the existing environment. General ecology is primarily concerned with an emphasis on the word, WHAT. What is the organism (animal, plant, microorganism) and what is its function in its environment? As an example, take the largest vertebrate and its function in its environment. The largest (ever) vertebrate on earth is *Balaenoptera musculus*, the elusive blue whale. Understandably, studies on its functions in the environment of the deep blue sea have been rather limited. From the environmental biology realm of general ecology, this elusive organism is increasingly seen from the perspective of the blue whale's life-essential environmental factors; i.e., the attention is more focused upon the blue whale's specific temporal environments (habitats), and correlated to any impacts upon the whale. When the whale is at the surface, the environmental impact of exploitation by man is well-known; over-harvesting wild populations always causes a problem. Little else is known of other significant environmental aspects upon the blue whale's life cycle. Nothing is known of the crucial environment (habitat) in which newborn whales enter the world. This essential winter habitat must be somewhere in lower latitudes. Wikipedia notes that in the North Atlantic, two groups are recognized. A Western Atlantic group is found off Greenland, Nova Scotia and the Gulf of St. Lawrence. An Eastern Atlantic group can be spotted near the Azores in the spring and near Iceland in the summer. Some Antarctic blue whales approach the eastern South Atlantic coast in winter, and occasionally their vocalizations are heard off Peru, Western Australia, and the North Indian Ocean. Nursing young have been seen in large numbers during southern summers and mid-March (fall) off Chile's Gulf of Corcovado. Whales need to spend time in a location that provides adequate food for this largest-ever animal. A crucial blue whale habitat is one of abundant krill (a shrimp-like crustacean); it's the whale's primary food source. Large populations of krill (and whales) can be found during spring and summer in the fertile far northern and southern oceans. Following spring break-up (ice), krill populations become one of the ocean's most prevalent macro-zooplankton food sources. This krill-rich, near-polar spring and summer marine environment is essential for the whale to acquire enough food to attain its immense size; it must also provide enough stored nourishment in females for developing (unborn) whales, including an adequate supply of milk. This krill-rich environment is now in sharp decline; krill density has dropped 80% in as many years. The changing conditions of life are highly relevant to the blue whale's continued existence. Krill harvesting for humans, livestock, and fish farming was over 500,000 tons/year in 1982; it still averages around 100,000 tons/year. Over-harvesting wild populations always causes a problem. Blue whale survival is mainly dependent upon maintaining blue whale habitat integrity.

Environmental biology is very much concerned with the word, WHERE. Where is the largest invertebrate found? It is likely to be found in the great expanse of water with the highest oxygen saturation in the marine environment; cold water holds more oxygen. The coldest water on earth spills out into the surrounding water from a shelf under the Antarctic ice. The circumpolar current waters surrounding Antarctica are nearly as cold; the water below the surface is constantly mixed by strong winds that blend the water below to extreme depths. The only thermal stratification found here is in the saltiest, deepest and coldest waters at the very bottom, spilling-out from under the Antarctic ice. The result of all this is a significant vertical mixing in the greatest expanse of highly- oxygenated waters on earth. As depth increases, the partial pressure of oxygen dissolved in the seawater also increases. Invertebrates attain larger size as environmental oxygen availability increases. The largest invertebrate should be found where oxygen reaches it's highest environmental availability, primarily due to this limiting environmental condition of life. It is unlikely to be found anywhere else. There is an improved scientific understanding of life if the focus is on the selectivity of the environment itself. While *Mesonychoteuthis hamiltoni*, the largest of all giant/colossal squids, is certainly important, it is the selectivity of the rich oxygen environment that enables the attainment of greatest size; the role of the environment is no less important than the role of genetics. Both profoundly impact all life, everywhere.

Location is extremely important to an environmental biologist. Not only does location specify climate and substrate, which delineates primary production, the location has a range of variation for other essential conditions that affect the life within it, as in the above example of the colossal squid. The location of an environment makes a great deal of difference to the biology within it. The significance of location is primarily addressed in biogeography, the analysis of spacial distributions of organisms. Alfred Russell Wallace, the father of biogeography, is also co-founder of the theory of natural selection. Wallace contributed much to an understanding of the geographic distribution of species; his work laid the foundation for further environmental impact studies. Between 1913 and 1924, Richard Hesse wrote *Tiergeographie auf oekologischer Grundlage*, which was translated into English by Hesse, Allee and Schmidt as *Ecological Animal Geography* in 1937. "Zoogeography is the scientific study of animal life with reference the distribution of animals on the earth and the mutual influence of the environment and animals upon each other." This is the classic text on environmental considerations for animal life. "Zoogeography corresponds to phytogeography and with it forms the single science of biogeography.... Phytogeography has been the subject of research for a much longer period and has accordingly been much more intensively studied in special fields. Modern work in plant geography continues in both historical and ecological aspects. Among plants, the relations with the total environment are much more direct and obvious than among animals. The capacity for motion from place to place makes animals to a degree independent of their environment; the majority of them are at least able to move towards water, food, or warmth in new localities, and thus they become exposed to new conditions. The formation of spores or seeds, effectively protected from unfavorable influences and easily distributed passively, favors the wide distribution of plants, enabling them to cross barriers with less difficulty, so that limitations of the distribution of plants in accordance with their ancestral history are much less evident than among animals."

This is not a text on biogeography; there are many excellent texts on biogeography. And, they are not all that different from other texts on ecology. Hesse's *Ecological Animal Geography* made a superb text for the study of animal ecology. When it came to seeing the impact of the environment upon the distribution of organism's within it, Hesse's text was outstanding; it is, almost, the classic text of environmental zoology. Even paleobiogeography, which goes a step beyond biogeography to earlier times, and includes paleogeographic data and considerations of plate tectonics, only supplements the WHEN and WHERE of environmental biology. Although biogeography does address the WHY and HOW of spatial distribution patterns, biogeography really does not address WHY or HOW environmental variability impacts the structural and functional biology of the life within it. The location affects the conditions of life; the focus of biogeography is very much upon on the location itself. Biogeography primarily addressees the spatial distribution changes that are a product of the environmental variation; but, it does not focus that much beyond the location, examining the specific conditions of life (and any changes in these conditions) as a primary focus. And while there is some overlap, environmental biology's emphasis is less on the geography and more on the changing conditions of life within the environment itself; and, as just noted, environmental biology addresses the environmentally-related mechanisms of biological change (WHY and HOW these variable conditions change the lives confined within this environment). Just as ecology includes biogeography, ecology encompasses environmental biology. Biogeography and environmental biology are similar, yet different, topics; each one complements the other. Environmental biology is the study of the conditions of life (or conditions of existence) and the impact of these conditions upon and within the life it contains. As important as location is, it is the variable conditions within the location that make all the difference for the impact of not only WHEN and WHERE, but WHY and HOW the biology subsequently changes.

Just as ecology includes environmental biology, environmental biology also includes environmental science. Both disciplines (environmental biology and environmental science) focus primarily on changing environments and ongoing biological changes within them, the former—on life in general, and the latter—on humanity. Environmental biology is an applied general ecology; it is primarily concerned with the environment and environmentally-induced biological change. Environmental science is applied human ecology (environmental applications of human ecology-usually for non-majors); it may include human history, human biogeography, human economics and anything else humans do that changes the environment. Although there will be some discussion of environmental science material, this will not be an environmental science text; there are numerous texts on environmental science. Environmental biology is gaining an increasing global awareness; the perspectives developed in this text could to be useful. The discussion of humanity and the environment promises to be of great interest and importance to every living being. Humanity may have already exceeded the carrying capacity of the land. What will happen next? This is not a gloom and doom environmental book, leading to frustration from the futility of inaction; it offers options that are both positive and very realistic. It is not supportive of environmental extremism and associated terrorism. There is good news and there is news that is not good. The environment played a significant role in the evolution of mankind and will continue to play a significant role in mankind's future. The environmental carrying capacity of mankind on earth has recently increased and enabled a tenfold population increase in the past century or so. The favorable environmental conditions that enabled this population increase will not last forever. As human populations continue to increase further and the conditions of all life worsen, the earth's environmental carrying capacity could once again return to the lower levels seen throughout all but the most recent historical time.

1

Overview

The conditions of life include everything that affects an organism during its lifetime, the circumstances of biological existence, or the environments effects upon an organism. The effect may be subtle; it may be severe. No environment is permanent; change is inevitable. Habitat destruction eliminates environments. The loss of habitat is the loss of the (necessary) conditions of (existence for) life; the loss of habitat is the primary cause of extinction. Habitat destruction occurs throughout time, in varying degrees of severity. The degree in severity of extinction seems to parallel the degree of habitat destruction. The conditions of life are not limited to the present time; environmental biology has applications to any time in the history of life on earth. An understanding of changes in types of life over time (evolution) is a basic component of environmental biology; past environments have influenced the evolution of today's biology. Environmental biology is concerned with location; specifically, upon the conditions of life within this location, and on the impact these conditions have upon and within the life it contains. It is logical to believe that both the environmental conditions of life and the survival of the fittest competitors (within a species) played strong roles in the evolutionary development of all life on earth. The environment is the gatekeeper for genetic success. If natural selection is the key to the "preservation of favorable variations and the rejection of injurious variations," that key must unlock the environmental gate if life is to pass into the next time, i.e., continue living in a newly-created environment. Environmental creation includes and follows an environmental extinction event; the destructive process triggers the creation of new environments, with new opportunities for the survivors.

Environmental creation can also occur in a more steady-state environment; it may be subtle. Something is very special about environmental creation; it enables the selective forces of the environment to interact with life's genetic potential.

To date, changes in types of life over time have been primarily focused on the genetic mechanisms of evolution, namely natural selection, preservation of favorable genetic variations, an individual's norm-of-reaction, population variation, and mutation. Natural selection was first seen as a preservation of favored traits in a "struggle for life" process, a competition with others. The evolutionary effect of environmental influence on an organism seems to be greater than fatal competition selection alone. It seems as if any natural process with selective evolutionary impact could be considered as part of a natural selection process. Natural selection (survival of the fittest) is strongly connected to life's primary limiting factor, the environmental conditions of life, which often includes competition as part of ecological community dynamics. Natural selection is the result of an environmental selection process; natural selection is a resulting genetic bias. Natural selection is the preservation of genotype variations in expressed phenotypes, modified primarily by an environmental selection process. That is, the phenotype is an environmentally-integrated expression of the genome's natural selection genetic bias, a display from the inherited genotype norm-of-reaction, and it is connected to different previously-encountered environments. Ultimate understanding of evolution and environmental biology cannot be achieved unless we seek to better understand both natural selection genetic bias and the conditions of life. Selective forces in the environment are variable, and are as important to life as genetic variation. Environmental variation changes the form of environmental selection over time; this causes interrupted changes in types of life. There is a biological pattern from environmental variability in evolution; it is an interrupted, environmentally-advantageous role repetition. And there is a biological pattern to genetic variability in evolution; it is inheritability of environmentally-advantageous traits.

Almost all variation in expression comes from the individual's plastic range of expressing its norm-of-reaction. This is environment-sensitive. Life is situational and directional, and life's course is determined by both genetics and the existing environmental conditions. Environmental selective forces and existing natural selection genetic bias may even interact to produce something new. Environmental stress may lead to an adaptation response in the developing offspring of survivors. Here the environment does not directly control genetic expression, but it does influence the initiation of genetic expressions, stabilization of genetic expressions, and survival of the individual. This non-random environmental influence may help form an advantageous environmental integration-into-phenotype design. Early developmental adaptations are environmentally-advantageous changes and are inherited. It is environmental adaptation that allows "the preservation of favored races in the struggle for life." Assuming that genetic variation with improved survival potential exists, the creation of an isolated, changing environment must occur to initiate this life-changing phenomenon.

The conditions of life give life meaning; the dynamic interaction of heredity and the environment produce the biodiversity in life. The environment indirectly changes the phenotype. An adaptation is an environmentally-integrated inheritable phenotype, with an environmental edge. The environment indirectly changes the genotype. Natural selection genetic bias is the genotype preservation of successful phenotypic adaptations to previously-experienced environments. Nearly all phenotype variation and genotype variation has been influenced by the interaction between environmental selection and natural selection genetic bias. Adaptations are new phenotype variations, modifications of form, function, behavior, etc., that are supported by environmental selection; they are intimately involved with (often adverse) environmental selection. Adaptations are constructed from components of previously-preserved developmental pathways that are stabilized genotype norm-of-reaction adaptation pathways to previously-encountered environments. Said otherwise, adaptations are new phenotypic expressions, made from newly-modified developmental pathways that utilized previous pathways in the genome (natural selection genotype bias) as raw material. A new environmentally-stabilized adaptation is an environmentally-integrated expression of the genotype, even if it began as a random phenotype expression resulting from environmental stress. Random exploratory cellular behavior can be stabilized by environmental feedback and functionally-integrated into genetic expression hierarchy. Environmental selection will then continue to stabilize, or destabilize, the developing organism. Successful new phenotypes pass advantageous adaptations to offspring. A new environmental adaptation allows the continued preservation of previously-preserved environmental adaptations (natural selection genetic bias); adaptation is the basis of environmental integration-into-phenotype design. Nearly all adaptations are environmental adaptations. Evolution is a change in phenotypic adaptations, in response to variable environmental conditions, over time.

In simple life forms, speciation loses relevance, but may be a result of genetic change in expression of the phenotype in a changing, isolated environment. If this change has survival advantage, it may become established and predominate in the vegetative-reproducing population. In complex life, the origin of species may also involve genetic change in expression of the phenotype to a changing, isolated environment. A complex-life breeding population could include different phenotypes, including any new phenotype change with survival advantage (an adaptation). This adaptation (change) may become established and predominate in the sexually-reproducing population. In complex animal life, speciation is increasingly the result of sexual selection, a behavior phenotype, which may itself be constrained by isolating factors in the environment. The origin of species through natural selection may not be the driving force of evolution. Speciation is an isolating mechanism (for a population). The driving force of evolution may be the product of early developmental organism interactions between environmental selection and genetic variation in a changing, isolated environment. In a word, the driving force of evolution is adaptation. Gene pool isolating mechanisms, including sexual selection, results in the origin of species, which genetically supports preferable adaptations.

There are two scenarios to consider that play a strong role in environmental biology. Both scenarios begin with adverse environmental change (environmental selection) in a confined area. With adverse environmental change, by far the most common result is extinction of any population confined to this area. Sometimes the adverse environmental selection process does not cause extinction of the severely-stressed isolated population. In this case, a second scenario involves three adaptive options for survival early in an organism's development. Any one of the three options could allow survival in an adverse newly-created environment; timing is everything. The 1st option is an increased rate of mutation and an immediate expression of a phenotype, which fortuitously has survival value in the stressful environment. This can be easily understood, but is likely to be a very rare occurrence, even in the seldom-seen process of successful environmental adaptation. Most mutations are silent, or if expressed, fatal. Less rarely, a 2nd option, a pre-existing phenotype structure (or function) allows an organism to quickly adapt (improve interaction) to the hostile new environment. Because there is little room for neutral (phenotypic) variation in structure or function in life, this "pre-adaptive" approach would still be a less-common route to adaptation. When adaptation does occur, the seldom-seen process of successful environmental adaptation usually involves a 3rd option, a life-saving, environmentally-generated and environmentally-stabilized, norm-of-reaction plastic change in genetic expression, which makes a difference in the survival of some population members. In this last case, there is a genome-related response (utilizing the plastic range of existing genotype variation) that results in a new genetic expression (new random phenotype) in response to the adverse environmental challenge. If this structure (or function) allows an organism to improve interaction with the hostile new environment, (environmental and genetic) stabilization of the new phenotype (adaptation) occurs. This last case doesn't involve a genome (DNA) change per se, only genome expression of phenotype change (modulation of gene expression- epigenome change); this may even include expressing a mutation that occurred at an earlier time. The expression of an earlier mutation as a result of environmental change is slightly different from how the cell normally varies the phenotype under frequently-occurring conditions. It utilizes the plastic range of expression to form an expression from outside the norm-of-reaction (but still utilizes epigenome environmental sensitivity), and seems to require oppressive signals of extreme environmental stress that do seem to be necessary for generating extremes of random phenotype expression. But other than these differences, in the usual plastic range of epigenetic sensitivity, (far less intense) environmental signals from frequently-encountered environments use the same core mechanisms to develop the inheritable norm-of-reaction phenotypes (from natural selection genetic bias). Small changes in expressed variation are more commonly associated with small environmental variations. Major adaptive change parallels major environmental change.

At every level of cellular organization, environmental differences generate changes in the phenotype displayed, without any change in the genes themselves. Cells change their properties many times during the transition from a single fertilized cell to the billions of cells that make up an adult organism. They do this in an environment of surrounding cells, moving and changing in a pattern that is constant from one individual to the next, changing very slowly over evolutionary time. Complex life's environments are increasingly internalized into genetic hierarchy; but at the same time, phenotype formation is simplified when random exploratory cell function is stabilized by environmental signals. It is the environmentally-sensitive physiologic stabilization of random exploratory cellular processes that mostly forms the phenotype. Developmental compartments are isolated environments. Environmental creation is achieved through local differences along chemical gradients produced by developmental or regulatory genes; changing their activity changes the environment of structural or locally-functional genes. Cells receive information from their environment and translate that information into one of the thousands of different potential responses from the differential activation of genes. Subtle alterations in the timing, location, and levels of protein synthesis have considerable consequences at the molecular and organism level. Protein production by local genes can be modified by controlling expression intensity of developmental or regulatory genes (transcription modulation via allosteric gene switches). And protein production by local, developmental and/or regulatory genes can be modified by changing the genes that are expressed (modulation of gene expression via allosteric epigenome switches). Epigenetics invalidates the mindset of genetically-determined evolution.

The genome of living organisms functions in formation of norm-of-reaction phenotypes. It has components of varied environmentally-integrated expressions in accordance to environmental changes over time.

The epigenome is very much affected by the environment and the epigenome controls genome expression. The indirect influence of the environment can also initiate and stabilize a new change in expression of the epigenome. During environmental stress, the adaptive process utilizes existing genome pathways to norm-of-reaction phenotypes as raw material to produce a new environmentally-integrated more-fit individual. The adaptation process is dependent on an environmental integration into genetic expression early in development, and the inheritable transfer of this to offspring. While it may be true that the environment doesn't directly instruct a gene on how to vary its expression, the influence of the environment (causing plastic change in expression of the epigenome) should be acknowledged. The epigenome, and through it, even the genome, have an indirect heritable connection to environmental variation. Variation is not in genetics alone. Environmental selection may initiate new directional change (from the genetic potential) in the surviving progeny of a heredity-environment marriage. Signals from the external and internal environment may initiate flexible, regulated, modifications in epigenome expression during early development. The epigenome can change, according to an individual's environment, and this change is passed from generation to generation. Knowing random phenotype structure or function can be indirectly stabilized by environmental feedback, and that this cellular-level stabilization can then indirectly stabilize the random phenotype expression from the genotype (need epigenome expression integration with environmental feedback signals); this doubly-indirect environmental influence should also be acknowledged. This feedback may even temporarily stabilize a non-adaptive phenotype. Only adaptive phenotypes prosper (positively supported by the selective conditions of life) in environments most favorable to that adaptive phenotype. Only adaptive phenotypes manage in environments that are more limiting to non-adaptive phenotypes. Only adaptive phenotypes survive extreme environments that are lethal to non-adaptive phenotypes. The survivors incorporate this new adaptation within their genetic bias of natural selection (the preservation of favorable variations and the rejection of injurious variations). The generation of shared character states in survivors is an indirect genetic response to environmental challenge. Epigenetic changes involve environmental modifications of gene expression. Instinctive behavior is a heritable phenotypic trait (adaptation) that is environmentally-integrated. (This is the *ne plus ultra* example for adaptive environmental feedback transfer to heritable mechanisms; environmental feedback stabilized neural circuitry that was linked to adaptive-genetic pathways that form the phenotype.) The actual mechanism of change (for example epigenome expression modification through signals from the environment), will lie in the biochemistry of genetic mechanisms.

Even though epigenome adaptation occurs as a biochemical process, it can occur at different levels of hierarchy within an individual organism. Environmental selection may also operate at different levels of hierarchy within an individual organism; however, survival only relates to the individual organism. Genetic isolation, genetic potential for survival, and the time to respond to a changing environment, before species extinction occurs, are contained in the recipe for adaptive evolutionary success. Environmental selection acts on life at all times; it is especially active during times of environmental extinction events (habitat destruction). The primary cause of species extinction is an environmental extinction event. Environmental extinction events (catastrophic and/or constant) seem to be more significant for evolutionary change than intraspecies constant competition for new and different jobs in a stable environment. The environment is not passive; it is anything but stable, and it plays a role in evolution that is essential for creating natural selection genetic bias of the phenotype and continuing the ongoing adaptive process. Evolution is genetic success in response to environmental conditions.

As individuals of an environmentally well-adapted population increase in number, competition for resources increasingly becomes more of a limiting environmental factor. All biologic and physical changes modify the degree of an individual's environmental compatibility, which continually varies the outcome of environmental selection. Quantitatively, environmental selection varies directly with the sum and/or intensity of environmental limiting factors; the population's numbers vary inversely with the sum and/or intensity of environmental limiting factors. Biodiversity is directly proportional to environmental stability; i.e., species richness is directly proportional to environmental stability over time; or, habitat continuity over time stabilizes and enhances biodiversity. Biodiversity is inversely proportional to a location's environmental extremes and directly proportional to habitat diversity; i.e., species richness is directly proportional to favorable conditions and habitat diversity. Without some form of population control or means of dispersal to unpopulated areas, increasing numbers beyond the carrying capacity of a finite habitat will lead to environmental destruction and an inevitable population crash.

Human overpopulation is causing a significant increase in environmental destruction, environmental extinctions and loss of biodiversity today. Irresponsible overpopulation not only threatens the future of humanity, it threatens the conditions of too many other lives; humans will suffer as a result of this. The conditions of life will control the destiny of humanity. The environment ultimately determines the standard at which human overpopulation occurs. The worst example of an unsustainable human population growth tragedy is yet to come; it will certainly come and it will happen at the location of the highest population growth on the planet. Global warming is inevitable, and it is not just due to greenhouse gas emissions, which are increasingly spiraling out-of-control as presently overpopulated countries raise their standard of living. Reflecting on the high probability of worsening global warming, followed by another Little Ice Age, Great Retreat of Agriculture, or whatever it will be called, dooms any large human populations in the not-that-distant future. At issue is a choice between the quantity of human life and the quality of human life. A quality human life requires a quality environment. Overpopulation and environmental degradation create an environment that selects against humanity. The environment cannot provide for humanity if humanity does not provide for the environment; it is in the best interest of all to become more environmentally-aware. Survival tomorrow primarily depends upon increased consideration of the environment, today, as well as the education and empowerment of all women, today.

2

Retrospective

Thomas Malthus wrote *An Essay on the Principle of Population* in 1798. In it he noted that human beings reproduce at a rate far outstripping food supply; their population grows until resources are exhausted. He said that there must be positive checks such as wars, disease and famines to keep them from overpopulating the world. This by its very nature involved a "struggle for existence." Specifically, he said: "The power of population is indefinitely greater than the power in the earth to produce subsistence for man. Population, when unchecked, increases in a geometrical ratio. Subsistence increases only in an arithmetical ratio. A slight acquaintance with numbers will shew the immensity of the first power in comparison of the second. By that law of our nature which makes food necessary to the life of man, the effects of these two unequal powers must be kept equal. This implies a strong and constantly operating check on population from the difficulty of subsistence. This difficulty must fall somewhere and must necessarily be severely felt by a large portion of mankind. Through the animal and vegetable kingdoms, nature has scattered the seeds of life abroad with the most profuse and liberal hand. She has been comparatively sparing in the room and the nourishment necessary to rear them. The germs of existence contained in this spot of earth, with ample food, and ample room to expand in, would fill millions of worlds in the course of a few thousand years. Necessity, that imperious all pervading law of nature, restrains them within the prescribed bounds. The race of plants and the race of animals shrink under this great restrictive law. And the race of man cannot, by any efforts of reason, escape from it. Among plants and animals its effects are waste of seed, sickness, and premature death.... We will suppose the means of subsistence in any country just equal to the easy support of its inhabitants. The constant effort towards population, which is found to act even in the most vicious societies, increases the number of people before the means of subsistence are increased. The food therefore which before supported seven millions must now be divided among seven millions and a half, or eight millions. The poor consequently must live much worse, and many of them be reduced to severe distress. The number of laborers also being above the proportion of the work in the market, the price of labour must tend toward a decrease, while the price of provisions would at the same time tend to rise. The laborer therefore must work harder to earn the same as he did before."

Using insight extending far beyond a man's lifetime, Charles Darwin carried this "struggle for existence" concept even further, to include all life and all time. In 1859 he published *The Origin of Species by Means of Natural Selection, or the Preservation of Favored Races in the Struggle for Life*. Darwin gave evolution credibility when he correctly recognized that competition between organic beings led to the preservation of favorable variations and the rejection of injurious variations, and this was the key to explaining evolutionary change. Darwin drew comparisons from man's selection of traits in types of domestic animals (artificial or human selection) to a natural selection of traits in types of all living creatures, present and past. This theory of natural selection's inherited preservation of favorable traits, as the explanation for the origin of species (different types of life), has been regarded by many as the greatest scientific discovery of all time; perhaps it is.

Evolution isn't as new as some believe; Aristotle thought of natural selection as a cause for the creation of different types of life and evolution, but dismissed it in favor of essentialism, a world of teleology and immutable types (Jeffrey Levinton, *Genetics, Paleontology and Macroevolution*). In philosophy, teleology is the doctrine that attempts to explain the universe in terms of ends or final causes. Teleology is based on the proposition that the universe has design and purpose. In Aristotelian philosophy, the explanation of, or justification for, a phenomenon or process is to be found not only in the immediate purpose or cause, but also in the "final cause," the reason for which the phenomenon exists or was created. In Christian theology, teleology represents a basic argument for the existence of God, in that the order and efficiency of the natural world seems not to be accidental. If the world design is intelligent, an ultimate designer must exist. Teleologists oppose mechanistic interpretations of the universe that rely solely on organic development or natural causation. The powerful impact of Charles Darwin's theories of evolution, which hold that species develop by natural selection, has greatly reduced the influence of traditional teleological arguments. Nonetheless, such arguments were still advanced by many during the upsurge of creationist sentiment in the early 1980s (Microsoft-teleology). Today, points of view are still strongly divided between the emotionally-conditioned religious stories of creation and the scientific community's theory of evolution. Some people try to find a way to incorporate both points of view. Some scientists don't accept Darwin's theory of evolution; these scientists do not include the biologists. The religious views predominate worldwide; most people still reject the concept of evolution in favor of whatever story of creation is told by their religion.

Both the spontaneous generation of living forms from inanimate objects as well as Jean Baptiste de Lamarck's doctrine of the gradual transmutation of one species into another had a following before Darwin published *The Origin of Species*. Lamarck came to the conclusion, "that none of the animals and plants now existing were primordial creations, but were all derived from pre-existing forms, which, after they may have gone on for indefinite ages reproducing their like, had, at length, by the influence of alterations in climate and in the animate world, been made to vary gradually, and adapt themselves to new circumstances, some of them deviating, in the course of ages, so far from their original type as to have claims to be regarded as new species." In order to explain (after an indefinite lapse of ages) how so many of the lowest grades of animal or plant still existed, he imagined that the germs or rudiments of living things, which he called monads, were continually coming into the world; he believed that there were different kinds of these monads for each primary division of the animal and vegetable kingdoms. This last hypothesis does not seem essentially different from the old doctrine of equivocal or spontaneous generation. Charles Lyell (a geologist from the time of Lamarck and Darwin), argued against the sudden destruction of vast multitudes of species, followed by the abrupt ushering into the world of new batches of plants and animals. Lyell opposed the idea of catastrophism (the belief that catastrophes caused a lot of change to the earth in a short amount of time); he claimed that the earth had been formed by the same kinds of processes that we see today. Lyell "whole-heartedly" followed Scottish geologist James Hutton in support for his "Uniformitarian Doctrine" (that assumed that the earth is run on perfect principles), which being perfect, know no change.

Regarding the catastrophic destruction of life, Darwin held views similar to those of Lyell. Darwin knew "many groups, formerly most extensively developed, have now become extinct." Darwin believed that species go extinct gradually; "each area is already fully stocked with inhabitants, it follows that as each selected and favored form increases in number, so will the less favored forms decrease and become rare. Rarity, geology tells us, is the precursor to extinction." Darwin believed changes through geological time were intimately connected to natural selection; extinction resulted from the fatal competition of natural selection. Darwin believed that the fiercest competition was between individuals of the same species. Darwin (Laws of Variation chapter) believed that "species of the same genus descended from a single parent." In his chapter on Natural Selection, Darwin maintained that "natural selection necessarily acts by the selected form having some advantage in the struggle for life over other forms; there will be a constant tendency in the improved descendants of any one species to supplant and exterminate (in each stage of descent) their predecessors and original parent." Darwin believed evolution to be in a direction towards higher forms of life (taxa). Darwin clearly recognized that geographical isolation played a role in evolution. In this role, geographical isolation is an environmental description, a created condition of life.

Darwin said: "isolation is an important element in the process of natural selection. In a confined or isolated area, if not very large, the organic and inorganic conditions of life will generally be in a great deal uniform; so that natural selection will tend to modify all of the individuals of a varying species throughout the area in the same manner in relation to the same conditions. Intercrosses, also, with the individuals of the same species, which otherwise would have inhabited the surrounding and differently circumstanced districts, will be prevented. But isolation probably acts more efficiently in checking the immigration of better adapted organisms, after any physical change, such as of climate or elevation of the land etc.; and thus new places in the natural economy of the country are left open for the old inhabitants to struggle for, and become adapted to, through modifications in their structure and function. Lastly, isolation, by checking immigration and consequently competition, will give time for any new variety to be slowly improved; and this may sometimes be involved in the production of new species." At the same time, Darwin believed that isolation of large populations provided the greatest genetic variation for the creation of new species.

Darwin understood that environmental limiting factors of climate and food affected populations, discussing this in his chapter on Struggle for Existence. "The action of climate seems at first sight to be quite independent of the struggle for existence; but in so far as climate chiefly acts in reducing food, it brings on the most severe struggles between individuals.... Even when climate, for instance extreme cold, acts directly, it will be the least vigorous, or those which have got least food through the advancing winter, which will suffer most." In Darwin's chapter on Laws of Variation, he stated: "How much direct effect of climate, food and etc. produces on any being is extremely doubtful. My impression is that effect is extremely small in the case of animals, but perhaps rather more in that of plants. We may at least, safely conclude that such influences cannot have produced the many striking and complex co-adaptations of structure (NOTE-adaptation = variation) between one organic being and another, which we see everywhere throughout nature. Some little influence may be attributed to climate, food and etc.... such considerations such as these incline me to lay very little weight on the direct action of the conditions of life." Natural selection, the (inherited) preservation of favorable variations and the rejection of injurious variations, (inheritance of favorable traits), was strongly contrasted against the conditions of life (everything that affects an organism during its lifetime, the circumstances of biological existence, or the environments effects upon an organism).

Few people have been misinterpreted or misquoted more than Darwin. It is common to freely change (the context of) the wording on Darwin's ideas of natural selection, using the natural selection and the conditions of life in roles inconsistent with his original writings; both have been blended completely into shades of each other. This does not help the understanding of Darwin's intent. It is helpful to return to the original work to understand what Darwin really said. There are countless other examples of biological phenomena getting interchanged with (wording related to) the conditions of life (environment). It is often done in such a way that the particular wording can be very confusing to a reader. This is a problematic obstacle to an understanding of evolution and environmental biology; this obstacle only perpetuates public problems with evolutionary theory.

In 1997, in *Guns, Germs and Steel,* Jared Diamond made a very strong case for environmental differences being the primary determinant in the evolution of mankind's societies. In a short version of history, he noted that "only a few areas of the world developed food production independently, at different times (depending upon available environmental resources). From these areas, neighboring hunter-gatherers quickly learned food production or were replaced by invading food producers, at different times. People with a head start on food production gained a head start on the path leading to productive societies that displaced other societies. People of areas ecologically suitable for food production that neither evolved nor acquired agriculture, remained as hunter-gatherers until the modern world finally swept upon them. The result was a long series of collisions between haves and have-nots." Although some might see this as a struggle for survival of the fittest, Diamond's point was that fitness had little to do with it; it was simply a matter of available resources in the environment. According to Diamond, the environment played the primary role in mankind's development and Western Civilization's evolution into world power. The environment still plays a primary role in the fate of human societies; climate determines food production, the most essential of all production (environmental differences in food resources gave some cultures historical advantages over others). The environment determines resource availability for other production (productive societies displace other societies). The environment is the base of a productive society's economy. The environment harbors disease; this is perhaps the major limiting factor for mankind (other than food—recall Malthus). The environment provides air and water, two resources even more important to us (for short-term survival) than food. Man's presence excessively disturbs the environment. The environment can be degraded by overpopulation; there are waste, energy, topsoil and wildlife concerns. Environmental degradation (leads to war, disease and famine) causes society degradation; it is the ultimate population check. **It is logical to believe that both the environmental conditions of life and the survival of the fittest played strong roles in the development of mankind's societies.**

In 2005, Diamond published *Collapse*, which deals with the fall (as opposed to the rise) of societies. He listed five factors: ecological collapse (damage that people inadvertently inflict on their environment); climate change (climate dominates the environment); hostile neighbors; decreased support by friendly neighbors; and finally, society's response to the listed environmental issue (or combinations of these environmental issues). All five are population-related and/or resource-related, which makes all five environment-related. As previously discussed with the rise of societies in *Guns, Germs & Steel*, environmentally-connected roles of agricultural-based technology and disease always apply. However, *Collapse* insightfully explores the over-utilization of resources and the consequences of overpopulation in the isolated, changing environment (along with other environmental issues). Recall that Malthus stated that "there must be positive checks such as wars, disease and famines (limiting environmental conditions) to keep people from overrunning the world, and that this by its very nature involved a struggle for existence." **Survival of the fittest (or competition) likely plays a role in the fate of man's future societies, but it assuredly plays no greater role than that of the environment (the conditions of life).**

Is it possible that environmental conditions might have played a more significant role, regarding the evolution of all life over all time, than presently perceived by the scientific community? Can Diamond's environmental approach to man's development be extended to apply to all life for all time, as Darwin did with Malthus' "struggle for existence?" After all, Malthus did emphasize the importance of population checks (limiting environmental conditions). So, why couldn't the environment have played an equally significant role in evolution? **Environmental biology is the study of the conditions of life (or conditions of existence) and the impact of these conditions upon and within the life it contains.** Science has provided some insight on the conditions of life over geologic time. If the conditions of life did play a major role in evolution, what was the role and how did it function? Darwin said far more than that he had an idea that there was some vague biologic mechanism behind his view on evolution. In order to form a new species from a common ancestor, he proposed the mechanism of natural selection (an <u>inherited preservation of favorable traits</u>) as the basis for a change. **It is logical to believe that both the environmental conditions of life and the survival of the fittest played strong roles in the evolutionary development of all life on earth.**

3

Beyond Ancient History

Environmental history tells us the age of universe is 13.7 billion (thousand-million) years (radiation signature); the age of solar system is 5 billion years; the age of earth is nearly 4.6 billion years. The earth was formed from many collisions of solar system debris in the earth's orbital area. While the earth was still forming, 4.5bya (billion years ago), a planet the size of mars smashed into the hot molten earth, throwing enough debris into orbit to form our large moon; this stabilized the earth's spin and stabilizes climate today. The crust then began to cool. The age of the earth's first rocks is greater than 4 billion years; the age of first life on earth may nearly be that old. Rocks associated with undersea hydrothermal activity, formed 3.8bya (billion years ago) contain cell-like structures; ocean sediment layers from near the same time period (>3.83bya) contain extensive deposits of fossilized carbon, further supporting the existence of widespread microscopic ocean life by this time in earth history. (The types of life will be discussed in the chapter—Common Ground.) Find water anywhere on earth and you will find life on earth. Some chemically-powered bacteria, chemobacteria (or chemosynthetic bacteria), live under extreme conditions; they are known as extremophiles. Chemobacteria extremophiles are found in undersea thermal vents (235°F or 113°C), hot springs, salty seas, icy caves, polar ice (12°F or –10°C), alkaline dump sites, acidic old mines, underground rivers and particularly deep underground in water-filled porous rock formations, which can be up to greater than one kilometer below the ocean's bed and up to greater than 10 km below land surface. These deep rock extremophiles live in water from just below the earth's surface down to the hot, lava-like, liquid mantle; they may be the most abundant life form on earth. Bacteria have always been the preponderance of biomass on earth; 70% of the bacteria on the planet today live below ground (soil aerobes, anaerobes and chemobacteria). Most of the subsurface bacteria are the deep-rock extremophiles; there are millions in a single teaspoon. Chemobacteria extremophiles that live deep in the earth only need water to live in and heat of the earth (energy) to reduce (or oxidize) rocks and minerals. Hydrogen (H_2) gas is released as water seeps through rock (produced from iron oxide reacting with water); molecular hydrogen can be reacted with carbon (C), oxygen (O) or sulfur (S) to sustain deep-rock extremophile respiration. Another rock type, limestone extremophiles, chemically reduce the rock they live in with their very slow rates of metabolism, releasing carbon dioxide (CO_2) and forming caves. Sea-vent extremophiles draw thermal energy from below the ocean floor to make carbohydrates; these chemobacteria can create carbohydrates capturing close-by carbon in a chemosynthetic cousin of the carbon cycle. For example, hot water gushing from deep-sea vents is laden with hydrogen-sulfur and hydrogen-iron compounds. Extremophiles near the vents combine these molecules with the carbon and oxygen molecules dissolved in sea water. These reactions form larger molecules, carbohydrates (from the H/O/C atoms).

Scientists believe life on earth may have begun in the ocean-bottom environment of a thermal vent (very hot—800°F, mineral rich, reducing environment). Life evolved. Photosynthetic blue-green "algae," which are primitive bacteria (cyanobacteria), are still abundant today, made earth's oxygen atmosphere and supercharged respiration 2.2bya (billion years ago). Today these cyanobacteria can be easily observed as thick mats covering the bottom in shallow waters (e.g. *Oscillatoria*). Cyanobacteria are the most abundant organisms in the ocean. Cyanobacteria have recently found in the most barren area of Antarctica, where no other life has been found; similar observations have been made in the high Arctic. They live just below the surface of rocks (within the rock-like chemosynthetic extremophiles). Cyanobacteria favor carbonate substrates, residing in limestone ($CaCO_3$) is a widespread pattern. Stromatolites are carbonate rock-dwelling cyanobacteria that make-up the oldest fossil structures, known as stromatoliths. Stromatoliths are large columnar, calcium carbonate rock-layered structures; some have been found to be up to 3.6 billion years old. These were originally formed largely through in-situ precipitation of layers (within the cyanobacterium) during Archean (2.5–3.6bya) and older Proterozoic (2.2–2.5bya) times; however, younger Proterozoic (0.55–1.0bya) stromatoliths grew largely through the accretion of carbonate sediments, enabled by an oxygenating environment. Stromatoliths still exist today in environments salty enough to deter grazers from feeding upon them, such as those found in Australia's shallow tidal flats or The Dead Sea. Complex organisms are relatively new on earth. Sponges are a possible animal ancestor from over a billion years ago.

In Peter Ward and Donald Brownlee's *The Life and Death of Planet Earth*, we learn that oxygen began accumulating in earth's atmosphere

2.2bya (billion years ago). The arrival of free oxygen (O_2) changed the environment. The atmosphere and ocean turned blue in color, respectively changing a Mars-orange sky and vitriol-green sea. There may have been some kind of a bacterial mass extinction caused by this new toxic O_2 waste, but there may have been no extinction at all. A large die-back is not really an extinction; chemosynthetic bacteria still live in extremophile environments. Around rift zone sea vents, it is common to find enough free iron today to form significant quantities of black ferrous sulfide (FeS_2) from hydrogen sulfide (H_2S). Also, sulfate-reducing bacteria (sulfide-producing bacteria) thrive today in the black layer below moist soils, water-covered sediments and in black water columns containing hydrogen sulfide. Above this sulfate-reducing layer is the omnipresent red-brown soil, or brown sediment layer, or water-column cloud of sulfate-fixing bacteria (where oxygen is available). Above this layer is the green soil, surface, or water column zone of photosynthesis that contains algae and/or photosynthetic bacteria.

Cyanobacteria survived the freezing of earth's ocean surface when the world became a snowball, like Europa (Jupiter's moon), 2.3bya. This snowball (worldwide glaciation) occurred as the oceans were losing their last iron salts (iron enhances photosynthesis). Atmospheric oxygen levels were near 1%. Prior to forming an atmosphere containing significant amounts of free oxygen, any available atmospheric oxygen (O_2) would have first oxidized atmospheric methane (CH_4), producing carbon dioxide (CO_2) and water (H_2O). Methane (CH_4) is a greenhouse gas, about 25 times (25×) as effective in this role as carbon dioxide (CO_2). Losing atmospheric methane (CH_4) would have had a chilling effect, even with earth's originally-high atmospheric carbon dioxide levels. Back then, respiration was primarily anaerobic. There were at least two more snowballs (worldwide glaciations) between 580–750mya (million years ago), as photosynthesis from cyanobacteria once again decreased atmospheric greenhouse gas levels (cooling the surface). The snowball's ice cover decreased photosynthesis and kept any CO_2 in the atmosphere, and that helped with warming and melting the snowball. Volcanic CO_2 may have helped reverse the more recent snowballs. The main source of volcanic CO_2 today is from extensive deposits of seabed limestone that are continually buried at the edges of continents, in tectonic plate subduction zones. As the limestone ($CaCO_3$) plunges deep into the earth, CO_2 is released and vented to the surface, often in violent eruptions. Upon eruption, volcanoes at first enhance global cooling; they also spill greenhouse gasses (water vapor + CO_2) into the atmosphere, eventually warming the surface. There wasn't much seafloor limestone to do significant cooling or warming 2.3bya, and there was still far less than today's limestone deposits back 580–750mya. (Super-continent Rodinia formed 1.1bya and subsequently broke-up 750mya.) During the time following the more recent snowballs, oxygen (O_2) levels began to surpass the atmospheric levels of today (collagen needs O_2 and complex animal life needs collagen) , grazers of the seafloor shallow-water green mats arrived, along with predators, to off-gas CO_2 (from aerobic respiration) and help to prevent future snowball formation. Complex life became larger and more abundant. The conditions of life have been anything but constant, or similar to the conditions of our present time. In Peter Ward and Donald Brownlee's *The Life and Death of Planet Earth*, we learn that the sun is getting hotter; it is 30% brighter today than it was 4.6bya. At the time of the first snowball (2.3bya), when the glaciers froze the oceans all the way to the equator, the sun's intensity was 6% weaker than it is today.

4

The Atmosphere

The atmosphere (troposphere) forms a thin coat of gases (mostly nitrogen and oxygen) 10km (kilometers) in thickness around the massive earth. The atmosphere of Mars and Venus is mostly carbon dioxide (CO_2). Mars (average temperature –63°C) once had enough greenhouse CO_2 gas in the atmosphere to maintain large amounts of liquid water on the surface; losing its magnetic core resulted in losing nearly all of its atmosphere and water. Abundant CO_2 greenhouse gas and closer sun proximity makes the surface of Venus hot enough to melt lead; Venus is a life-hostile environment. Life locked-up earth's carbon dioxide early-on, removing it from earth's atmosphere. Had it not been for life locking-up the CO_2, the earth would also have a CO_2 atmosphere (like Mars or Venus). The earth's early atmosphere was much like Jupiter and Saturn, a reducing atmosphere of mostly hydrogen (H_2) and helium (He); both were lost early in its history. Methane (CH_4) and ammonia (NH_3) were present in smaller amounts (Saturn's moon Titan—atmosphere is 90% N_2 and 10% CH_4), but when the earth's H_2 and He were lost, that is what remained. All of these are abundant in the solar system and are (or made of) all the lightest elements of low chemical valence. The most common compound in the universe was also present, as atmospheric water vapor. Much of the Earth's water had an extraterrestrial source (outer rim of the asteroid belt or comets from Kuiper belt), coming from early impacts in our planet's history. The water from these impacts formed the oceans. Most of the water on the planet today is chemically-bound in hydrated rocks (like granite). In addition to the greenhouse-potent methane (CH_4) gas, carbon dioxide (CO_2) levels were initially quite high, greater than 100× (times) the levels of today (Ward and Brownlee). Global warming was significant. Methane (CH_4) was likely the primary atmospheric greenhouse gas (25× as potent as CO_2), keeping ocean water in liquid form at this time of lower sun intensity. There was likely insufficient atmospheric CO_2 to prevent the snowball (worldwide glaciation) of 2.3bya, once a threshold amount of methane (CH_4) in the atmosphere was lost to oxidation (to CO_2). Yet, there must eventually be enough atmospheric CO_2 (or other greenhouse gases— or something else) to re-warm the oceans once again. The problem with an increasing CO_2 eventuality is that there was likely an inadequate sea-bed limestone source for plate tectonics to generate adequate volcanic CO_2 emissions 2.3bya, and perhaps even 580–750mya. Recall that stromatoliths were likely the only source 2.3bya; significant carbonate accretion did not begin before 550mya. At the time of the first snowball (2.3bya), the in-situ layering within stromatolite cyanobacteria produced some limestone (on the anaerobic sea floor) but it was unlikely it produced anywhere near as much limestone as is produced today. Other than very limited CO_2 source possibilities at areas of volcanic activity, it was also possible that some quantities of out-gassed CO_2 could have resulted from reactions of buried methane (and other organics) with relatively oxidized silicates of the crust. Under the snowball's thick ice, photosynthesis would sharply decrease (atmospheric CO_2 reduction blocked), and the organic decomposition from the die-back, combined with the respiration of the survivors, would eventually increase atmospheric methane (CH_4) levels (+ possibly CO_2 levels). At the time of the first snowball (2.3bya), the continents were perhaps only half as large. Also at this time, due to above-ground continental ice significantly lowering the sea level, the sea level was the lowest in the history of the wet and cool earth; this would expose massive amounts of frozen methane and anaerobic bottom sediments worldwide. The frozen methane (CH_4) deposits on the sea bottom would have been far more prevalent than that seen today, perhaps even suggestive of the limestone sea-floor deposits of today; carbon sedimentation was still anaerobic. The oceans at that time were much shallower than the oceans of today. When it was no longer held by hydrostatic pressure (widespread continental ice lowered the sea level), seasonal temperatures would have allowed massive volumes of exposed and shallow water hydrated frozen methane to escape into the atmosphere. This would increasingly warm the atmosphere, which would release additional methane gas (CH_4). Today, either cold temperatures of 1–2 degrees Celsius or Centigrade (34–36°F), or the ocean deep (400 m+ with high pressures and slightly warmer waters) are needed for the frozen hydrated methane crystals to form; importantly, two thirds of the ocean depth is greater than 1.5 km, often having temperatures of minus one degree Celsius or Centigrade (30°F). There are extensive ice-methane deposits in the ocean basins, as well as in the shallow Arctic waters (e.g. in Alaska's Prudhoe Bay), where temperatures even near the surface keep it frozen. Frozen hydrated methane (CH_4-{20}H_2O) deposits form when methanogenic extremophile bacteria (deep in marine sediments) generate methane, which rises through the sediment and, if the temperatures and pressures are right, forms ice-balls dispersed in other sediments, or even larger masses of this "gas-ice." (Frozen ice-methane can even be found in permafrost, formed in seasonal wetlands—20% of land surface is tundra). Methanogenic bacteria (anaerobic extremophile) in deep marine sediments convert (reduce) carbon dioxide into methane and water, while wetland-generated methane is more of a byproduct of (non-extremophile anaerobic) fermentation. Today the largest fermentation sources of atmospheric methane are wetlands and rice fields, agricultural fertilizers, digestion of ruminants and termites, waste disposal sites, and the gas produced by sewage treatment plants. In addition to the methane released with the snowball, it is conceivable that both the atmospheric CH_4 and exposed sediments may have slowly oxidized (2.3bya atmosphere was perhaps 1% O_2) to form more atmospheric CO_2.

Increasing amounts of free atmospheric oxygen appeared after the first snowball melted 2.2bya. The oxygen (O_2) was a by-product of photosynthesis (via cyanobacteria or blue-green algae); some oxygen (O_2) was also produced from photolysis of water vapor. Aerobic photosynthesis, involving the loss of atmospheric carbon dioxide (CO_2), the breakdown of water (H_2O), and the release of oxygen (O_2), is likely a chemosynthetic bacteria variation of an anaerobic photosynthetic depletor of atmospheric CO_2 seen today in estuaries (discussed later). Iron (Fe) precipitation took some time before the earth's reducing atmosphere became an oxidizing atmosphere 2.2bya (Ward and Brownlee). Free oxygen (O_2), produced by cyanobacteria, would have first rusted iron (Fe) in surrounding seas; this required the production of 20 times the oxygen (O_2) compared to that seen in the atmosphere today. Additionally, just as the bulk of the earth's water was bound in the rocks, oxygen increasingly became a major component of sedimentary, metamorphic and igneous rocks; about half the atoms in any rock found today are oxygen atoms. Magma (lava) is rich in silicates (SiO_4=). Sedimentary rocks (limestone [$CaCO_3$], sandstone [SiO_2] and shale [Al_2O_3–$2SiO_2$]) make up 5% of the earth's crust, but cover 75% of earth's surface. Minerals precipitate from sediments and bind sediment particles together; the most common cements are calcite ($CaCO_3$) and quartz (SiO_2), but dolomite (Ca-$Mg[CO_3]_2$), and iron oxides (Fe_2O_3 and Fe_3O_4) or hydroxide (FeO-OH-H_2O) also bind particles in sedimentary rocks. As the iron (Fe) went out of solution 2.2bya, the oxygen (O_2) was then free to act on aqueous sulfides, aqueous ammonia and atmospheric methane. Oxygen (O_2) oxidized methane (CH_4) to carbon dioxide (CO_2). Ammonia (NH_3) was trapped in the water vapor (H_2O) and life off-gassed nitrogen (N_2) into the atmosphere. As free oxygen (O_2) became available 2.2bya, there may have been enough nitrogen oxides formed to act as a greenhouse gas and contribute to global warming at that time. Nitrogen (N_2) forms 80% of today's atmosphere; this indicates very high levels of ammonia (NH_3) in earth's early atmosphere. Some amounts of nitrous oxide (N_2O) may have formed in the process of oxidation of ammonia (NH_3) to nitrogen (N_2). When ammonia (NH_3) is applied to moist soil today in agriculture, small amounts of nitrous oxide (N_2O) are formed and released into the atmosphere. Nitrous oxide (N_2O) emissions range from 0.5 to 4 pounds (0.2 kg–2 kg) per acre per year, depending on moisture and temperature. Although relatively small amounts are released, it is a particularly potent greenhouse gas, having 300× the greenhouse gas potency of CO_2. Nitrous oxide lasts 150 years in the atmosphere; current levels are up 20% over pre-industrial levels. Any burning of methane (CH_4) in the atmosphere would also generate N_2O, NO and NO_2 (from N_2 in the atmosphere); this process also occurs today in fossil-fueled electric power plants and automobile fuel consumption.

A carbon dioxide molecule (CO_2) only lasts 10 years in the atmosphere today; the carbon cycle moves it around; the O_2 stays with the carbon atom. Rain and surface water dissolve CO_2 gas (according to CO_2 atmospheric partial pressure). The vast majority of CO_2 added to the atmosphere will eventually be absorbed by the oceans and become bicarbonate (HCO_{3-}) ion, but the process takes close to 100 years, primarily because most seawater rarely comes near the surface, where this exchange occurs. The ocean waters hold 50 times the carbon as the atmosphere (Ward and Brownlee). Cyanobacteria, like today's top primary producer, *Prochlorococcus marinus*, locked-up carbon dioxide (CO_2) into the ocean's food chain. Photosynthesis has dropped atmospheric carbon dioxide (CO_2) levels to earth-historical lows, those of our presently ongoing ice age (Ward and Brownlee). As oxygen levels in the atmosphere and oceans approached today's levels, skeletons of different life forms first appeared almost 550mya (million years ago). These skeletons are an adaptation to protect against predation; these skeletons are also the earth's primary CO_2 lock-up. The majority of skeletons (and sedimentary rock) are made of calcium carbonate ($CaCO_3$- also note oxygen atoms). The skeletons form a marine snowfall on the newly-formed ocean floor; it will become thousands of feet thick on older portions of the ocean floor and over the continental shelves. By far, the greatest quantity of the earth's carbon dioxide (CO_2) is locked-up as limestone ($CaCO_3$) from ocean deposition. Ocean sediments are mostly (>50%) marine limestone and land-sourced clay (Sverdrup-*The Oceans*). In sedimentary shale (calcareous, bituminous, carbonaceous, fossiliferous and oil shale), significant carbon lock-up amounts are between limestone levels and the following lesser lock-ups. Most of the lesser organic sediments lock-up as frozen hydrated methane (CH_4) in permafrost, on continental shelves and on ocean basins. Other lesser organic lock-ups, like oil from plankton (lipid storage vacuoles), comprise 1% of marine sediments (Sverdrup). All lock-ups combine to keep atmospheric carbon at very low levels. Until tectonic plate subduction releases the bulk of earth's CO_2 (locked-up in limestone) as volcanic CO_2 emissions, the locked-up ocean sediment carbon is trapped for some time. Presently, a hundredfold greater release of carbon is rapidly returning to the atmosphere as a by-product of fossil fuel combustion. And global warming is heating the sea water, slowing ocean uptake of CO_2. As waters warm, some shallow water deposits are beginning to bubble CH_4. Melting permafrost is also beginning to release CH_4.

Nevertheless, atmospheric carbon has been in a long steady decline (Ward and Brownlee). Ten percent of all buried carbon metamorphoses into inert graphite and leaves the carbon cycle; it will never again return as atmospheric CO_2. Recall that early atmosphere CO_2 levels were over 100× those of today (380ppm or 0.38%). Around 550mya (million years ago), just before the Cambrian explosion of the animal green mat grazers and predators, CO_2 levels were about 15× higher than present day levels (Ward and Brownlee). Grazers and predators caused a change; around 450mya CO_2 levels were 20× higher than present day levels (Ward and Brownlee). CO_2 levels in the atmosphere dropped sharply with the rise of vascular land plants 400mya (Ward and Brownlee). This coming ashore allowed plants to colonize increasingly growing continents; therefore, photosynthesis increased. The soil then also became a carbon reservoir; surface soil holds twice the carbon of the atmosphere (Ward and Brownlee). By 300mya, wet and warm fernlike forests (up to 50 m tall) dominated the northern hemisphere. The forests (swamps) partially decomposed (anaerobically) to form extensive deposits of coal. Large amounts of carbon were buried both in the land and at sea. Mangrove swamps continue to bury carbon today; the rate of new land formation matches the present sea level increases of two millimeters/year and stabilizes tropical shorelines. Grasses, today's most abundant complex-life plant, first appeared 35mya and began to bury significant amounts of carbon into the wet soil; polar ice caps formed. Very large amounts of carbon are being continually locked-up over time on the edge of continents as shallow-sea sediments of limestone and shale. The extensive shallow-sea limestone deposits on the floating continents have excluded much of their carbon-limestone content from further carbon cycle activity (CO_2 return to atmosphere). For the most part, this continental lock-up (e.g. sedimentary rocks—limestone and shale—of the entire Mississippi drainage) is permanent (continents have doubled in size, and drifted from different positions on the earth surface—due to plate tectonic activity). The silicate-carbonate geochemical cycle, involving physical weathering of rock and plant root acids (that give rock weathering a 4–10× boost) does break down continental rock. But, the carbonates released from the continental silicate-carbonate geochemical cycle (physical weathering and plant root acids) enter streams, bypassing the atmosphere, and further lower atmospheric CO_2 by enhancing additional CO_2 oceanic lock-up (Ward and Brownlee). Less than two per cent of the carbon dioxide dissolved in the ocean is present as a gas. The rest is in the form of either bicarbonate (HCO_3-) or carbonate (CO_3=) ions; carbon dioxide reacts quickly with water to form these two ions. The balance between carbon dioxide, bicarbonate and carbonate is controlled partly by water temperature and pressure (cold soda under pressure holds more CO_2) but also by biological activity that combines carbonate with calcium to build shells and skeletons. The silicate-carbonate geochemical cycle was boosted even further by the recent (geological) collision of India and Asia (> 15mya), forming the still-rising Himalayan mountain range. This single event created the thickest crust on the planet today, and greatly contributed to our existing 2.5 million-year ice age (Ward and Brownlee). Ice ages occur as the greenhouse effect of CO_2 diminishes (Ward and Brownlee). Greenhouse gas loss will lead to the next snowball earth, which will eventually melt in the distant future, as the sun continually increases in intensity.

Atmospheric CO_2 reached an earth-history all-time low (160ppm) when the present ice age began 2.5mya (Ward and Brownlee). Were it not for volcanism from plate tectonics, the CO_2 locked-up long ago in sea sediments would have been lost forever (Ward and Brownlee). Plants are continuing to cause CO_2 to disappear from the atmosphere; the level is still near an earth-historical low (Ward and Brownlee), now 380ppm, up from 280ppm only 100ya (years ago). The glaciations of ice will return. In the distant future, plant life will die from lack of CO_2 (Ward and Brownlee). The grazers, browsers, etc. will not survive. Extinction events are always tougher on animals as they move up along the food chain. Top predators will be among the first to go. Plate tectonic volcanism will not be able to keep pace with the steady decline of atmospheric carbon (Ward and Brownlee). Bacteria will still survive long after we are gone (Ward and Brownlee).

5

Milankovitch Cycle(s)

While it may seem as if the sun rotates around the earth, the spinning earth rotates once a day, to create the illusion. Although it also seems as if the sun moves from the northern hemisphere to the southern hemisphere as it crosses the equator for the fall equinox, this illusion is due to the yearly orbit of the earth around the sun. Due to a tilt in the earth's axis of rotation, there are yearly changes in sun intensity for different latitudes. When the spinning earth is at a certain place in its orbit around the sun, the northern hemisphere is tilted (closer) toward the sun and it experiences summer. Six months later, when the earth is on the opposite side of it's orbit around the sun, the northern hemisphere is tilted (farther) away from the sun and it experiences winter; the seasons are reversed for the southern hemisphere. The angle of tilt periodically changes in a repeating cycle over thousands of years. Small differences in sun intensity and duration of exposure make a big difference on the earth's surface. The spinning earth also wobbles some (like a spinning top), over a very long period of time; this wobble is known as precession. Precession is giving the northern lands more summer sun exposure today; i.e., the North Pole is aimed closer to the sun in the present direction of wobble. This means that the North Pole is presently getting more summer sun than it has in the historical past. In addition, the earth's orbit around the sun is rarely round; most of the time it is shaped like an oval (ellipse). The periodic change in the orbit's oval shape covers an even greater period of time than that of the wobble. Over time, all of these cyclic variations change the distribution and intensity of sunlight reaching the earth's surface.

All of the planets have elliptical orbits around the sun. The earth's orbit around the sun is said to be variably elliptical. Over a 100,000 year periodic cycle (Milankovitch Cycle), the earth's orbit (oval-shaped ellipse) around an off-center sun shortens into a shape forming more of a circle and then returns back into an elliptical shape. As the orbit changes shape to be more round, this distance-variability effect increasingly becomes greater than it might seem. This distance-difference is increasingly significant because illumination (heat from the sun) varies inversely with the square if the distance. For example, reducing a distance by half would increase illumination intensity by four times over the previous level for that position in the orbit. Today the earth is approaching, but hasn't yet reached, an orbit that is as round and as continually close to the sun as it will get. The increasingly rounding yearly ellipse now places the earth slightly farther away from the off-center sun in July (apogee) and closer to the off-center sun in January (perigee). The closest extreme (perigee) now occurs in the southern hemisphere (the northern winter), due to the tilt, that is, seasonal position in the orbit. Presently, a 3% difference in distance means that the overall earth's surface now experiences a 6% increased distance-effect difference in solar energy in January compared to July. The date (time of year) for this closest point extreme doesn't remain fixed over very long periods of time; it slowly progresses (moves later) through the year. The earth now reaches this closest point to the off-center sun, in early January, only about two weeks after the December solstice. In the future, the closest ellipse extreme dates will be March, July, September and then again in January. In 3000 years (3% of 1,00,000), the earth's elliptical orbit around the sun will be as round and close as it ever gets, providing maximum year-round constant illumination and heat intensity to our planet, somewhat like an Alaskan summer. So, the earth will certainly get warmer, no matter what, but the global warming will not last forever, the glaciers will eventually be back. Today, continental glacial advance would become significant following a 10% decrease in illumination. In 50,000 years, when the earth's orbit around the sun is most elliptical, the amount of solar energy received at the furthest extreme would be around 20–30% less than at the closest extreme.

Including the additional variable of precession in the above, recall that over a 23,000 year cycle, the earth's axis wobbles like an out-of-balance spinning top, causing it to vary in the direction of true north, relative to the position of the North Star (direction the North Pole points into space). Because the direction of the earth's axis determines when the seasons will occur, precession will cause a particular season (for example, winter in the northern hemisphere) to occur at a slightly different place in the earth's orbit around the sun from year to year. Five to six thousand years from now, the precession effect will balance the summer and winter distance and the orbital ellipse will still be very close to round (= to today, or 3% off-round in orbital distance variability in extremes). This was also the case five to six thousand years ago (warmer than today), with an orbital ellipse pattern of 9% off-cycle maximum roundness. While the upcoming short-term climate pattern looks like a planetary global warming forecast, the ice will return, someday.

Twelve thousand years from now, summer will occur near the closest elliptical peak extreme in the southern hemisphere, but at a 6% more off-cycle roundness (ellipse) than that of today (apogee 9% off-cycle roundness). Illumination of the northern hemisphere will decrease (should increase cooling), even though the summer sun will still be as close to earth as it was 5000–6000 years ago. Twelve thousand years ago, precession also focused the sunshine on the southern hemisphere, and the northern hemisphere summer occurred at the furthest extreme of a 12% more off-round elliptical cycle (than today); glaciers advanced to a moderate degree.

The axis of earth's rotation tilts between an angle of 21.5 to 24.5 degrees to the sun, more or less at a 40,000 year interval. Because of the periodic variations of this angle, the severity of the earth's seasons changes as the tilt increases. Said otherwise, the more the tilt, the more the climate extremes; recall that this tilt basically accounts for the summer and winter seasons as the earth orbits the sun. This third variable in the Milankovitch Cycles combines effect with the other two (precession and ellipse change) and has a major impact on climate extremes. Today the axial tilt is in the middle of its range; at present, the earth's axial tilt is about 23 degrees. When the axial tilt towards the sun decreases, the sun's solar radiation is more evenly distributed between winter and summer. However, less tilt also increases the difference in radiation received between the equatorial and polar regions (favors glaciers). Integrate this with the peak and minimum extremes of the yearly ellipse as it slowly changes summer and winter intensity on the 100,000 year cycle. With precession, summer will occur at a slightly different time relative to the closeness of the peak over the 100,000 year exposure. Previously noted, all variables change the distribution and intensity of sunlight reaching the earth's surface.

6

Climate

Atmospheric gases and the Milankovitch Cycle changes have greatly influenced global climate and the conditions of life. Climate dominates the environment. Atmospheric greenhouse gases, notably low carbon dioxide levels, play a strong role in climate determination. In the familiar yearly cycle, a photosynthetic summer on northern lands decreases atmospheric CO_2 and winter respiration increases atmospheric CO_2. In like manner, a long-term cooling from the freezing effect of an entire ice age of 2.5 million years would tend to increase atmospheric CH_4 and CO_2 (like the snowball earth), due to rain reduction, an ice cap on the water surface, decreased photosynthesis (on land and under ice), continued respiration and continued volcanism, while interim periods of photosynthesis during interglacial periods (within the ice age) would once again decrease atmospheric CO_2, bringing back the cold. Bands of sedimentary rock (from ocean cores) match the Milankovitch Cycles of periodic cold spells associated with glacial expansion every 100,000 years or so. The 23,000 year precession cycles are visible in continental rocks over most of the geologic time that complex life has existed on earth. Glacier cores from Greenland and Lake Vostok in Antarctica also confirm the cycles, with even greater accuracy. Air trapped in the ice reveals that atmospheric CO_2 levels today are 30% higher than cyclic highs of the past 400,000 years. Interestingly, the past high levels of atmospheric CO_2 have followed, not preceded, 100,000 year cycles of warm spells (organic decomposition is enhanced, increasing both atmospheric CH_4 and CO_2; warm water holds less carbonation-gas laws of Boyle, Charles, Dalton and Gay-Lussac = oceans release more carbon during global warming). Conversely, 100,000 year cycles of low temperatures seem to have caused a drop in atmospheric CO_2 (organic decomposition was put on ice; cold water holds more carbonation and other gasses like O_2 or H_2S- gas laws of Boyle, Charles, Dalton and Gay-Lussac = oceanic plankton take up more carbon during global cooling).

This Milankovitch-related pattern seems to indicate a strong role for the Milankovitch cycle for not only determining climate, but on temporarily influencing atmospheric carbon dioxide (CO_2) as well. This Milankovitch-related increased CO_2 pattern directly correlates with a satellite-surprise source of methane over forested areas today.

Milankovitch Cycle-related global warming increases atmospheric carbon both from the land and sea. The world's living land vegetation emits between 10–30% of all annual global emissions of CH_4 in a year; even dried leaves and grass contribute another amount equal to 10% of the living plant total (the anaerobic leaf/grass breakdown by ground bacteria to produce $CH_4 + H_2S$ is documented). The CH_4 emission in living plants occurs in the presence of O_2; the chemistry of CH_4 production is presently unidentified. However, it could be connected to unidentified pathways related to photorespiration (e.g. lost carbon skeletons). The rate of methane production increases drastically when plants are exposed to the sun; significant emissions occur at temperatures above 30 degrees C (Centigrade). The largest portion (2/3rds of the 10–30%) originates from tropical areas, the location of the warmest plant biomass. This means that the rain forest today not only decreases atmospheric CO_2, but produces a potent (CH_4) greenhouse gas as well. The CH_4 eventually oxidizes to CO_2. Both CH_4 and CO_2 contribute to global warming. This CH_4 generation could significantly offset the land plant role of photosynthesis in decreasing atmospheric CO_2 if temperatures continue to rise; the ultimate carbon sink is the ocean. The remaining 70–90% of atmospheric CH_4 comes from anthropogenic emissions (especially agricultural cultivation, but also from landfills and incomplete combustion- primarily fire; all of these are responsible for the well-documented increase in atmospheric methane since pre-industrial times), as well as from bacterial fermentation sources in wet, low-O_2 environments (swamps, marshes, rice paddies, buried organics, sewerage systems and animal digestive systems, e.g. termites and other animals—including humans); ocean CH_4 emissions are low (presently- but increasing-e.g. off CA coast).

Twelve thousand years from now, precession will expose the southern oceans to more sunlight; the axial tilt will be lower, resulting in cooler northern summers and winters. Northern continents could again be in an ice age period of expanding glaciers; but because of sun proximity, probably not as severe as that seen 12,000ya (years ago). At a still later time (36,000yr), the Milankovitch Cycle's elliptical orbit will cause a summer sun to be increasingly distant from the earth. In 60,000 years, even more significant glacial expansions, comparable to that seen over the Wisconsin Ice Period 10–100,000ya will have returned. These significant glacier expansions will have a profound effect on the conditions of life. The ecological specialists that are specifically adapted to particular roles in unique environments will not do as well as ecological generalists, as the glaciers again move southward, covering continents with over a mile-deep sheet of ice. And while it will not cause a mass extinction of nearly all life on earth, it could be a notable/small extinction event. A loss of numerous species, even genera, could be expected; it has happened before, many times. At higher latitudes, it will devastate human populations of North America, Europe and Asia. Much of this area is now the area of greatest food production. The land is fertile because of past glaciations. For mankind, it will be a massive die-back. Assuming the population distribution of today, the great majority of humanity would not survive another Wisconsin-type glacial advance. Even with the atmospheric greenhouse levels of today, it would take only a 10% decrease in sun illumination to bring back the continental ice. That may or may not happen in 12,000 years, but will likely happen in 36,000 and/or 60,000 years. There is more than just one simple cause and effect when it comes to climate.

The earth has been in an ice age for 2.5my (million years); it is still in this ice age, during an interglacial period. The low CO_2 atmosphere-related ice age of the last 2.5my had periods of glaciers covering 30% of the land (partial snowball); today, it only covers 10%. Ice increases when the earth's orbit is stretched out in northern summers. Glaciers need land (today land is mostly north of the equator; the southern ocean is land-free at latitudes above the roaring 40's), low atmospheric CO_2, and low sun intensity in order to increase in size. As CO_2 levels continue their long-tern decline (temporarily arrested), and the Milankovitch cycles decrease sunshine in northern latitudes, the northern glaciations will eventually return with an ever-increasing intensity, never before seen by mankind. Concern for global warming is only a very recent awareness; before this, cold has been always perceived as the enemy of civilization. It is helpful to look at the history of climate over the most recent portion of the ongoing ice age. The earth is living in a warm interglacial period within a major ice age that started 2.5mya. Before this warm interglacial period began, the Wisconsin (Wurm in Europe) glacial period (within the 2½my ice age) was the coldest experienced by mankind; it occurred during the late Pleistocene, 10–100ka (thousand years ago). There was a warm interglacial period before the Wisconsin glacial period, the Sangamonian interglacial, which lasted 20,000 years. Before this interglacial warm period, the Illinoian glacial period in the late middle Pleistocene (120–200ka.), extended to the border of, and formed, the Ohio river in North America; it also lasted 80,000 years (as long as the Wisconsin glacial period). The interglacial warm period before the Illinoian is known as the Yarmouthian interglacial, and it only lasted 10,000 years. Prior to that was an over half-million year glacial period known as the Kansan, during the middle Pleistocene (0.13–0.7mya). The interglacial warm period before this is known as the very short Aftonian interglacial (< 1000 years). The very long (nearly a million years) Nebraskan glacial period takes the ice age as far back as the early Pleistocene (1.65–0.7mya). It formed the Missouri River. Our present ice age, containing 100,000 year cycles within it, began a million years before that, and was primarily caused by the all-time low level of atmospheric CO_2. This ice age allowed some increase in atmospheric CO_2, similar to the snowball earth, or a winter season effect. Recall that the freezing effect of an entire ice age of 2½ million years would tend to increase atmospheric CO_2, both from respiration and volcanism, while interim periods of photosynthesis during interglacial periods within an ice age would once again decrease atmospheric CO_2. Fossil fuel combustion is unquestionably boosting atmospheric CO_2, but the ice age CO_2 atmospheric increase, recent (Milankovitch Cycle related) global warming release of CO_2 from the oceans, and CH_4 from the rain forest and grasslands must also be taken into account. Although today's atmospheric CO_2 concentrations were exceeded during earlier geological epochs, present carbon dioxide levels are higher now than at any time during the past 2.5 million years, and at the same time, looking at scales longer than 50mya, lower than at any other time in earth's history. There are advantages and disadvantages to the present slightly elevated levels of atmospheric CO_2.

It does appear as if the earth is now in a present cycle of warming; it was even warmer around five thousand years ago. At that time the summer sun seemed even closer to northern lands (precession extremes—March and September). At the present time, the sun's radiation has increased 0.05% per decade since the 1970's. A recent increase in atmospheric CO_2 greenhouse gas (381ppm, up from 280ppm only 100 years ago) has caused global warming concern (atmospheric CO_2 levels today are 30% higher than cyclic highs of the past 400,000 years); atmospheric methane levels have doubled, almost tripled (to1.5ppm), in the past 200 years and nitrous oxide is up also 20% over the past 200 years. Atmospheric release of chlorofluorocarbons (like Freon) not only destroys the ozone layer, chlorofluorocarbons are also potent and long-lasting greenhouse gases. Previously to present time, wildfires and other fires caused by man had already slightly elevated CH_4 and CO_2 levels for over 10,000 years; a prolonged spike in atmospheric CO_2 is evident approximating the time of mankind's entry into Australia 40,000ya (years ago); this may or may not be coincidental. A prolonged spike in atmospheric CO_2 is really dramatic with the expansion of mankind in the Americas 12,000ya. Global warming from mankind and fire may have begun long ago. Although not counted as fossil fuel emission, burning of the rain forest produces approximately 4 billion metric tons of carbon yearly, equivalent to over 14% of all yearly human fossil-fuel emissions (28 billion metric tons CO_2 in 2005). And of the 28B metric tons calculated emissions, 8.7B metric tons, or 30% of these

CO_2 emissions, are released from burning to clear areas for subsistence agriculture or burning wood for cooking and/or heat. These all total nearly 13B metric tons (44% of the worldwide 28B). Also uncounted underground coal fires in China and elsewhere nearly match the CO_2 production of the forest fires. Global warming can change the environment, impacting the conditions of life to varying degrees.

The temperature of the Atlantic basin has slightly increased in the last half-century. Additionally, the Gulfstream has recently diverted 30% of its flow southward towards Africa, reducing the amount of oxygen-rich cold salt water sinking into the North Atlantic basin. Global warming could be the cause of this Gulfstream shift. The Gulfstream is an ocean river (100× Amazon River) surface flow of evaporated salty, warm tropical (Caribbean) water to Western Europe. This water is as clear as any water can be, yet it appears to the eye as hauntingly ink-dark blue (light is lost into dark depths below). Western Europe receives almost 1/3rd as much heat from the Gulfstream as it does from solar radiation. The waters off Europe are now cooling; hurricane-forming waters off Africa are now warming. The result of the above Gulfstream diversion could be an increase in Atlantic basin hurricane numbers and/or intensity.

In most latitudes, the surface layers and deep layers in the ocean do not mix, as they are separated by a thermocline layer of sharp temperature change that is approximately 1km deep; there may be lesser thermoclines above this level. Warm water is lighter (expands into more space) and it "floats" on top of cold water, down to the thermocline, the level at which wind cannot mix the layers. Ice also floats in cold water (water tetrahedron shift @ 4°C makes still colder H_2O lighter than H_2O @ 4°C = H_2O freezes on top); however, unlike pure water, seawater becomes denser as it becomes colder, right down to its freezing point of –1.9 degrees C, when it reaches its maximum density. Water freezes as fresh water; the saltier water residue sinks. The Deep Ocean Conveyor (Thermohaline Circulation) is a deep oceanic system of massive slow-moving currents that account for all ocean circulation (90%) below the top surface layer (top is only 10%). Previously noted, surface currents, like the warm Gulfstream that carries heat to Europe are very important to climate stability; but, perhaps even more so is its contribution to The Deep Ocean Conveyor (keeps tropic waters, and the tropics, cool). As the Arctic cold causes sinking of the Gulfstream's very salty water, there is significant Gulfstream contribution to the Atlantic basin portion of the Deep Ocean Conveyor (now reduced by 30%) ; this is the primary source of O_2 for the deep Atlantic basin. Limited turnover (mixing) of cold aerated Arctic waters also provides oxygen and nutrients throughout the depths of local waters. Global warming of the salty surface water in Polar Regions reduces the sinking of cold salty water and leads to anoxia in the water below.

There is also evidence of a slowing for this portion of the conveyor system. Not surprisingly, there is now evidence of decreased oxygen levels in the waters of the deep Atlantic basin. The world's coldest, saltiest water flows out from beneath the Antarctic ice; ocean salt concentrates in the water that does not freeze. The primary source of cold water for most of the world's Deep Ocean Conveyor is cold salty water from Antarctica. Turnover (mixing) in waters of the Antarctic Circumpolar Current also introduces the major source of oxygen in the Pacific deep ocean layers. Deep ocean concentrations of 6ml O_2 per liter seem to be decreasing worldwide. In general, most fish species will have a difficult life if O_2 levels become lower than 5–6ml O_2 per liter of H_2O.

Tim Flannery, in *The Weather Makers*, notes that as global warming increases, this increases the El Nino effect (frequency and intensity) of increased drought in former monsoon-supported areas of the Asian rain forest (on the wet side of the Asian mountain rain shadow: Pacific equatorial moisture that normally produced rain on land, moves eastward, back out to sea and stays there). The drought has led to fires, crop failures, and other major drought-related problems off Southeastern Asia. Flannery notes that Australia, Sub-Saharan Africa and even the Southwestern United States are also affected by this global warming-related drought-producer. Global warming has shifted the jet stream further towards the North and South Poles. Flannery further notes that many species are shifting to higher latitude (cooler) range, presumably maintaining temperature-consistent habitats; similar relocations of temperature-sensitive species to higher mountain elevations are also discussed. Continued global warming will drive these temperature-sensitive species to extinction; many will either be unable to move quickly enough, or will run out of suitable latitude or mountain elevation habitat. As global warming continues, for every degree Fahrenheit the Earth warms, the earth will experience an additional one percent of increased rainfall (Flannery). This rainfall will not be evenly distributed; previously discussed, there will be marked changes from previous patterns. In general, rainfall will increase more with increasing latitude, melting northern glacier ice and diluting the salty ocean water with the increased rainwater itself. That will increase sea level and further destabilize the Deep Ocean Conveyor. And that, like global warming of surface waters in the Polar Regions, will lead to increased deep-water anoxia (stagnation).

Global warming increases atmospheric water vapor; atmospheric water vapor level plays the greatest role (95% of greenhouse gas effect-trapping long-wave heat radiation) in greenhouse gas warming, one far greater than the role of atmospheric CO_2 (CH_4 is 3rd). However, once clouds form, sunlight is blocked, and clouds cause a local global cooling. Cloud moisture forms on pollen and sea-salt particles; cloudiness can be increased by more than 50% from particulate pollution. Warm air holds more moisture (up to 6% of all atmospheric gases); water is held in the atmosphere by Earth's gravity and a cap of cold dry air, 10km above the surface, near the boundary of the ozone-rich stratosphere. Global warming may be offset more than 50% by air pollution particulates and sulfur oxides from fossil fuel burn. Sulfur oxide production has subdued global warming, enough to even question the existence of global warming. In addition to the fossil fuel sulfur oxide source, there is another significant sulfur oxide source. As atmospheric CO_2 increases, and the earth warms, this induces an increase in the productivity of marine phytoplankton, specifically dinoflagellates (an algae), which results in greater production of oceanic dimethyl sulfide (DMS) and its release to the atmosphere. DMS is then oxidized to sulfur oxides and carbon dioxide (in the air). Sulfur oxides (e.g., SO_2) plus water (and O_2) produce sulfuric acid (H_2SO_4) cloud droplet nuclei. Cloud particle nuclei form clouds in the atmosphere and these clouds cool the planet (reflect sunlight like a mirror). DMS emissions drop off as the surface water cools. This significant marine mechanism, combined with widespread fossil fuel burning (coal and diesel-cars-planes), have produced enough cloud droplet nuclei to decrease the sunlight reaching most areas of the earth's surface by 1.5% per decade since the 1970's; some areas of sunlight dimming from pollution and DMS have run higher than 10%. There has recently been a great effort to reduce anthropogenic (man-made) sulfur oxide emissions, due to their highly toxic properties. As this is accomplished in developed countries, undeveloped countries are increasing their consumption of fossil fuels and surpassing earlier levels of the planet's pollution. Volcanic eruptions can cause an even more significant global dimming, due to the ash and sulfur (H_2S and SO_2) released into the atmosphere.

The global cooling effect of sulfur oxide (SO_2) lasts longer in the atmosphere than any other additional short-term global cooling, such as that which might arise from any source of volcanic ash or atmospheric dust (e.g. drought-related African dust now reaches America). However, atmospheric sulfur oxides (SO_2) still rain-out faster than atmospheric CO_2 levels can decrease. Recall that today, the entire atmospheric content of CO_2 is consumed and replaced every ten years as the carbon cycle moves it around. The vast majority of CO_2 added to the atmosphere will eventually be absorbed by the oceans and become bicarbonate (HCO_3-) ion, but this process takes on the order of a hundred years, primarily because most seawater rarely comes near the surface. Also recall that the ocean waters hold 50 times as much carbon as the atmosphere (Ward and Brownlee). The major carbon sinks: ocean sediments (shallow and deep limestone, shale, ice-methane and fossil fuels like petroleum), ocean water (+ life within), landlocked limestone, dolomite and shale, organic land sediments (fossil fuels like coal and peat), and soil (including life upon, within and below), hold most of the Earth's carbon (and CO_2). Importantly, the global dimming of SO_2 can offset the global warming from CO_2, and the recent combination of SO_2 global cooling and CO_2 global warming has produced a short-term climate stability; however, there is a long-term net effect of increased CO_2 buildup and global warming (DMS even produces CO_2). In addition to an eventual net increase of atmospheric carbon dioxide and global warming from burning fossil fuels, sulfur oxide production from fossil fuel burn has a highly significant down side; neither airborne sulfur dioxide nor sulfuric acid rain is at all complex life-friendly.

7

Major Environmental Extinction Events

When it comes to sulfuric acid rain, it is hard to beat the nasty conditions of life during the Cretaceous extinction. Around 65mya (million years ago) an asteroid (metal and sulfide-rich carbonaceous chondrite)10km (kilometers) in diameter struck the Yucatan peninsula of Mexico, and the surrounding shallow sea, at an angle of 30 degrees from the southeast. It is difficult to relate to the magnitude of destruction caused by a 10km asteroid like the one that killed the dinosaurs. When it struck earth, the opposite end was still at 35,000 feet, the altitude of today's commercial jet traffic. It drove deep in the earth, vaporizing rock in the impact area and throwing this vaporized rock half-way to the moon. It left a crater 180km wide. The vaporized rock sprayed North America, covering it in fire within a matter of seconds. The largest wave the earth has ever seen followed immediately afterward, and flooded all of North America, along with most of South America; this would have extinguished the American fires, leaving large amounts of scorched dead plant remains. The hot rock thrown into space then returned to earth, showering down fireballs (spherules) worldwide. Fire cooked much of the earth's entire surface. Extended Deccan volcanic continental basalt floods (>1M km^3) had already covered India on the opposite side of the planet; both combined to make most of the earth's surface like an oven. The impact area was carbonate/sulfate rock (less than 0.5% of earth's surface has this combination). This impact, combined with the ongoing half-million-years of volcanic activity, rained sulfuric acid (H_2SO_4) with unprecedented intensely; the amount of sulfuric acid (battery acid) deposited on the surface over a short period is unsurpassed in the history of complex life on earth. Additionally, superheated atmospheric nitrogen and vaporized carbonate rock added significant amounts of nitric (HNO_3) and carbonic (H_2CO_3) acid to the noxious downpour. Both terrestrial and marine environments suffered the results of the most severe acid rain. And all this was only the beginning of a very bad time for life on earth.

Over the long term, the impact combined effects with the previous problem of ongoing volcanism and created the largest dust/ash cloud in the history of the earth; it spread globally. The atmospheric dust blocked sunlight; there were years of darkness. It was darker than the darkest night. Photosynthesis stopped—everywhere. The food chain collapsed; survivors of the immediate impact (+ fallout) died of starvation. Sulfuric acid cloud droplet nuclei continued to reflect sunlight away from the earth, long after the dust settled. Global temperatures dropped as much as 10 degrees C (Centigrade). Many species were unable to survive the global freezing, which lasted for many more years.

The ice age at the end of the Cretaceous caused a drop in sea level (150 meters +), exposing large areas of anaerobic sediments and methane hydrate from the continental shelf sea floor. The anaerobic sediments, exposed to the atmosphere, oxidized to CO_2. The icy methane hydrate was depressurized, and the greenhouse-potent methane (CH_4) gas released into the atmosphere combined greenhouse gas effects with the CO_2 to trap returning sunlight as infrared heat. The methane (CH_4) eventually oxidized to CO_2, and combined with other sediment-generated CO_2, and existing high levels of volcanic CO_2 from the continental basalt flood covering India. Global warming followed the sun's return and the atmospheric increase in greenhouse CH_4 and CO_2, which increased temperatures at least 5 degrees C above pre-event levels, and killed even more of the survivors. The sudden rise in sea level (up to 300 m) that followed might have carried deeper anoxic waters further onto the continental shelves (Jeffrey Levinton, *Genetics, Paleontology and Macroevolution*); by this time, the majority of the extinction had already happened.

Overall, marine life suffered the most damage; the aquatic environment is more fragile for the complex life forms within it. In the oceans, the prolonged darkness, cold and toxins devastated the plant life. The cold broke thermocline stratification and the anaerobic bottom ocean water turned over, exposing aerobic life in the oxygenated waters above to the anoxic waters below; the anaerobic bottom water also contained toxic hydrogen sulfide (H_2S). The rich chalk biotas of the upper Cretaceous gave way to miserably sparse clays (Jeffrey Levinton, *Genetics, Paleontology and Macroevolution*). The fossils best preserved had hard shells; mollusk survival was correlated with increased geographic range, indicating a habitat-related survival pattern (Levinton—chapter on Patterns of Diversity, Origination and Extinction). This habitat-related survival pattern was also repeated by increased survival of diatoms with benthic-resting stages (Levinton). Groups associated strongly (Foraminifera, coccolithophores) or weakly (ammonites) with the water column suffered the most, whereas benthic forms (e.g. bivalve mollusks) generally suffered less. Foraminifera species with specialized morphologies and larger size were eliminated, and simpler morphologies were favored (Levinton reference—Keller). Generally, marine sediment processors and shallow-water herbivore species suffered more extinctions than did omnivores or selective deposit feeders (Levinton reference—Smith and Jeffery). Members of food webs less dependent upon plant material (marine deposit feeders and scavengers) suffered less than strict herbivores (Levinton reference—Sheehan). While the darkness, cold, and toxic water environment early in the extinction event damaged marine life most, global warming added to the death and destruction, seemingly to more adversely affect the life on land.

It was a starving time on land; there was little to eat. Plant life on land shriveled in the prolonged darkness, froze, and then baked; 90% of all plants with leaves died. Also, 70% of all land animals died. Notably, large animals that ate the plants died; these large herbivores (grazers) were accompanied by the death of large predators, which even seem to go first. This eradicated the dinosaurs; closely-related terrestrial birds of the Cretaceous also went extinct. Freshwater biotas seem to have emerged relatively unscathed (Levinton). There is evidence from China (the side opposite impact and some distance from Deccan traps) that smaller birds associated with aquatic habitats survived the extinction; the only dinosaur-like life that did survive the extinction event were these fresh-water habitat aquatic birds. Fresh water in-stream inhabitants also suffered less. The conditions of life seem to have made a difference in survival. Small burrowing, insectivorous warm-blooded mammals (with high metabolism) also survived. Medium-sized cold-blooded reptiles, like crocodiles, survived, but the large crocs did not. There was a worldwide fern "spike" found in the fossil record; ferns flourish when all other plants perish. Most life and most types of life (85%) on earth died; life on earth was never the same again. Although Darwin believed that species gradually go extinct, the fossil record does not support this. This was a sudden, massive extinction event.

The conditions of life were even worse at the end of the Permian. The greatest threat to complex life on earth happened around 250mya (just over halfway through the time of complex animal and plant life existence). The earth's greatest-ever volcanic activity occurred in Siberia. Continental basalt floods covered nearly all of Siberia (> 1.6M km^3) and began to liberate large amounts of ash, sulfur dioxide (SO_2), carbon dioxide (CO_2), and iron-rich basalt. Continental basalt floods are often associated with extra-terrestrial impacts. There is some evidence for an extra-terrestrial impact in an area near Australia 250mya. Yet, the ocean floor surrounding

Australia provides only inconclusive findings for this time in earth history. Some shock quartz is present, but iridium (commonly found with asteroid impacts), is not to be found. Nevertheless, the presence of iridium is not necessarily required to document an extra-terrestrial impact.

Interestingly, there does seem to be a magma-filled crater in eastern Antarctica; the 500 km-wide crater lies hidden 2 km beneath the East Antarctic Ice Sheet. Its size and location, in the Wilkes Land region of East Antarctica, south of Australia, suggests that by creating the tectonic rift that pushed Australia northward, it could have begun the breakup of the Permian Gondwanaland super-continent. Geological events of this type often happen slowly. Ralph von Frese, a professor of geological sciences at Ohio State University and Laramie Potts, a postdoctoral researcher in geological sciences, led the team that discovered the crater. Von Frese noted that: "Approximately 100 million years ago, Australia split from the ancient Gondwanaland super-continent and began drifting north, pushed away by the expansion of a rift valley into the eastern Indian Ocean. The rift cuts directly through the crater; the impact may have helped the rift to form." Von Frese and Potts used gravity fluctuations measured by NASA's GRACE satellites to peer beneath Antarctica's icy surface, and found a 320 km-wide plug of mantle material, a mass concentration ("mascon" in geological terms) that had risen up into the Earth's crust. "If I saw this same mascon signal on the moon, I'd expect to see a crater around it," von Frese said. "And when we looked at the ice-probing airborne radar, there it was. There are at least 20 impact craters this size or larger on the moon, so it is not surprising to find one here." He continued: "Based on what we know about the geologic history of the region, this Wilkes Land mascon formed recently by geologic standards, probably about 250 million years ago." "The active geology of the Earth likely scrubbed its surface clean of many more." Von Frese and Potts do admit that such signals are open to interpretation. Even with radar and gravity measurements, scientists are only just beginning to understand what's happening inside the planet. Still, von Frese believes that circumstances of the radar and mascon signals support their interpretation. "We compared two completely different data sets taken under different conditions, and they matched up." The Wilkes Land meteor could have been up to 50 km in diameter, 5 times greater in diameter than the dinosaur-killer of 65mya. The Wilkes Land crater is more than twice the size of the Chicxulub crater in the Yucatan peninsula, and far deeper.

For whatever the reason, there was an unprecedented, massive outpouring of molten lava on the opposite side of earth, covering Siberia, much like the 3 km-thick ice sheet covered it during the last recent glacial period. Peter Ward, in *Gorgon*, tells a fascinating story; Ward realized that something other than an impact, or following an impact, may have caused the actual extinction. Just as iron locked-up oxygen prior to 2.2bya, the iron-rich continental basalt flood would also begin to rust, consuming most of the atmospheric oxygen. This would have caused serious problems for all O_2-acclimated aerobic complex life on earth; previously, Permian atmospheric O_2 levels had been close to 30%. Prolonged Siberian volcanic emissions of atmospheric ash, H_2S and SO_2 caused a further decrease of atmospheric O_2. H_2S oxidized to SO_2. Prolonged relative darkness and a prolonged ice age that was even more severe than the one following the Cretaceous extinction, caused an even more significant sea level drop (200 meters—the lowest level in earth's history since the snowball episodes). This then exposed massive areas of anaerobic organic-rich continental shelf sediments (like those found in the Black sea today), which also oxidized (rusted), consuming even more O_2, and made more CO_2, adding to very large volcanic excesses of atmospheric CO_2, which were likely higher than CO_2 levels of the Cretaceous extinction. The aerobic life in the oceans was doubly poisoned; H_2S is not aerobic-life friendly. Prolonged emission of CO_2 and SO_2 combine with water (and O_2) to make acid; low atmospheric O_2 created severe marine anoxia. Aerobic life needs oxygen for respiration. Under the cold temperatures, the ocean thermocline disappeared and the anoxic hydrogen sulfide-rich layers below mixed with surface layers. This only increased the previously-mentioned marine acidic-toxic-anoxic problem. In this toxic/anoxic marine environment (high H_2S-H_2SO_4-H_2CO_3-HNO_3; low O_2), photosynthesis did not effectively decrease CO_2 and replace O_2 levels. Photosynthesis was impaired by the atmospheric dust, sulfates (sun-block + acid rain), and cold; this impacted both the marine and terrestrial environments. Atmospheric O_2 dropped to 10% and stayed there; the iron-rich basalt floods continued to rust and maintain low oxygen levels. The prolonged anoxia had a profound effect on aerobic life.

After the ash and atmospheric sulfates (acid rain) finally settled out, the sun finally reappeared and the excess CO_2 buildup began global warming (uncompensated—to a degree limited by lock-ups). Recall that methane (CH_4) is a much more potent greenhouse gas than CO_2. Frozen hydrated methane, on the extensively exposed and shallow water areas of the continental shelves, melted, entered the atmosphere, warmed the atmosphere, and then slowly oxidized to form more CO_2. The world (Pangea) then got hotter and drier as a result of massive greenhouse gas emissions; global temperature warmed again to Permian levels and then may have risen still another 10 degrees C. The warmed ocean water continued to release any methane (CH_4) not held in place by hydrostatic pressure. Sea level rose 400 meters, from the 200 m drop. Land plants (not flooded) then died from the oppressive heat and drought. The food chain fell apart; many land and marine animals were asphyxiated during the continuing conditions of high temperatures and anoxia. This was the hottest extreme that the earth had experienced in the history of complex life existence on the planet; this is still the hottest extreme that the earth has experienced in the history of complex life's existence on the planet. Temperatures at the poles rose as much as 25 degrees C. More than 95% of aquatic species and 75% of land species died 250mya; these types were never seen again. Mass extinctions like the Permian environmental disaster had little to do with competition for resources. It was a time when nearly all complex life on earth died.

Considering the conditions of life during the Permian extinction, there were so many extremes, it is surprising that there were any survivors. A better understanding of this might be gained by describing a scenario of just one point in time during the prolonged process. The onset of the marine extinction will provide insight. Examination of a tropical reef ecosystem, just before the marine extinction, would reveal a similarity to the biodiversity of a coral reef today; it would just look different. Today, 90% of the species are found in the tropics; although the tropics may have reached higher latitudes, this is likely similar to the pattern of that time. If the ocean was unmixed, it would resemble the Black Sea of the mid 20th Century, with rich life in the aerated water above an ocean thermocline that was likely around one kilometer (km) deep. Below the Black Sea thermocline lay anaerobic water with hydrogen sulfide gas in solution; H_2S is not at all friendly to most aerobic complex life. As the extinction began, the atmosphere filled with volcanic H_2S, SO_2 and ash. The prolonged relative darkness, acid rain and cold would have killed some land plants, which mainly bordered the rivers, along with some animal life. That would introduce nutrient pollution in the aerated upper surface layer of the ocean, dropping oxygen levels and creating an increasingly expanding dead zone in the formerly productive upper layer above the thermocline. This is similar to what has happened in the Black Sea today. Like today, numbers and biodiversity would increasingly suffer a significant setback. Volcanic dust and sulfates sharply reduced marine photosynthesis; it was a time of prolonged relative darkness and prolonged cold. Primary production all but ceased. All of this continued to change the chemistry of the surface water. The environmentally-sensitive corals and the photosynthetic plant producers, the base of the community, would have been the first to die. The herbivores and their predators would have quickly followed. Resident aerobic reef life would have also experienced prolonged acid rain (snow, sleet, etc.), unprecedented in the history of complex life. Acid rain is tough on aquatic ecosystems; pH changes of this magnitude in the ocean environment had never been experienced by complex life. Marine organisms could not make carbonate shells for protection from predation. By this time, atmospheric oxygen began approaching 10%, less than half of today's level. This atmospheric problem added to the already low oxygen partial pressure in the water and made aerobic survival increasingly more difficult. By the time the warm tropical water got as cold as water can get, most of the tropical reef sea life had already been immobilized from the cold and died from this, or the increasing anoxia, or the increasingly toxic mixtures in the water. Their death only added to the decomposition-induced oxygen debt of the water. The cold then led to thermocline breakdown; oxygen-starved waters above the thermocline mixed with the totally anoxic, sulfide-rich water below. Any traces of oxygen in the upper layers immediately disappeared, for a very long time. Recently-enhanced quantities of hydrogen sulfide from below poisoned whatever aerobic life that had survived to that point. The water's pH then dropped even lower (acidity worsened). The cold introduced an ice age that led to a sea level drop of around 200 meters (Levinton *Genetics, Paleontology & Macroevolution*—chapter on Patterns of Diversity, Origination & Extinction). It was not a good day for complex aerobic aquatic life.

The end-Permian environment was extremely hostile to life; most all complex life forms just died. Over a period of about 80,000 years, there was first some extinction on land; this occurred over a period of 40,000 years (cold). The land extinction was then followed by an abrupt, massive, marine extinction that killed virtually almost all life in the ocean (thermocline breakdown). The marine extinction was then followed by a prolonged (40,000y), second, more massive land extinction that occurred over the remaining period of time (global warming). Between the marine extinction and the second land extinction, large quantities of biological carbon-12 were liberated into the atmosphere; this was likely from frozen methane. In these anaerobic conditions, Ward notes that H_2S was likely to be present in the atmosphere as well, possibly in significant quantities; this gas is also highly toxic to aerobic life on land. Ward noted that the predominant land plants were fungi, characteristic of plant degradation. Plants had succumbed to acid rain, global ice age cooling, (partial) runaway global warming and extreme dryness; but, some made it, possibly through seed/spore resistance. Animal life could not meet the extreme demands of the climate, food chain breakdown and the anoxic, likely toxic, environment. For a while, life all but disappeared. Marine sediments from this time are undisturbed (the perfectly laminated layers are barren), indicating a lack of plant activity and burrowing. Yet, life endured; it's likely that cyanobacteria were responsible for getting photosynthesis production and CO_2 lock-ups functional once again. This reduced atmospheric (and aquatic) CO_2 and limited any further greenhouse warming, causing a cooling to more normal temperatures. After some time, nearly another 100,000 years, limited types of survivors increased in number (like an arctic summer) and later diversified into new species, fitting into new niches.

What kind of plant or animal life could survive such an environmental ordeal? Could the preservation of favorable variations and the rejection of injurious variations have played any role at all? Certainly some generalist scavengers of small size could survive on land, if they could get enough oxygen (small size helps). Atmospheric oxygen partial pressure was near our survival limit today (which is 10%, without supplemental oxygen; even @ 15%, supplemental O_2 is needed to work); any exertion would be extremely difficult (Ward—*Gorgon*). *Lystrosarus*, a Permian mammal-like reptile the size of a sheep, somehow survived. Following the Permian extinction, Triassic fossils are associated with an abundance of *Lystrosarus* fossils. Peter Ward, still continuing the story in *Gorgon*, noted that Greg Retallack had presented a paper suggesting that "*Lystrosarus*, virtually alone among the mammal-like reptiles, had an enormous rib cage and most probably large lungs within it." It was the one of the few candidates able to function in the relatively anoxic environment. This sounds very much like the preservation of favorable variations to me. Was the Permian extinction just another example of natural selection (survival of the fittest) playing the solo role, as always, throughout time?

Objectively, it is very difficult to believe that "the conditions of life" weren't the dominant overriding factors involved with the extinction and evolution of life during (and following) the Permian and Cretaceous extinctions. Natural selection genetic advantage was surely involved as the key to survival in the struggle for existence, but the environment was an oppressive gatekeeper and nearly all life found their passage to the future blocked and locked. No complex life survived the most severe environmental extremes. Most life could not adapt to areas where even the least severe environmental extremes were encountered. Most life (>90%) could not pass into the next time. It helps to have the right key. If natural selection (the preservation of favorable variations and the rejection of injurious variations) is contrasted to the conditions of life (everything that affects an organism during its lifetime, the circumstances of existence, the environment), the conditions of life do seem to be very important in the process of evolution.

8

Extinction Events Initiate Change

Major environmental (mass) extinction events have occurred more than twice. Paleontologists define (name) periods between major extinctions by representative animals and plants in the fossil record. In every mass extinction event, the meek do inherit the earth; small scavengers survive. Survivors adapt and evolve. In the following adaptive radiation, grazers (herbivores) get bigger and/or faster to escape predation. Predators get bigger and/or faster to become more deadly. The arms race continues until the next mass extinction event; then, the large herbivores and top predators die, as the others died in the previous extinction. The specialists are also gone (again). And, once again, the meek do inherit the earth; a new and different cycle of life forms begins.

Following the Permian extinction, rust-related anoxia continued for another 50my (million years); atmospheric oxygen remained below 15% (Ward—*Gorgon*), equivalent to an altitude of 15,000 feet or more today. During this anoxic time known as the Triassic, ancestral dinosaurs (+ mammals) evolved. Dinosaurs likely had lungs like a bird, or vice versa. Birds can fly over Mt. Everest (30,000 feet). They can do this because their peculiar lungs have better ventilation than a mammal lung; their lungs have numerous air sacs extending throughout their body, even into their bones. The air-sac system is around 3 times as efficient for low atmospheric O_2 utilization. The hemoglobin of birds is also better-adapted to function in low oxygen environments. No doubt natural selection genetic change keyed-in these adaptations for the anoxic conditions of the environment. Nevertheless, the conditions of life played an equally significant role. At the end of the Triassic, there was another extinction event; it was another profound mass extinction. Conditions of life at the end-Triassic extinction were once again devastating for the mammal-like reptiles and the newly-arrived mammals (Ward—*Gorgon*). Several asteroids struck earth 214mya; there was extensive volcanism. Once again, atmospheric oxygen fell to 10%; it may have dropped even a little lower than end-Permian levels. This time the abundant *Lystrosarus* succumbed to the profound environmental extinction event (Ward— *Gorgon*). Lasting nearly 40my (million years) is an achievement for any genera that existed during this prolonged period of anoxia. No species (or genera) lasts forever; over 99% of all species that ever lived are extinct. Few species last over 20my; average species endurance is 1–2my. During the Triassic extinction, 35% of all animal families were lost; yet, in all animal families, only 25% of the land species were lost. The marine environment (again) suffered most; 80% of the marine species went extinct. Ocean life went through another marine toxic-anoxia episode. Other environmental changes occurred. The supercontinent of Pangea began to split apart, so rainfall on land increased; both helped dinosaur evolution. The Pangea breakup isolated dinosaur populations, limiting migration to more favorable environments, and laid the groundwork for adaptations to occur in the created changing, isolated, environments. Increased rainfall greatly expanded the amount of plant life (food) available; atmospheric oxygen levels rose sharply in the Jurassic. The environmental change allowed oxygen-efficient dinosaurs to reach their largest size in the late Jurassic. Any size increase first required increased oxygen availability, which allowed herbivores to increase in size and escape predation. Then, predators also had to increase in size. One thing for sure, there is a major environmental role, in addition to a natural selection survival of the fittest competitor, involved in the evolution of complex life on earth.

There seems to be about a 200my pattern of the worst extinction events. The first complex life extinction of the Precambrian, a total-snowball glaciation occurring about 650mya, was of such severity that almost all complex life micro-organisms were completely wiped out; it also affected stromatolites and organic-walled microfossils. Multicellular complex life, as we know it, has only existed about 550my. The most severe multicellular complex life mass extinctions occurred 440mya, 250mya and 65mya. There also seems to be roughly a 50my or 100my pattern of large extinction events; this pattern of events ranges from mass extinctions (> 50% loss of species) to moderate (significant) extinctions. The Cambrian Period (510–544mya) was dominated by marine arthropods. The Cambrian extinction (510mya), when 50% of all animal families were lost, had a climate change and sea level change, suggestive of an ice age. The Ordovician Period (440–510mya) was dominated by marine mollusks. The 2-pulse Ordovician extinction (440mya) had 85% of all marine species go extinct; there was a climate change (ice age, followed by global warming) and sea level changes. In the Silurian Period (408–440mya), bugs (arthropods) were back. Scorpions and plants went ashore; land plants began to increase atmospheric oxygen. The Devonian Period (360–408mya) was the age of fishes; vertebrates gained prominence and invaded the land. The first fern-like trees appeared alongside streams. In the Devonian extinction (360mya), 30% of all animal families were lost and 70% of all species (mostly shallow marine) went extinct; there was an ice age. Bugs blossomed again in the Carboniferous Period (286–360mya); some land arthropods were huge. Forested swamps of giant horsetails, mosses, and ferns made most of the coal found today. The Permian Period (248–286mya) belonged to the reptiles. Conifers covered the land. Previously discussed, the Permian extinction (248mya) had over 50% of all animal families and over 90% of all species lost; there was a massive asteroid, extensive volcanism and their toxic/climatic squeals; it was strongly marine toxic and anoxic, and much the same on land. During the Triassic Period (208–248mya), 10% atmospheric O_2 supported dinosaurs, but kept mammals small. The Triassic extinction (214mya) lost 35% of all animal families. The moderate extinction in the early Jurassic (183mya) was mostly aquatic (toxic/anoxic); there is evidence of frozen methane release into the atmosphere and subsequent global warming; continents increasingly separated (promoted rainfall, plant growth, population isolation); volcanism was prominent; and, increasing atmospheric O_2 levels were advantageous to the dinosaurs. At the close of the Jurassic (144mya), as flowering plants (and pollinating insects) suddenly dominated the landscape, there was another impact-related moderate extinction event (O_2 levels temporarily dropped); the biggest-ever dinosaurs disappeared. Cretaceous armored herbivores and super-predators (e.g., *Tyrannosaurus rex*) developed from smaller Jurassic ancestors. Nocturnal mammals stayed small, but became brainy. Previously discussed, the massive Cretaceous extinction (65mya) had an 85% species lost; a large asteroid, extensive volcanism, and very toxic/climatic episodes killed the dinosaurs, except for the ones that still live on as birds. **The degree in severity of extinction seems to parallel the degree of habitat destruction.**

9

Chicken Little Was Right

The Canadian Shield (greatest expanse of ancient rock on the surface) has over 50 extraterrestrial impact craters. One is a massive impact crater (200 km in diameter) that struck the Sudbury diamond mine area 1.85bya (billion years ago, or 1,850mya). Another impact crater was caused by an impact event 214mya, in present-day Manicouagan, Canada. The impactor was a 5 km meteor; the crater diameter is 100 km. It is believed to be one of several responsible for the mass-extinction found at the Triassic-Jurassic boundary layer. There was another impact event in Morokweng, South Africa, 145mya. The crater diameter is 70 km; it is believed to be responsible for the moderate (significant) extinction found at Jurassic-Cretaceous boundary layer. Around 74 million years ago (9 my before the killer asteroid ending the Cretaceous struck earth 65mya), a 2 km meteorite created a 37 km diameter crater in Iowa (Manson). This is the largest intact, on-shore meteorite crater in the United States. Everything within 200 km of the point of impact was instantly engulfed in flames. Trees were toppled 300 km away and standing vegetation was devastated within 600 km of the impact site. Most animals over 1000 km away did not survive the Manson impact, and dinosaurs over 1300 km away would have been knocked off their feet. The impact caused a world climate cool-down that had far more than a local effect; sea-level dropped (ice period?) and there was a global warming that followed. The climate of North America became more arid. Dinosaur diversity, particularly in this area, may have never quite recovered; there is evidence of depressed biodiversity (in this area) when the mass extinction occurred 9 million years later. This Manson impact was the first of three final, fatal upheavals affecting the dinosaurs' conditions of life. Some (Keller) believe there may have been more than one killer asteroid at the time of the dinosaur extinction 9 million years later. Even if there wasn't, the Deccan traps (65.5mya) occurring (geologically) not long after the Manson impact on the opposite side of the earth, followed by the Yucatan impact (65mya), left no conditions for dinosaur life to continue past that time. These three occurrences changed conditions of life on earth for many lives. Small burrowing mammals and aquatic birds survived.

There have been other extra-terrestrial impacts since the time of dinosaur destruction. Lesser (less-significant) mammalian and marine extinction events of 55mya (Paleocene extinction) and 34mya (Eocene extinction) are associated with a local iridium deposition, indicative of an extra-terrestrial impact. Iridium is uncommon on the earth's surface, but commonly associated with extra-terrestrial impact. Also associated with the iridium deposits is evidence of global cooling, sea level change and global warming; both of these lesser extinction events demonstrate less environmental change than seen in the moderate (significant) Jurassic extinctions, and demonstrate far less environmental change than that seen in the Cretaceous or Triassic mass extinctions.

There may have been as many as four Cambrian extinctions; recall 50% of all families were lost, equaling the benchmark number of Permian family loss. Was the environment as harsh during the Cambrian extinction(s) as during the Permian and Cretaceous extinctions? The major animal Phyla were well established by then; any changes that occurred in later extinctions were below this taxonomic level (see chapter: Cambrian Explosion? in Levinton's previously mentioned text). The environment of the two-pulse Ordovician extinction was supposedly even harsher than the Cambrian extinction environment; it is rated as the second-worst extinction of all known extinctions. Recall that the Cambrian and Ordovician extinctions were associated with sea level changes (suggestive of an ice age and global warming); large complex life was still confined to the marine environment. Extra-terrestrial bombardment seems to be the most likely cause of the Cambrian and Ordovician extinctions.

The location of the impact site may make a difference. The earth may have taken a fairly recent non-vital hit. Evidence, if correct, points to a space rock some 5km across having crashed into the Ross Sea near Antarctica around 3mya. This is 10% the diameter of the end-Permian impactor that created the 500 km-wide crater in eastern Antarctica 250mya, midway between the Manson Iowa impactor (74mya) and the Cretaceous extinction impactor (65mya), and comparable in size to the Triassic-Jurassic boundary layer impactor of 214mya. The depression is 250 km in diameter and 850 km deep; this was a major impact. Yet, there wasn't much worldwide environmental change, even if there is evidence of Antarctic ice sheet melting 3mya, and paradoxically, a major world climate cooling at the same time. There was no mass extinction, moderate (significant), or even a lesser (less-significant) event at that time. (Elizabeth Vrba, in a Turnover-Pulse Hypothesis relating climate change to extinction and speciation, did detect an increase in speciation among bovine mammals 2¾–2½ mya, but there doesn't seem to be notable evidence for even a small climate-based extinction 3mya.) The Isthmus of Panama formed 3mya, connecting the Americas and shutting-down a permanent El-Nino, as well as deflecting warm waters towards the poles. The Antarctic impact may or may not have contributed to our ice age that began 2½mya. Atmospheric CO_2 reached the all-time low 2½ mya (160ppm). Earlier glacial periods in our 2½my ice age were more prolonged. There were some extinctions during the 2½my ice age, but they occurred later, in the Pleistocene (close to last 1½ my). And they seemed to be climate-related. Impact latitude does seem to make a difference. If this is true, consider the implications of having the massive Permian Antarctic impactor striking the earth at lower latitude.

Our solar system orbits the galaxy center (of massive black holes) once about every 220my. Our location is within a spiral arm (of many stars) that rotates around our galaxy center, about midway out on the plane of our galaxy. Our star (sun) vibrates up and down every 52my through a cloud of similarly-vibrating stars (in our spiral arm area), which results in increased numbers and increased velocity of the stars in the denser center positions. This pattern places our sun (solar system) in a traffic-jammed crowd of rapidly moving stars every 26my; we are presently out-of-center in this pattern. In 13my, our solar system will return to the more dangerous high-speed traffic area of accelerating stars in greater numbers. This speeding traffic jam of stars can cause disturbances in the asteroid belt between Mars and Jupiter; asteroids displaced from previous orbits could be displaced in a direction towards Earth. Three billion (3 thousand-million) years ago, Jupiter and Saturn aligned enough to send a large number of these asteroids into Earth and the Moon. Recall that the Earth was formed from many collisions of solar system debris in the Earth's orbital area. As the solar system formed the planets, the asteroids were formed by an interference of Jupiter's huge mass, which prevented any proto-planetary bodies from growing larger than 1,000 km. Several thousand of these bodies now form a class known as near-earth asteroids; their orbits cross or come close to the Earth's orbit around the Sun. All asteroids over 5 km diameter have been located and 80% of the asteroids over 1km have also been located and tracked. As many as a thousand asteroids may be at least 1km in diameter, perhaps near the size of the impactors from 55mya and 35mya. A one km (1000 m) asteroid impact is quite capable of causing a lesser extinction event. Asteroid impactors a few times this size really devastate the Earth, causing an environmental disturbance (habitat destruction) of the greatest extremes. Additionally, there are innumerable asteroids and comets beyond Pluto, in remote regions of our solar system; these may present an additional problem.

Some paleontologists see a 26my cycle of extinctions, with varying intensity. If a plot is made from the rate of genus level extinctions, as a function of time over the past 250my, a statistically significant periodicity of 26my is found; the same pattern occurs for the extinctions of entire families. This suggests an extinction event regularity, in what might be assumed to be a totally random process of environmental fluctuation. Yet there is no 26 million year interval between 74mya, 65mya, 55mya, and 34mya (Germany had a small impact 15mya). Considering the impact of 3mya, it even looks more like there could be more of a 10 million year interval of asteroid impacts, some of little consequence (due to lack of size, +/or location of impact). While there may indeed be a 26my statistically-significant correlation between extraterrestrial impacts and extinction events, the earth is certainly not a 26-million-year scheduled stop for wayward asteroids.

Impacts need not be limited to asteroids (or meteorites—small earthly remains of an asteroid). Multiple comet impacts may occur; on July 16–22, 1994, over twenty fragments of comet Shoemaker-Levy 9 collided with the planet Jupiter. Unimagined damage to Jupiter was observed; the damage was severe and long-lasting. It has undoubtedly happened here. Astronomical, geological and palaeontological evidence is consistent with a causal connection between comet showers, clusters of impact events and extinctions (another statistical correlation). Sometimes impacts do only have a local effect. In Tunguska, Siberia, on June 30, 1908, an asteroid or comet (estimated to be 80 m diameter) traveling at supersonic speeds entered earth's atmosphere and exploded (vaporized in atmosphere) over an unpopulated region. The explosion was heard over 1000 km away; millions of tons of dust blasted into the atmosphere and spread worldwide. Researchers visited the site in 1927 and found that, except for few trees standing in a small area of the blast center, every tree within the area of nearly 2000 square kilometers area was knocked down. Still, it is unlikely that such a local event as this would be of evolutionary significance. It only supports the belief in continued extra-terrestrial bombardment, which varies in size and intensity. About a hundred tons of space debris, including comet particles, falls to earth each day. Thankfully, large body extra-terrestrial collisions with the earth are relatively rare.

10

Fire Down Below

Volcanic activity at the surface can affect the conditions of life. The earth's crust has a big long continuous crack in it; it's an undersea crack, and most of the volcanic activity on earth occurs here. Volcanic activity leaks out from this fracture, forming a ridge; this is an underwater mountain range, and in comparison, it dwarfs terrestrial mountains. This continuous ridge is in the ocean basins; it's often found in middle of a basin, but it can be near a continent. It is 60,000 km long, 1000 km broad and is over 2000 m tall. Here the basalt ocean crust is created from rising magma plumes and volcanic activity spreading outward. Associated with the ridge is tectonic plate divergence, which creates a rift zone (valley) in the midline. The seafloor spreads laterally outward, and becomes covered with increasingly thicker sediments. The basalt ocean crust is heavier than a continent's granite crust; both "float" on the earth's hot plastic outer mantle. If the ocean crust and continent crust collide, subduction occurs; i.e., the heavier basalt seafloor crust slides under the floating continental crust and rejoins (melts into) the hot plastic mantle below. In these subduction areas, continental volcanoes release water, carbon dioxide, sulfur (including H_2S and SO_2) and nitrogen. In other areas, rising mantle plumes can also cause volcanic activity in the earth's crust above isolated "hot spots."

In Iceland, the Mid-Atlantic Rift is visible at the surface. The environmental impacts of a very small flood basalt eruption (565 km^2- size of Long Island, NY) in 1783 are notable; this is our only modern-time flood basalt reference. Ben Franklin (American Ambassador in Paris) described 1784 as "a year without a summer." Ash from the eruption blacked out the sky and crops failed across Europe. In a relatively brief period of volcanic activity, toxins emitted by the fissure eruption damaged and destroyed vegetation from the Arctic Ocean to the Mediterranean. Air pollution (ash and SO_2) was so intense that human health was affected and the national death rate increased dramatically in both England and

France. The previously-discussed flood basalt flows associated with the Permian and Cretaceous extinctions were far, far larger than this.

Antarctic ice cores reveal that volcanic activity has tripled in the last 2000 years. Much of the earth's subduction-related volcanic activity on land borders the Pacific Ocean, in a "ring of fire." Tambora was the largest volcanic eruption in historic time. Tambora (Indonesia) erupted in 1815 (150km³ of ash) and resulted in another "year without a summer." Aerosols from the Tambora eruption blocked out sunlight and reduced global temperatures by 3 degrees C (6°F). It snowed during every month of summer in both Europe and America. Famine (e.g. India) was widespread because of crop failures. Krakatoa, a comparatively mild to moderate eruption (25km³ of ash) in 1883, dropped global temperatures 1 degree F for 5 years, temporarily matching the magnitude of the following century's (20th century) global warming rise of ½ degree C (1°F). The "Little Ice Age" (1200–1900AD) cold was associated with a very high frequency of Krakatoa-level volcanic eruptions. Even recently, Pinatabo erupted in the Philippines in 1991 (25km³); temperatures again dropped ½ degree C (1°F), temporarily masking global warming at that time.

A single volcano can cause even greater changes in the conditions of life. A notable world climate change occurred around 75,000 years ago, when "super volcano" Mt. Toba in Sumatra (Indonesia) erupted. At that time, 800 cubic kilometers of ash (Mt. St. Helens had 1.2km³) caused a "volcanic winter" that lasted 6 years. The volcanic winter was followed by, for mankind, 1000 years of the coldest time in the present ice age on record (during Wisconsin ice period). Over 90% of sunlight was blocked; plant growth stopped. Temperatures dropped as much as 15 degrees C (30°F). Rainfall ceased. From an estimated near-million humans on earth, genetic studies suggest that less than 10,000 of our ancestors survived. The survivors (with the greatest genetic diversity today) were located in equatorial East or Southeast Africa. Ocean currents (Easterly Equatorial Countercurrent) and upwelling atmospheric conditions along the equator dispersed the ash cloud to higher latitudes. While 15 cm of ash fell on southern India, equatorial Africa's more remote location would have received far less. Later on, as temperatures fell sharply, the equatorial proximity contributed to moderating the severe cold environmental conditions. Even though the glaciers never approached the equator, ice age cold and drought challenged our survival. Lake Malawi (East Africa-550km long and 700m deep), the third largest lake in Africa and ninth largest on earth, was reduced to a couple of pools no more than 10 km across and 200m deep. Only a prolonged continent-wide drought could have had this effect. "Survival of the fittest" produced a new kind of man for the upcoming great leap forward (What doesn't kill you...). Did this near-extinction of mankind play a role in making *Homo sapiens* even wiser?

Modern man left Africa once again, heading north and east. Neanderthals and the hobbit-like *Homo erectus* also survived. The near-extinction of man was undoubtedly accompanied by other notable extinctions. Climate dominates the environment.

Yellowstone is another "super volcano." Yellowstone erupted 630,000 years ago, 1.3mya and 2.1mya; is it about due for another eruption? Nearly 400 cubic kilometers of debris were liberated at the last eruption (½ the size of the Mt. Toba eruption). A similar eruption today would immediately kill all complex life within 100km. Thousands of people would die as the superheated cloud of dust, sulfuric, nitric and hydrochloric acid sweeps across Middle America. The volcanic dust would spread across all of America, burying the Midwest. Inhaling the dust with broken-glass sharpness, Marie's disease killed extensive numbers of prehistoric life in far less extensive past volcanic eruptions. Silicosis-induced lung cancer would continue far into the future. Billions of people would die in the volcanic winter and famine that would last several years. The U.S. would not recover anytime soon. The entire world would not recover anytime soon. While it will not cause a mass extinction, or even a lesser extinction event, it could still cause a notable (small) extinction event. Like an ice age Milankovitch Cycle, a loss of numerous species, even genera, could be expected. In order to erupt, Yellowstone's magma chamber needs to be 50% filled with molten rock; its now estimated at 10–30%.

Some of the greatest volcanism the world has seen in the past half-billion years occurred at the end of the Permian and Cretaceous periods; the floods of basalt lava respectively covered Siberia (2M km^2) in an area equal to the size of the United States, and covered most of India (1.5M km^2). Curiously, neither of these massive flood basalt flow areas is located in an ocean basin ridge spreading area. It is possible that the end-Permian impact in Antarctica is associated with the volcanic activity of the Siberian traps? Could the Antarctic impact crack the crust at the opposite pole of the earth and cause Siberia to be covered with lava at the end of the Permian? The end-Cretaceous basalt flows that covered India (Deccan traps) occurred (65.5mya), following the Manson, Iowa impact of 74mya, on the opposite side of the earth. Rocks from the time of the Manson impact show intense volcanic activity. This impact and basalt flood happened before the Yucatan impact of 65mya; did the former impact promote this unusual basalt flood? In both cases, the earth actually split apart and covered entire continental areas with massive flows of lava. Flood basalt eruptions that cover such large areas of the planet are exceedingly rare. Could it be shock wave convergence on the opposite side of the earth? To some, this may seem like a stretch, but is no more far-fetched than the chances of increased occurrences of earthquakes or volcanic eruptions in areas directly under, and following, a solar eclipse, which does seem to happen. Jupiter's moon's Io and Europa demonstrate possible exaggerated examples. Io demonstrates massive volcanic eruptions as it is turning it inside out at a rate of 1cm every 3000 years. Europa's gravitational flux causes extensive fracturing of the 25 km of surface ice as the massive watery tides rise to this surface. The solar system demonstrates even more evidence of shock-wave crust fracture in the following examples. Jupiter's moon, Callisto is covered with impact craters; Valhalla Crater is surrounded by concentric ripples that extend over 1/3rd of the way around that moon; similar ripple patterns are seen around all of Callisto's numerous large impact craters. Saturn's moon, Mimas, is scarred by an impact crater that covers nearly 1/4th of its surface; stress marks on the opposite side indicate that the impact nearly destroyed it. Saturn's moon, Tethys, is girdled by Ithasca Chiasmata, a huge crack on the opposite side of impact crater Odysseus that covers over 2/3rds of the circumference. Uranus' moon, Titania, is completely covered with impact craters and a web of interconnected valleys (chiasmata). Both Siberia and India were near the exact opposite side of the earth at the time of each impact (highly uncommon coincidence?). Could there have been more than one impact, in the area of eruption itself? While dual opposite strikes are possible in both instances, this degree of coincidence would seem to be even more unlikely. There does seem to be a strong association between extra-terrestrial impacts and volcanism. Recall the impact in Canada, 214 million years ago; the impactor was a 5 km meteor; the crater diameter is 100 km. It is believed to be one of several responsible for initiating a mass extinction found at the Triassic-Jurassic boundary layer. Following this, a massive rift (that approximates the mid-Atlantic rift today) poured forth another massive basalt flow around 200mya (similar in size to the Siberian traps); the supercontinent of Pangea split apart and the Atlantic Ocean began to form. Additionally, high levels of volcanism are associated with impacts of 55mya and 34mya, and their respective Paleocene and Eocene extinctions (Levinton—chapter on Patterns of Diversity, Origination and Extinction). There are also data that imply comet/meteorite impacts correlate with Yellowstone eruptions 1.3 and 2.1 million years ago. The impact that caused the Yellowstone "hot spot" probably occurred about 17.5mya (Burchard).

11

Constant Extinction

It is difficult to find examples of extinction due to (fatal competition) natural selection alone. Charles Darwin observed that "neither the strongest nor the most intelligent necessarily survive; survival or evolution depends on response to changing conditions." In other words, species extinction occurs when previous adaptations are no longer suitable to a changed environment. There seems to be a variable pattern of either 10my, or 26my, or 50my, or 100my, or 200my extraterrestrial-based extinction events. Differences in the severity of environmental change from the impact-related event are associated with corresponding differences in the severity of extinction. The case has been made for a constant pattern of impact-related major/massive extinction events, moderate/significant extinction events, and lesser/less-significant extinction events. The case has also been made for volcanism-related environmental change extinctions. These events also range in severity of eruption, with massive eruptions causing massive extinctions, to more frequent (constantly-occurring) eruption events causing noticeable, but small extinction events. Environments are habitats. **The degree in severity of extinction seems to parallel the degree of habitat destruction.**

Scientists have found rather strong correlations between relatively recent climate change, extinction and biological speciation. In a turnover-pulse hypothesis, Elizabeth Vrba (*Macroevolution, Diversity, Disparity, Contigency*) suggested a decade ago that extinction is climate-based (Milankovitch cycle or tectonic-related) and extinction promotes greater possibilities for speciation. Climate dominates the environment.

Notable changes in climate cause noticeable changes in environments. The ongoing ice age of the last 2½my was primarily caused by the photosynthetic depression of atmospheric greenhouse gases and biological carbon sink lock-up. At the present time, the Milankovitch Cycle plays the other major role in determining present climate conditions. Recall that the Milankovitch Cycle will play a major role in an upcoming notable icy extinction. **The extent of the extinction will parallel the degree of habitat destruction.** Climate change can be seen as a prolonged trend of increasingly-severe global cooling, which intensified 35mya with the rise of grasses and the corresponding drop in greenhouse gasses (notably CO_2). Recall the corresponding extra-terrestrial impact evidence from that time (34mya), as well as the evidence of a noticeable extinction event (Eocene extinction); ice formed at the Earth's Polar Regions. There was an abrupt cooling 23mya, likely related to decreasing atmospheric levels of CO_2 (cool wet grass rapidly buries carbon); this was associated with a climate-related noticeable extinction event (Oligocene). The India-Asia collision of 15mya became increasingly significant and added to the long-term trend of global cooling. Africa, previously covered by tropical rain forest, became much cooler and drier by 8mya. Global cooling was even more significant 5mya, a time of peak glaciation and another noticeable extinction event (Miocene). Global cooling was most significant 2½mya, when the present ice age began. Earlier in this 2½my ice age, the extensive glaciation periods could remain for almost a million years. An advancing continental sheet of ice 3 km thick is a major habitat change; it wipes the slate clean of complex life in any area it occupies. It profoundly changes peripheral environments. Nearby katabatic wind speeds may reach 200–300 km/hr. Glacial advancement crushes and freezes anything that stands in the path; melting can cause unimaginable flooding. Extensive glaciation profoundly affects sea level and available atmospheric moisture over the entire planet. With advancing glaciation, climate zones often shift faster than key residents (e.g. plants) can accommodate. Tropical environments are desiccated and repeatedly decimated (10% loss of area). Habitat destruction destroys resident populations; homeless populations cannot survive. **Nothing else is more damaging to a population than the destruction of its habitat.** The most significant life change observed during the Pleistocene ice age of the past 1½ million years was species extinction (Levinton). Even in aquatic environments, shallow-water tropical invertebrates succumbed to the global cooling. The glaciations ended in the late Pleistocene (12000ya) with a final sudden cold spell and glacial advance. During the Pleistocene ice had periodically covered around 30% of the earth's surface (partial snowball) with each 100,000yr Milankovitch cycle. The earth's icy cap now only covers 10% of the surface, and it seems to be shrinking. The pattern that emerges here, and in at least some other intervals, is that there are more numerous, but sporatic, climate-based (noticeable but small) extinctions spaced from less than one to a few million years apart (up to about 10 million years). These climate-based small extinctions have strongly shaped the course of life up to the present. Although not as severe as the world-changing mass, moderate or lesser extinction events, sporadic, small, climate-based extinction events constantly occur, with variable intensities. No environment is permanent; stable environments only last so long. **The degree in severity of extinction seems to more or less parallel the degree of habitat destruction.**

Constant extinction involves more than climate change. Shortly after the Pleistocene glaciations ended, the largest carnivorous predators, like the short-faced bear, saber-tooth cats, and dire wolves, which only ate the largest mammals (mammoths, mastodons, stag-moose, camels, horses, and other large mammals—over 100 species in North and South America), went extinct in a (mammal-record) short period of time. Any evidence of an extra-terrestrial impact is inconclusive; however, a cold spell and evidence of widespread fires in North America do allow for this possibility. Still, Native Americans used fire frequently and extensively, modifying ecosystems to maximize human hunting and/or agricultural advantage (Mann, 1491). Widespread fires would have caused the destruction of existing environments and the creation of new environments; the largest herbivores (giant sloth-browser, mastodon-browser, mammoth-grazer of longer grasses or sedges) were vulnerable. Other grazers were attracted to fresh new growth, controlling grass-eating herbivore distribution. And as the habitat lost cover, the natural predator ambush would have been greatly compromised, making specialized predators the most vulnerable. The extinction dates match man's arrival; Europe, Asia, and Africa were less affected. Man's environmental impact upon arrival in Australia 40,000ya is another dramatic example that illustrates a similar disappearance of megafauna. Relatively speaking, there has been a stable climate since the last period of Wisconsin glaciation (ending just10,000ya). The climatic conditions of The Little Ice Age (1200–1900 AD) were irregular, but still, unusually stable. This climatic irregularity suppressed human populations, keeping human numbers well below one billion. It is said that there is supposedly a baseline "natural" extinction rate of one species per decade. Humans caused 2–3 species per decade to go extinct in the time period from 1600–1850 AD. Human populations increased to even greater numbers as The Little Ice Age ended; the past century ended with a moderate warming. This very recent moderate warming has created a climate optimum that has allowed human populations to dramatically increase over the past 100 years or so (10× over previous 500 years). Even stable environments experience ongoing population expansions and extinctions. Although it may take a very long time for development of a new species to occur, it doesn't take very long, in our present time, to witness daily extinctions of numerous species. Humans and regional human population expansions are now pushing up to 20,000 species per year into extinction (mostly tropical—90% of all species are tropical). Habitat destruction, exotic species introduction, genetic assimilation, pollution, disease, and inappropriate fishing or hunting of wild populations are all human-related causes of extinction today. The number of humans is still increasing, as new areas of wild habitats are occupied and/or modified. This, in turn, only increases the extinction of organisms that formerly occupied these habitats. This daily extinction is most certainly a form of constant extinction. **Today's daily eradication of environments is an accelerated form of the very low-level constant extinctions that occur throughout time,** even in "stable environments." The primary cause of extinction today is habitat degradation (destruction of ecosystems), which is itself caused by human overpopulation. The adaptive radiation following these very recent extinctions of environments is a continually-expanding human adaptive radiation.

12

Ecological Succession

Ecosystems are comprised of living (biological community) and non-living (physical environment) components. In a stable environment, constancy in the conditions of life promotes constancy in the life contained within. A persisting constancy of populations in an ecosystem's biological community forms a recognizable pattern that is known as a climax community; these climax community populations are relatively unchanging, as long as the conditions of life are unchanging. Ecological succession is related to a change in the conditions of life, an interruption of the constant conditions. Ecological succession is primarily focused on plant life, the primary productive (and structural) base of the (animal-inclusive) biological community. Complex plant life seems to be more environment-dependent than complex animal life; however, the animal life is dependent on the plant life. In a given location, the plant life is mostly dependent upon the climate and substrate; it also must reach that location if it is to grow there.

When a small portion of an environment is disturbed, a new environment is created. The newly-changed environment will then slowly return to increasingly look like the environment that existed before the disturbance occurred. The return will begin with a pioneer (plant) species and transform itself through a series of stages of ecological succession, finally ending up as a climax (plant and animal) community that resembles the climax (plant and animal) community it replaced. This is the most common type of ecological succession. Following any environmental (habitat) disturbance, it can be seen nearly everywhere; it is known as secondary ecological succession.

If the environmental disturbance is large enough to cause the extinction of a species, which was limited to this area, that particular population can no longer re-populate the climax community. The climax community will then be different, less diverse, and likely less well-balanced (stable); the difference depends upon that species' function in that ecosystem's web of life. This example of secondary succession, which created a new climax community following the environmental extinction of a species, is not an uncommon occurrence throughout geologic time. It is far too common of an occurrence today in the tropics.

Primary succession involves extensive environmental destruction; it is far less common than secondary succession. It begins with a total lack of organisms and bare mineral surfaces, or only water. Primary succession can begin on bare rock, lava flows, volcanic ash, pure sand, or standing water; it can take a very long time. Such conditions occur following extensive lava flows or when glaciers scrape away organisms and soil, leaving extensive areas of bare rock, only sand, or glacial meltwater. For example, the pioneer community of bare rocks is made up of lichens (algae and fungi). The resultant climax community depends upon the climate and substrate, as well as the ability of life to reach the location, so it can grow and reproduce.

13

Extinction Event Significance

In every habitat (residence), each species has a niche (job). It is uncommon to find open opportunities in life; environmental vacancies generally do not exist. This applies to intra-specific and inter-specific opportunities. The greatest competition for those best qualified for the job exists within the same species doing the same job, but this greatest competition understandably does not lead to extinction of the species; it leads to survival of the fittest. Recall the thoughts of Thomas Malthus and Charles Darwin. Recall that organisms reproduce at a rate to create an excessive number of offspring. Excess numbers of individuals lead to a shortage of resources, resulting in a "struggle for existence" (competition for the inadequate resources) and "survival of the fittest" (a natural selection of more-fit types). Darwin focused on the mechanism of evolution in the steady-state environment, assuming only minor environmental influences. In this steady-state environment, jobs are very hard to find and there just aren't enough jobs available for all to make a living. There are almost no opportunities for filling existing job positions, or finding any new types of jobs. Life and death levels of competition will decide which individual continues in any job; only the strong survive. Job positions in a stable environment do not change very much.

No environment is permanent; change is inevitable. The primary cause of extinction is an environmental elimination. Habitat destruction eliminates environments; habitat destruction occurs throughout time, in varying degrees of severity. Ecological extinction events (habitat destruction) create new environments, with new job openings. The opportunities for candidates increase with the size of the newly-created environment. Mass extinctions (with massive destruction of habitats) create a massive vacuum, with innumerable jobs waiting to be filled. If a mass extinction event has killed most candidate types, an entirely new environment with almost unlimited opportunities awaits the survivors. A moderate, but significant, ecological extinction event will kill fewer types of individuals, so the extent of job openings in the new environment will be fewer than in a mass extinction. Nevertheless, it will be greatly significant to many new types of life radiating into new jobs. Natural selection genetic bias provides life the genetic variation to adapt to a newly-created environment. Constant rate extinction, related to local environmental disturbances (perhaps only a single population's habitat loss), occurs when acquired adaptations are no longer relevant to survival of a species that is confined to a changing environment. Even a single extinction of a species could be important to an organism looking for work. A local extinction of a species, when other population members still exist elsewhere, is referred to as an extirpation. It is also referred to as an opportunity (for others).

Is environmental elimination a prerequisite to evolutionary change? Must it occur to provide a new environment for the radiation of new types? In a chapter on Patterns of Diversity, Origination and Extinction, Levinton concludes that: "In most cases of potentially competing groups, extinction seems to have preceded replacement." Catastrophic mass extinctions (habitat holocaust—like asteroid impact) lead to mass job openings in the new environment. Mass extinctions usually favor the survival of ecological generalists (Levinton). Constant rate extinctions (local habitat loss—like an environmental disturbance causing a single species extinction) also may lead to a job opening in a new environment. Extinction is usually related to loss and degradation of habitat rather than to particular properties of individuals (Levinton). Radiations of superior invading species have strong and variable effects, but do not necessarily drive local species to extinction (Levinton). Convergent evolution results in different species competing in one location for the same job; however, extinction does not seem to be the ubiquitous pattern (Levinton). Throughout time, both environmental and species constancy is the rule (Levinton). Natural selection evolutionary change, from competition alone, occurs between species; but it primarily occurs within the species (Darwin—*Origin of Species*).

Looking at the pattern of evolution, it seems as if the greatest extinction events trigger the greatest opportunity among survivors competing for new jobs, via the environmental connection. Greater environmental changes cause greater extinctions, and greater opportunities for change in survivors. Stable periods of time do seem to involve constant competition with a more gradual evolution of changes; however, these changes occur mainly with constant environmental changes (that still create new job opportunities in the changed environment). With the habitat destruction seen today, the constant-rate environmental extinctions are constantly increasing.

Understanding the cause of extinction helps understanding the significance of extinction, and any related implications to environmental biology. **Extinction is primarily due to the destruction of a population's environment; it is nearly impossible to find any extinction without an environmental connection.** Levinton, in the chapter Patterns of Diversity, Origination and Extinction, cautions that: "The evidence on extinction depresses us into realizing that we have no formula that can relate a degree of environmental change to percent extinction." Not everything needs a formula. Organism range of distribution in and out of the affected area would be a deciding factor in any extinction. Biodiversity (variation) within the species as well as individual organism adaptability are definitely wild cards. How can a formula account for variation in geologic time, variation of location (polar: Antarctic- 5 km impactor 3mya vs. sub-tropical: Yucatan 10 km impactor 65mya), and variation within the location, i.e., existing environmental conditions (including habitat diversity and habitat stability)? Size matters, but other variables also matter. Perhaps seeing a parallel relationship between the loss of habitat and extinction is enough, without creating a "formula." Douglas Futuyma (*Evolutionary Biology*), in the chapter on The Evolution of Biological Diversity addresses the causes of extinction: "Extinction has been the fate of almost all the species that have ever lived, but little is known of its causes. Biologists agree that extinction is caused by failure to adapt to changes in the environment. Ecological studies of contemporary populations and species point to habitat destruction as by far the most frequent cause of extinction." If the authorities on ecology and evolution can agree that environmental extinction (habitat destruction) is the leading cause of extinction today, the case for nearly all extinctions, due to variable degrees of habitat destruction, may have been established in this text. If this is true, biodiversity would then be directly proportional to environmental stability; i.e., species richness is directly proportional to habitat stability over time. Time matters. And, as in real estate, location matters (e.g., arctic vs. tropic environmental variation). So then, biodiversity would be inversely proportional to a location's environmental extremes and directly proportional to habitat diversity; i.e., species richness is directly proportional to favorable conditions and habitat diversity. (The adaptability of a species depends a great deal upon the genetic variation or biodiversity within the species.) Understanding the cause of extinction may help the understanding of environmental biology even more than it helps understanding the significance of extinction.

14

Environmental Extinction and Environmental Creation

The case has been made for habitat destruction as the primary cause of extinction over all time. **The loss of habitat is the loss of the** (necessary) **conditions of** (existence for) **life within. The loss of habitat is the primary cause of extinction.** It is now up to the evolutionists and geneticists with their genetically-determined theories and complicated equations, species-based punctuated equilibrium, variable mutation rates, speciation rates, diversity equilibrium formulas, and even stabilizing or directional selection calculations to prove it otherwise. Limiting factors in the environment today constrain all life. A better understanding of both genetics and the environment is necessary to form the best overall perspective of evolutionary change.

Now arguably established, environmental extinction events are primarily due to habitat destruction; **the destructive process triggers the creation of new environments with opportunities for the survivors.** The destructive process initially creates an extremely stressful environment for life within it. The populations within will vary in ability to tolerate this stressful environment. As the stressful environment eventually moderates and returns to more favorable conditions, newer, more life-friendly environments are created. In a way, extinctions are similar to ecological succession; both involve the creation of a new environment.

Mass extinctions parallel primary succession; the scale is very large and the replacements are very limited. There is likely to be an abiotic time lag. Recall that the end-Permian environment was extremely hostile to life; most complex life was sterilized from the face of the earth. The predominant land plants were fungi, characteristic of plant degradation. Plants succumbed to acid rain, global ice age cooling, extreme dryness and a significant degree of runaway global warming; still, some survived. Animal life could not meet the extreme demands of the food chain breakdown and the toxic/anoxic environment, and, for a while, animal life all but disappeared. Some animals survived. Sediments from this time are undisturbed, indicating a lack of plant activity and burrowing (Ward—*Gorgon*). Eventually, conditions of life became more complex—life friendly. For the survivors, the newly-created environment offered a time of opportunity. There was rapid radiation of new life forms into new jobs. This time was then followed by a time of species stability. Species and environmental stability predominated until the next environmental extinction event, which created another new environment. **Environmental creation includes and follows environmental extinction events.**

In lesser and constant extinctions, if the environmental disturbance was large enough to cause the extinction of a species, which was limited to this area, that particular population can no longer re-populate the climax community. The resulting climax community will be different; the difference depends upon that species' function in that ecosystem's web of life. This example of secondary succession, which created a less diverse climax community following the environmental extinction of a species, is not an uncommon occurrence throughout geologic time. It is far too common of an occurrence today in the tropics. **Environmental creation can also occur in a more steady-state environment.**

15

Environmental Creation Significance

Is there a special significance to environmental creation? Soot created a special environment in Darwin's England that favored the survival of dark-colored moths over the survival of light-colored moths. Today the soot is no longer present; light-colored moths now suffer less from predation. Environmental creation does not necessarily involve a major environmental change. It only requires isolation and changing conditions of life. Recall Darwin's views on geographic isolation. They basically describe the creation of a special environment, surrounded by a barrier or barriers, which then leads to speciation over time; geographic isolation is a spatial barrier to dispersal. Time will also create a barrier and provide a temporal creation of an environment. Climate creates special environments, which can also be a barrier to dispersal. Even the substrate can create a special environment; it can be a barrier to dispersal. These environmental barriers create an isolation of populations. When populations are divided by a barrier, two distinctly different populations (2 separate species) may result from the combined influence of genetic change and environmental change within each isolated newly-created changing environment (due to the creation of separate environments that formerly held a single population). This phenomenon is known as vicariance, and it involves the development of distinct differences between individuals, varieties, or species.

Created environments can be very subtle. While the uninformed may think of the oceans as one environment, this is certainly not the case. In his chapter on Genetics, Speciation and Trans-specific Evolution, Levinton discusses the tremendous variety of mtDNA genotypes of maritime taxa (mitochondrial DNA analysis is easier to manage than nuclear DNA analysis) across the southern Florida biogeographic break. "The marine species barrier is probably explained in the main by the separation of the

Gulf Stream from the southeast coast of Florida as it moves northward." Levinton also cites similar "major biogeographic barriers, such as Point Conception, California, and Cape Cod, Massachusetts, are loci of temperature change and separations of currents that might isolate populations."

Environmental creation may create conditions of life that are either negative or positive for the particular forms of life within. Free oxygen supercharged respiration and made complex life possible as O_2 became increasingly available between 2.2bya and 55mya. Free atmospheric oxygen was approaching today's levels 550mya, when complex life began to get bigger. Atmospheric O_2 levels reached their highest levels (> 1/3rd of all atmospheric gases) for all time on earth 300mya. Bugs became bigger; i.e., land arthropods were less-limited in size. Massive millipedes grew to a length of over three meters (> 10'); they were a half-meter wide. This could not happen today; it is obviously genetically-possible, but it is not environmentally-possible. The bodies of spiders were as large as a man's head; the atmospheric O_2 levels of today (21%) do not permit such a physiological possibility. Cockroaches were everywhere, and some were really large (> 10" long and 5" wide). Predatory dragonflies, flying up into their new highly-oxygenated aerial environment, had wingspans exceeding two meters (> 7'). The first reptiles and the amniote egg also appeared at this time; the moisture-proof skin and egg allowed these vertebrates to truly become land animals. The high atmospheric O_2 levels comprised an important part of the warm and wet Carboniferous Period habitats. This warm and moist tropical environment of the Carboniferous Period (354–290mya), with the high oxygen levels (atmospheric O_2 levels were 40% higher than today), gave way to a cooler and drier Permian Period (290–250mya); reptiles continued to flourish. A slight drop in atmospheric oxygen changed the conditions of life for the large land arthropods. **Life goes on only if the specific conditions of life can be found.** Even though there was no mass extinction at the Carboniferous-Permian transition, the high-atmospheric-oxygen-dependent large land arthropods of the Carboniferous Period were never to be seen again. Smaller arthropods continue to exist in lower atmospheric oxygen conditions.

Geneticist Sean Carroll discusses the remarkable icefish in his introductory chapter of *The Making of the Fittest—DNA and the Ultimate Forensic Record of Evolution.* The South Pole region began to sharply cool 35mya; polar ice caps began to form. At present, this area has the coldest (−2°C or 30°F) and most highly-oxygenated water in all the oceans. Icefish ancestors adapted to these increasingly colder temperatures by producing an antifreeze in their blood 25–30mya. At these low temperatures, red blood cells do not circulate well (viscosity increases); the numbers of red cells in circulation were reduced between 15–20mya. Icefish ancestors totally lost their globin gene function between 7½–15mya. The icefish, with clear blood devoid of red cells, evolved 7½mya; they also lost myoglobin from their muscle cells 2–2½mya, becoming transparent. Carroll does an excellent job discussing many fascinating genetic changes. In the above atrophy of the globin gene example, Carroll seems to actually imply that the environment played an unexpected important role in the natural selection preservation of the icefish. It appears that gene expression requires environmental selection activity in order to prevent loss from disuse. Furthermore, should the water become significantly less cold and less oxygenated, this non-expression of the globin gene could become a fatal mutation. In the next chapter, Carroll emphasizes: "The important messages from… (environmentally-integrated organisms), which will be reinforced many times and in many ways in the upcoming chapters, are that mutations can be 'creative' and that the main limit to evolution is not so much what is mutationally possible, but what is ecologically necessary." **Evolution is genetic success in response to environmental conditions.**

Something is very special about environmental creation; it enables the selective forces of the environment to interact with life's genetic potential. The creation of a special new environment must occur to initiate an important life-changing phenomenon. The driving force of evolution seems to act only under the special conditions of life in an isolated changing environment. **During and following an extinction event, the specific environmental conditions of the newly-created isolated environment select whether or not any existing inheritable adaptations can meet the criteria for life to continue existing in that newly-created, isolated environment. Stable environments also change, and create isolated, changing environments on a smaller and more subtle scale.**

16

The Rest of the Story

If the environment plays such a prominent role in evolution, doesn't the environment deserve greater consideration? **The environment is not passive; it is anything but stable, and it plays a role in evolution that is essential for the process of natural selection.** The environment is the ecosystem, which includes all life inter-relationships in the community as well as existing physical conditions; these are the conditions of life. **The environment is the gatekeeper for genetic success. The gatekeeper of genetic success could be thought of as an environmental selector. If natural selection** (the preservation of favorable variations and the rejection of injurious variations—Darwin) **is the key to survival in the struggle for existence, that key must unlock the environmental gate if life is to pass into the next time,** i.e., continue living in the newly-created environment. Environmental selection strongly influences the biological candidates in the newly-created environment; it sets the ground rules. Environmental selection has the "final word" in any process of evolution. Environmental selection can be a negative limiting or positive supportive influence; it's the active environmental component that creates an advantage, or causes disadvantage and/or death in environmental extinctions. Just one limiting factor, e.g., no oxygen, can make a big difference. **The influence of environmental selection could be only one negative limiting or positive supportive factor, but environmental selection is often from a combination of factors.** Natural selection (preservation of more fit types) is arguably dependent on an environmental selection process; there may be no fit type for a given environment (e.g. no water), so fitness has the dependent position. Gradual change (through natural selection) will mainly occur when the landscape changes (Levinton). The degree of fitness (success) is environmentally-determined; the environment grades dependent adaptation candidates (pass-fail, or A/B/C/D/F). In other words, **environmental selection varies directly with the degree of an individual organism's adaptation to that particular**

environment. This would most favor a maximum adaptation to a specific environment (ongoing specialization to a specific environment is a predominant pattern of life).

Specialization to a narrow environmental role can be highly adaptive and selectively advantageous in a stable environment, at least over a short term. Organisms that live in constant environments normally have a more narrow range of viability than those organisms that endure more varied environments, and have evolved mechanisms to tolerate a wider range of environmental conditions. In more unstable environments (environment is highly variable), environmental selection selects for organisms with mechanisms that produce a wider range of response (selects for that range of variability), which favor environmental conditions. Environmental selection may not control the genetic source, but it influences the success and/or survival of the individual, providing an indirect control of genetic combinations. **Environmental selection strongly influences any natural selection process; environmental selection is a major component of both evolution and environmental biology.**

It is easier to find ecological (than it is performance) causes for success or extinction (Levinton). Habitat specialization may be a blind alley (Levinton—chapter on Genetics, Speciation and Transspecific Evolution). Most paleontologists have recorded the extensive changes in the fossil record as products of forces in the environment (Levinton—chapter on Patterns of Diversity, Origination and Extinction). The geologic record shows a pattern of extinctions followed by massive radiations; the replacements that follow are different types that seem to have the same functional roles as the types they replaced (Levinton). **These extinctions are environmental (non-competitive) extinctions; these massive radiations are environmentally-dependent radiations.** Stasis (lack of change) in a population seems to be the dominant pattern, until extinction (Levinton). Traits with high fitness can disappear in extinction events, simply due to bad luck (Levinton). Natural selection competition (in a stable environment) seems to have a bias for increase in body size over most time (Levinton). During extinction events, environmental selection does not favor increased body size; the reverse is true. It seems that (stable environment) competition, over most time, has a bias towards diversification into specialized roles. During extinction events, environmental selection has a bias towards ecological generalists (Levinton). The underlying variable in all of these cases is the environmental conditions of life. **Selective forces in the environment are variable, and seem to be as important as genetic variation.** There is usually an abiotic time lag following a major extinction event (Levinton), during which time existing adaptations cannot cope until the affected environment moderates. If competition and related fitness were the cause of extinction, there would be no abiotic time lag. Geologic time was previously noted as an environmental phenomenon, resulting in natural selection dependence on that environmental condition. What happens when adaptations needed for survival cannot match the pace of a more rapidly changing environment (e.g. rising sea level on a shrinking island)? Reproductive isolation (e.g. geographical barrier) and time (both environmental phenomena) are required if "natural selection" is to "generate a new species." The day will come when a natural selection of existing environmentally-fit forms will be inadequate to continue any life forms on our planet, due to future extreme environmental conditions (Ward and Brownlee). There is a strong environmental connection to evolution; the environment selects the directional path for life's random genetic variation. It seems as if natural selection's preservation (genetic bias) of better-adapted traits is primarily the result of an environmental selection process. Said otherwise, **natural selection is the result of an environmental selection process; natural selection is the resulting genetic bias.**

Elizabeth Vrba (*Macroevolution, Diversity, Disparity, Contigency*) suggested a decade ago that extinction is climate-based and extinction promotes greater possibilities for speciation (a species-based or habitat-based hypothesis, i.e., speciation follows extinction in lock-step). "The hypothesis of speciation pulses forms a part of the turnover pulse hypothesis…. All species are specific for particular habitats consisting of the physical and biotic resources needed for life and reproduction. Climatic changes result in removal of resources from parts of the species' geographic distributions and therefore in vicariance…. For speciation to occur, physical change must be strong enough to produce population isolation, but not so severe as to result in extinction; and the isolated phase must be of sufficiently long duration for the relevant evolutionary changes to occur. Speciation is visualized as involving sustained isolation or near-isolation of shrinking populations in which habitat resources are dwindling, competition increases, and consequently, strong natural selection promotes new adaptation to these changes." Vrba also notes that in developmental change, there is evidence for both genetic and environmental influence (West-Eberhard discussion upcoming). "The likelihood that all metazoan (complex animal life) speciation involves heterochrony (changes in timing during development) of one kind or another implies a close relationship… with… physical (environmental) change." In vertebrates, the differentiation of cells in regulative development depends mainly on interactions with other cells; this is an environmental control (discussed later). Furthermore, differentiated cells can, at times, de-differentiate into stem cells in altered environments; this flexibility further demonstrates the importance of environment.

Environmental space, and even more so, nutrients are the limiting factors in the tropics. Inter-specific competition is greatest at the equator; 90% of all species are tropical. There are many species, but not necessarily large numbers of each species. Species biodiversity in the tropics contrasts sharply with Antarctica and the Arctic, where limited (species) cold-tolerance permits large populations of thriving individuals, particularly as environmental conditions improve (summer); natural selection (preservation of more fit types) is environment-dependent, not independent. More than ½ of all known species of complex life live in the tropical rainforest; extinction does not occur unless habitat is lost. Through habitat choice, behavior decisions and other ecological issues, organisms alter their environment (Levinton); this allows some flexibility in environmental fitness. It affects local distribution today and it applies to all time. Convergent evolution results in different species competing in one location for a job; what else other than environmental selective forces would cause this? Selective environmental forces make convergent evolution ubiquitous (Diamond—*The Third Chimpanzee*). Predation, an environmental phenomenon strongly affecting herbivores, often plays a greater role in evolution than herbivore competition for limited resources. Competition between predators can be a strong limiting factor for predators; however, these predators are also controlled by their environmental status in the environmental food chain (top predators eat lesser predators). Other environmental limiting factors, such as parasitism and/or disease, can play a significant evolutionary role. Mutualism, such as that seen between tropical shallow corals and their symbiotic algae (zooanthellae) is a (positive) supportive environmental factor for both the coral and the algae (limiting factors are negative). The tropical coral animals are totally dependent upon the algae for nutrition; loss of this relationship becomes a negative limiting factor. The zooanthellae will leave this mutualistic relationship when environmental conditions are unfavorable. Plant evolution is even more strongly related to environmental factors of climate and substrate; plant evolution is also intricately related to herbivores and pollinator activity in the environment.

The conditions of life are the conditions of the ecosystem; the conditions of life include all interactions within both the biological community and the associated **physical environment**. Doesn't this make the conditions of life (environmental or ecological conditions) the final determining factor of natural selection? If intra-specific and inter-specific competition for resources, lifestyles, and space is the stuff of natural selection, isn't Darwin's natural selection of favorable variations deeply connected to, and totally dependent on, the conditions of life (environmental or ecological selection)? Was Darwin wrong to say that conditions of life were far less important to evolution than the fatal competition of natural selection? The conditions of life actually include the fatal competition that Darwin used in his concept of natural selection. There were times when Darwin noted that different variations were adaptive under different "conditions of life"—different habitats or habits. However, he felt that the pressure of competition among species would favor the use of different foods or habitats. Darwin was criticized unmercifully, by both the church and other scientists. It continues, still, even today. In later editions, Darwin reluctantly accepted a then-popular scientific view that the environment directly instructs the organism on how to vary, retreating (under pressure) from the belief that the generation of biologic variation was random with respect to environmental conditions. This caused another problem; the more influence environment has in shaping biologic variation, the less the important it becomes in changing biologic variation through natural selection (genetic preservation of favorable traits). He was more correct in his first edition; biologic variation (genetic inheritance) is not directly generated from an environmental origin. Yet, he may have been close to correct, even in these later editions. Environmental selection affects biologic (genetic) variation; but, it does not form it. Therefore, **the conditions of life, which includes any competition between individuals, have an ultimate influence on the success and/or survival of any biologic variation, even though these conditions do not directly create this variation.** Life has adapted to environments unimagined in Darwin's time; extreme environments limit the life within them to very few candidates. Natural selection today is seen as "a change in gene frequencies in a population, owing to fitness of phenotypes' reproduction or survival among the variants" (Levinton). Natural selection (preservation of favorable traits) is dependent upon environmental conditions which are, and always will be, the final determinant in the fate of all life. Environmental selection provides direction and support to the randomness of genetic variation; natural selection is the resulting genetic bias. **Natural selection is the genetic preservation of favorable environmentally-related traits.** Try to find life without water. Try to find life without an energy source (food). Very high levels of temperature or radioactivity can create a sterile environment. Change the environment (e.g. atmosphere) and see what happens. **Ultimate understanding of evolution, even the understanding of life itself, cannot be achieved unless we seek to understand present and past genetic variation** (natural selection genetic bias), **PLUS present and past environmental variation** (environmental selection—limiting and supporting factors), **interacting simultaneously and continuously.** A discussion of the genetic perspective will soon follow in upcoming chapters.

17

Environmentally and Genetically Determined Evolution

Life interacts with the environment. Life does not function in a vacuum. Life in outer space does not interact; it ceases to function. Every environment is unique, each in its own way. Douglas Futuyama's 3rd edition of *Evolutionary Biology* notes that: "A species is not genetically uniform over its geographic range.... At least some of these differences appear to be adaptive consequences of occupying different environments." This is strongly suggestive of a primary environmental role in evolution. The environment is one of the two primary interacting life-changing entities. **Heredity AND environment are co-dependent.** Natural selection is the genetic preservation of favorable environmentally-related traits (Darwin definition revision).

The process of natural (environmental) selection has been compared to a sieve; yet, a sieve is rigid. Futuyama also notes that: "The environment (the physical and biological factors that impinge upon members of a species, i.e., a population) is in a constant state of flux, varying on time scales ranging from hours to millions of years." This means that all environmental conditions, everywhere they exist, are undergoing constant change (variable). Environmental selection employs a gatekeeper; the gatekeeper is more judgmental (than a sieve), selectively grading the degree of an organism's compatibility to the existing environmental conditions. This grading system restricts each individual organism in its passage through time, creating ongoing restrictions, expressed as organism or population (species) limitations. Therefore, environmental selection is really a more discriminating and dynamic process, with different, changing landscapes, having major geographical to micro-environmental aspects; it is anything but rigid. Integrated into this discriminating, dynamic, process, are many different populations (different species), with each single population (members of a species) of genetically-different individual organisms also in a state of constant change. On occasion, genetic changes within a population (a species), even within an individual organism (genetic expression may be altered by environmental conditions), can be of great environmental advantage. In the marriage of heredity and environment, the gatekeeper has become part of the family. A developing organism with ancestors having previously poor ratings of environmental passage through time can upgrade to first class passage (children will also get a free upgrade). First class passage is dependent upon continued first-class support from the environment. Later, as individuals of an environmentally well-adapted population (species) increase in number, competition for resources increasingly becomes more of a limiting environmental factor. All biologic and physical changes modify the degree of an individual's environmental compatibility, which continually varies the outcome of environmental selection. Quantitatively, environmental selection at a given time varies directly with the sum and/or intensity of environmental limiting factors; the population number (of individuals) varies inversely with the sum and/or intensity of environmental limiting factors.

Environmentally AND genetically determined evolution (change in types of life over time**) results when life comes under the selective forces of an isolated, changing environment; life will adapt or not adapt to that unique environment.** The adaptation process is dependent on harmonious environmental integration into genetic expression, early in development, and the heritable transfer to offspring. The adaptation or any variability in adaptations must be supported in later more stable environments. Genetic isolation, genetic potential for survival and the time to respond to a changing environment before extinction occurs is the recipe for adaptive evolutionary success. Levinton, in the chapter on Patterns of Morphological Change in Fossil Lineages noted that "Morphological evolution occurs, only occasionally, when a population is forced into a marginal environment and is subjected to rapid directional evolution." Said otherwise, **environmentally AND genetically determined change in types of life over time involves an environmental creation, and if it is an extremely adverse environment, death (extinction) occurs; or, on rare occasion, survival results from the interaction of environmental variation** (environmental selection limiting factors**) AND genetic variation** (natural selection genetic bias). Environmental selection initiates a new directional change (from the genetic potential) in the surviving progeny of the heredity-environment marriage. The environment doesn't control genetic variation from mutation, but it does influence genetic expression and genetic survival.

Levinton's chapter on Patterns of Morphological Change in Fossil Lineages notes that: "Fidelity to the environment often supercedes the effects of strong directional selection to jump to a new adaptive mode. The enslavement of organisms to their shifting environments is the essence of the fossil record." In other words, the inability to adapt to a new environment is far more common than the ability to adapt to a new environment. Is this caused by an unchanging (adaptation) commitment to the pre-existing environment? Is change even a greater risk if previous prevailing environmental conditions return? Environmental extinction occurs when acquired adaptations are no longer relevant to survival of a species that is confined to an isolated changing environment (environmental creation). From the perspective of life, significant environmental change in an isolated environment causes an extinction; or, on rare occasion, the changing environment generates an adaptation response.

Levinton, in the chapter on Patterns of Morphological Change in Fossil Lineages, further notes that: "Long periods of time with only insignificant taxa change are associated with unchanging environments such as the aerobic benthic areas from the early Pleistocene to recent times." In other words, constant environments maintain a constancy in structures, functions and species richness; i.e., they do not undergo significant change or extinctions, they may undergo significant genetic change in this unchanging environment. This constancy pattern supports the environmentally AND genetically determined evolution model. There is even further support of the model. Levinton, in the chapter on Patterns of Morphological Change in Fossil Lineages, additionally notes that: "Even when specific taxa change (as a result of significant extinction), functional groups remain the same over long periods of time; predominant physical and biotic environmental constancy leaves little room for expectations of major directional change. This constancy may be behind the lack of common landscape phyletic change." In other words, **there is an interrupted, but general overall pattern of environmental selection similarity. That is, there is a pattern of environmental variability in the evolution of life; it's an interrupted, environmentally-advantageous role repetition. And there is a pattern of genetic variability in evolution; it is an inheritability of environmentally-advantageous traits.**

With environmentally AND genetically determined evolution, stable environments involve less evolutionary change than destabilized environments. This would mean that **environmental selection acts on life at all times; it is especially active during times of an environmental extinction event.** The creation of favorable new environmental conditions is often preceded by the severe environmental conditions of environmental extinction. An extinction event creates a very hostile environment for life. Times of mass extinction involve massive amounts of habitat loss over a short time. Most environmental extinctions are not mass extinctions, but are due to different degrees of habitat loss. Most habitat loss is a local phenomenon and may only reduce biodiversity within a species (population); however, at times, habitat loss is significant enough to eliminate species, even genera. Fatal competition mostly occurs between members of the same population (species); it primarily increases fitness of individuals within that population. Increased fitness does not lead to extinctions of species, but becomes more important as essential resources are depleted from the environment. **Variation in availability of an essential resource constrains the life within that environment; this essential resource in short supply becomes a limiting factor of environmental selection.**

18

Alone

Environmental biology is the study of the conditions of life and their impact on past, present and future organisms. Population isolation is environmental (creation—may be biological). Recall that Darwin clearly recognized the significance of geographic isolation in *The Origin of Species*. Speciation is an isolating mechanism. Species are mainly significant as a breeding population of individuals; i.e., individuals selectively breed only with members of the same species. Isolation restricts genetic flow. This isolation creates a unique environment for that population, which allows environmental selection to interact with genetic potential in that population. The environment that was created by isolation continues on as time passes. With environmentally AND genetically determined evolution, as environmental stress increases, life will adapt, or not adapt, to a changing unique environment, depending on the genetic variation available. Assuming no escape from an isolated changing environment, if the environmental stress exceeds the ability of an organism's normal range of response to manage an environmental challenge, organism survival comes into question. The survival outcome is dependent upon the product of an interaction between environmental variation and genetic variation. For the individual organism and its offspring, this interaction makes all the difference.

The origin of species through natural selection may not be the driving force of evolution. Speciation does not necessarily create significant change (Levinton). Genera of small body size have far more species than those of large body size. If speciation was the driving force in evolution, morphological evolution would be more prominent on small forms, but the reverse is so (Levinton). **The driving force of evolution may well be the product of early developmental organism interactions between environmental variation** (variable environmental selection) **AND genetic variation** (natural selection genetic bias)**, in a changing, isolated environment. The creation of a changing, isolated environment**

must occur to initiate this life-changing phenomenon. Environmental selective forces in the changing isolated environment and their influence on then-present genetic patterns (preserved genetic variation) join and produce the successful change, something new. **In a word, the driving force of the changes in types of life over time is** (genetic) **adaptation** (to an isolated changing, environment).

Could speciation be over-rated? In simple life forms, speciation itself seems to lose relevance. For simple life, fatal competition doesn't cause speciation; it doesn't even fit. Members of a vegetative-reproducing population do not vary from one another unless something occurs that leads to a change in gene expression. The origin of a simple life "species" may be due to a genetic change in expression of a (pheno) type in a changing, isolated environment. If this change in type has survival advantage (an adaptation) it will likely become established and predominate in a vegetative-reproducing population. **For complex animal life, the origin of species is only a behavior adaptation (sexual selection) to minimize genetic incompatibility in organism development and sexual reproduction.** Sexual reproduction within a species constrains (as well as it enhances) complex life's genetic variation, and this aids in formation of a potential adaptation. The conditions of life, including sexual selection, isolate breeding populations in time and space. Under increasingly stressful conditions of life in an isolated environment, organisms utilize existing genetic bias and somehow integrate an environmental influence into expression of a newly-designed (pheno)type. An environmentally-supported change in a developing organism (an adaptation) must also be supported by a breeding population for the change to continue. **Gene pool isolating mechanisms, including sexual selection, results in the origin of species, which genetically supports preferable adaptations.** Speciation in complex animal life is increasingly the result of sexual selection, an ancient, complex animal life behavior pheno(type); speciation is only indirectly associated with (not caused) by other isolating factors in the environment. A species is a sex club; it supports harmonious reproduction and development in complex animal life. Flowering plants have thrived because they are better at sexual reproduction than other complex-life plants. In flowering plants, their sexual reproduction is dependent upon intimacy with co-evolved pollinating insects, often a single species. It was different species of plants that were first described and placed into the taxonomic system; classification was based on differences between sexual components. Botany is fundamentally about sex. However, plant speciation itself can become unclear (discussed in epilogue). In the chapter, Domestication: Evolution in Human Hands (wheat species discussion), of *The Evolving World*, David Mindell states: "The process of hybridization among disparate forms and of increase in the number of chromosomes via duplication (sometimes of entire genomes) is a common method of speciation in plants, and the increase in genetic material allows for the evolution of novel traits and protein functions." **Speciation is the vehicle carrying the changes in types of life over time; however, speciation is far less significant than adaptation to a changing, isolated environment.** In complex-life, the formation of a new species can take a very long time. Identifying a newly-evolved species and documenting an environmental advantage that is only found with this new species would be coincidental. In complex life, the origin of a species may or may not involve a detectable change in expression of a (pheno)type in a changing, isolated environment. A complex life breeding population (gene pool) often includes different (pheno)types, including any new (pheno)type change with a survival advantage—an adaptation; this adaptation will likely become established so as to predominate in the sexually-reproducing population. At all times, environmental selection is the gatekeeper for survival of an adaptation. As early complex animal life gained variation in (pheno)types, a change in behavior (a phenotype change) included a sexual selection of preferable (pheno)types, which led to further genetic isolation. Sex complicates the conditions of life. Geographic isolation affects the variation available to sexual selection possibilities in the isolated gene pool. Fatal competition primarily increases fitness of individuals within the breeding population; this doesn't necessarily originate a new species. Natural selection competition for limited resources affects sexual selection, but it is not the same as sexual selection or competition for breeding rights. Natural selection is the (genetic) preservation of favorable variations and the (environmental) rejection of injurious variations (Darwin—but embellished). Any gene expression change in (pheno)type involves an interaction involving both environmental variation and genetic variation. For any of this to make sense, it will require a better understanding of the phenotype. This requires some knowledge of genetics.

19

Genetic Perspective

Genetics is the science of genes, heredity, and the variation of organisms. In the cells of complex life, genetic activity mainly occurs in the nucleus; it's the control center. At the time of cell division for growth or reproduction, genetic material coalesces into recognizable patterns called chromosomes. Duplicated chromosomes separate to carry genetic information on to newly-formed cells. At other times this genetic material is in an uncoiled configuration of fine filaments (chromatin), to permit enhanced biochemical interaction with its surroundings. The fine filaments of chromatin are made of DNA (deoxyribonucleic acid), itself a coiled threadlike double-stranded molecule that conveys genetic information based on an alphabet with four letters (nucleotide bases). A gene is a section of a chromosome (a linear double-stranded DNA segment) that alphabetically codes for (produces) a protein (polypeptide), using a DNA-similar molecule, single-stranded RNA (ribonucleic acid), in an intermediate step, to accomplish the protein processing. Proteins form the structure of life and do the work of life. A gene may cause the appearance of a trait; a gene may produce more than one trait (pleiotropy); or, several genes may add influence to produce a variable trait (polygenic effect). A genetic adaptation to the environment will often involve forming a new structure that is made from proteins; proteins are made in the ribosome, a two-part intracellular body, outside of the nucleus. As long as an organism is alive, its heredity reacts with the environment; the interaction determines what an organism is like at any given moment as well as what it will become at a future time. If genes make proteins, and the proteins form specific morphological (structural) adaptations to the environment, this anatomical structure can often be identified. The identification of an anatomical structure on an organism can provide a visual link to genetic function. Gregor Johann Mendel (July 20, 1822–January 6, 1884) was an Austrian Augustinian priest and scientist, often called "The Father of Genetics" for his study of inheritence of traits in pea plants. Mendel showed that the inheritance of traits follows particular laws, which were later named after him.

In the Gradualism chapter of *Developmental Plasticity and Evolution*, the world's authority on the book's subject, Mary Jane West-Eberhard notes that: "Fischer (a geneticist) promoted the belief in the primacy of selection over variation as the architect of design, beginning with the premise that all selectable variation is genetic, or mutational in origin. If all new selective variation is (genetic) and mutation is random with respect to adaptation, then selection is completely responsible for the direction of evolution and the origin of adaptive design, the only possible exceptions being the rare examples of environmentally-induced recurrent adaptive mutation and the cases when the range of available mutations may be said to limit the scope of selection. Since these two exceptions are usually regarded as negligible, selection (natural selection genetic bias) is still regarded as by many as being entirely responsible for adaptive design." This is a genetically-determined view of evolution and in her view, it is incomplete. I strongly support her in this. Within the limitations of this incomplete view, many creative and complicated processes have arisen to explain evolutionary change; many findings, particularly from paleontology, are inconsistent with a genetically-determined view. **The environment is also variable; it is also responsible for adaptive design. This fundamentally changes almost everything;** these changes will be introduced, slowly.

20

Genotype and Phenotype

The genotype (genome) is the organism's DNA sequence; the unique assembly of genes inside the nucleus makes the genotype. It is helpful to think of genes as a collection of special chemicals inside a cell nucleus that run the machinery of the cell; i.e., they assemble the molecules of living organisms, using simple raw chemical materials from the environment (such as amino acids) to form the structures (made of proteins) in the cell. There is a variety of building materials and construction plans. Even though the same building materials are often used, the results of the construction project do not necessarily look the same, or even work the same way.

The phenotype is what you see; it is the completed project. The phenotype is the anatomy, physiology, behavior, development, and the expressions of change in the body through life; some of this is inherited and some of this is dictated by the environment. The environment does not directly instruct in the formation of the phenotype; the environmental influence is indirect. Phenotypes are inherited genetic expressions that are indirect responses to environmental conditions; the greatest response occurs from environmental feedback early in life; i.e., during the early development of the organism. It is helpful to reference other definitions and other examples.

The phenotype is the "outward, physical manifestation" of the organism. These are the physical parts, the sum of the atoms, molecules, macromolecules, cells, structures, metabolism, energy utilization, tissues, organs, reflexes and behaviors; anything that is part of the observable structure, function or behavior of a living organism (Microsoft-phenotype). The phenotype of an individual organism is either its total physical appearance and constitution, or a specific manifestation of a trait, such as size or eye color, that varies between individuals. Phenotype is determined to some extent by genotype or by the identity of the alleles that an individual carries at one or more positions on the chromosomes.

Phenotypes are much easier to observe than genotypes; the red color of a flower is an example of a genetic trait. It doesn't take biochemical analysis or genetic sequencing to determine the color of a flower. Classical genetics uses phenotypes to deduce the functions of genes. Breeding experiments can then check these inferences. In this way, early geneticists were able to trace inheritance patterns without any knowledge whatsoever of molecular biology (Wikipedia—phenotype).

The genotype is the "internally coded, inheritable information" carried by all living organisms. This stored information is used as a "blueprint" or set of instructions for building and maintaining a living creature. These instructions are found within almost all cells (the "internal" part); they are written in a coded language (the genetic code); they are copied at the time of cell division or reproduction and are passed from one generation to the next ("inheritable"). These instructions are intimately involved with all aspects of the life of a cell or an organism. They control everything from the formation of protein macromolecules, to the regulation of metabolism and synthesis (Microsoft—genotype).

The relationship between the genotype and phenotype is a simple one; the genotype codes for the phenotype. The "internally coded, inheritable information", or genotype, carried by all living organisms, holds the critical instructions that are used and interpreted by the cellular machinery of the cells to produce the "outward, physical manifestation", or phenotype of the organism. Thus, all the physical parts, the molecules, macromolecules, cells and other structures, are built and maintained by cells following the instructions given by the genotype. As these physical structures begin to act and interact with one another they can produce larger and more complex phenomena such as metabolism, energy utilization, tissues, organs, reflexes and behaviors; anything that is part of the observable structure, function or behavior of a living organism (Microsoft—genotype/phenotype relationship).

So, the genotype (genome) is the basic heritable chemistry of life. It can be influenced by the surrounding environment to produce the structure of life; the interaction between the genotype and the environment expresses the phenotype. Said otherwise, the phenotype is an environmentally-integrated expression of the genotype. There can be a range of different environmentally-integrated phenotypes for the genotype to display; this is the genotype's "norm of reaction" (of genotype-programmed inheritable phenotype variations). Environmental influences early in an organism's development have the greatest impact in genotype expression of the phenotype. Differences in the environment during and even after development of an organism generate different environmentally-integrated genetic programs. These environmentally- integrated programs are stored in the heritable genetic hierarchy as the genetic component of phenotype patterns that are environmentally-coordinated for optimum adaptation to frequently-encountered environments. It's an efficient programming that orchestrates maximizing utilization of resources and responses. If a completely new and different phenotype (environmental adaptation) is formed, it can only change form or function by using the already-existing phenotype-forming pathways. **It is naïve to think that because a phenotype is inheritable, that genetics alone creates the phenotype.**

In sexually-reproducing organisms, germ cells for future generations separate early in development, as the body (somatic) cells of the phenotype continue to develop and continue to interact with the environment. Any changes to the body cells of the phenotype cannot be passed to offspring, unless they occur early enough in development to transfer this change to the forming germ cells. The body cells continue to interact with the environment throughout life, but environmentally-integrated changes in expressed phenotypes are on a decreasing scale of frequency after the structures form.

Heritable variation change of the genotype would have originally required mutational change of the genotype (DNA). Furthermore, mutation itself can only change what already exists in the genome. The environment cannot directly control a mutation; however, it can increase the rate of random genotype mutation. Direct mutational change is not the only way to change a genotype. When environmental selection acts on the genome's environmental expressions (phenotypes), positive or negative environmental selection of the phenotype changes the genome heritability, and affects the possibility of genome transfer to offspring. That is, genotype survival or non-survival occurs when the environment interacts with the existing phenotype (displayed by the genotype). Only phenotypes that survive to reproduce continue their genotype's hereditable variation. Said otherwise, only the genotype is inherited, but only the phenotype is environmentally selected for survival.

21

Alleles

For complex life, most of the genetic material is in the nucleus, most of the time as fine filaments of chromatin (a double-strand coil of deoxyribonucleic acid (DNA) that conveys genetic information based on an alphabet with four letters—building blocks). A gene is a section of a chromosome (a defined segment of chromatin, i.e., a linear double-stranded DNA segment) that alphabetically codes for (produces) a protein (polypeptide), using a single-stranded but similar molecule, RNA (ribonucleic acid), in an intermediate step, to accomplish this protein processing. A given gene occupies a particular location (locus) on a chromosome, like an address; locus may be a term even used for the gene itself. In the normal, uncoiled, working state, chromosomes exist in pairs (like a pair of shoes); these pairs match each other, but not exactly. Opposite genes on a pair of chromosomes may be quite different. Complex-life sexual reproduction produces one maternal chromosome, and one paternal chromosome (like a right and left shoe); the children of sexually reproducing parents inherit one chromosome from each parent, which then becomes present in all cells of the offspring. This creates four readily-available gene-locus combination possibilities in chromosome pairing, for each matching pair of chromosomes in the offspring.

An allele is any one of a number of alternative forms of the same <u>gene</u> (sometimes the term refers to a non-gene sequence) occupying a given <u>locus</u> (position) on a <u>chromosome</u>. An example is the gene for blossom color in many species of <u>flowers</u>—a single gene controls the color of the petals, but there may be several different versions of the gene. One version might result in red petals, while another might result in white petals. Many organisms are <u>diploid</u>, that is, they have paired <u>homologous</u> (similar) chromosomes in their <u>somatic (body) cells</u>, and thus contain two copies of each gene. An organism in which both copies of the gene are identical—that is, have the same allele—is said to be homozygous for that gene. An organism which has two different alleles of the gene is said to be heterozygous. Phenotypes associated with a certain allele can sometimes be dominant or recessive, but often they are neither. A dominant phenotype will be expressed when only one allele of its associated type is present, whereas a recessive phenotype will only be expressed when both alleles are of its associated type. However, there are exceptions to the way heterozygotes express themselves in the phenotype. One exception is incomplete dominance (sometimes called blending inheritance) when alleles blend their traits in the phenotype. An example of this would be seen if, when crossing snapdragons, flowers with incompletely-dominant "red" and "white" alleles for petal color—the resulting offspring would have pink petals. Another exception is co-dominance, where both alleles are active and both traits are expressed at the same time; for example, both red and white petals in the same bloom or red and white flowers on the same plant. Co-dominance is also apparent in human blood types. A person with one "A" blood type allele and one "B" blood type allele would result in a blood type of "AB". A wild type allele is an allele which is considered to be "normal" for the organism in question, as opposed to a mutant allele which is usually a relatively new modification (Wikipedia—allele).

22

Mutation

A mutation is a new change in the genetic material; mutations are formed by alterations in existing genes. Said otherwise, a mutation is a change in the DNA sequence of a gene; i.e., a mutation is a change in one or more of the nucleotide bases that make up the gene. Mutations in a gene's DNA sequence can alter the amino acid sequence of the protein encoded by the gene. A mutant allele is a modified allele, with a changed gene; i.e., mutations result in the formation of entirely new alleles. The change can be isolated in a small coding sequence, or major structural modifications in chromosomes, and even missing or extra chromosomes. Mutation can only change what already exists in the genome. The mutation, mixed with the recombination of existing parental allele variation, creates the possibility of new genetic expression of the new mutation, creating traits that never existed before (within a population's gene pool). In *The Making of the Fittest: DNA and the Ultimate Record of Evolution*, Sean Carroll emphasizes: "Given sufficient time, identical or equivalent mutations will arise repeatedly by chance, and their fate (preservation or elimination) will be determined by the (environmental) conditions of selection upon the traits they affect."

A formal definition of mutation is often helpful. In biology, mutations are permanent, sometimes transmissible (if the change is to a germ cell) changes to the genetic material (usually DNA or RNA) of a cell. Mutations can be caused by copying errors in the genetic material during cell division and by exposure to radiation, chemicals, or viruses, or can occur deliberately under cellular control during the processes such as meiosis (cell division for sexual reproduction) or "hypermutation" (rapid immune system response to an antigen). In multicellular organisms, mutations can be subdivided into germline mutations, which can be passed on to progeny, and somatic mutations, which (when accidental) often lead to the malfunction or death of a cell, and can cause cancer. Mutations are considered as the driving force of evolution, where less favorable (or deleterious) mutations are removed from the gene pool by natural (environmental) selection while more favorable (or beneficial) ones tend to accumulate. Neutral (no change in phenotype) mutations do not affect the organism's chances of survival in its natural environment and can accumulate over time, which might result in what is known as punctuated equilibrium, a disputed interpretation of the fossil record. Contrary to tales of science fiction, the overwhelming majority of mutations have no significant effect. Visible effects are especially rare, since DNA repair is able to reverse most changes before they become permanent mutations (Wikipedia—mutation).

Even though mutation is seen in the above as the driving force of evolution, it's mostly a dead end. Mutation is the driving force of genetic variation (genotype) within an individual and its offspring. From an individual's perspective, DNA mutations are as infrequent as winning the lottery. Only in a large population, over a period of many generations, does a DNA mutation occur (as in someone will win the lottery sometime). In a unique environment created by isolation, a changing environmental condition of life known as time allows genetic variation (from mutation) to occur within a population. Mutations that are not repaired by genetic moderation may or may not be expressed in the phenotype; the latter (no expression) is far more common. Mutation alone changes nothing in the evolution of new types, unless the mutation changes expression of the phenotype, and environmental selection interacts favorably with this phenotype (i.e., it has adaptive value). That is, the effect of a mutation that has an immediate influence on the phenotype displayed ranges from none to lethal; it is often lethal. Any mutation contribution to evolution is limited by environmental selection of an expressed phenotype. Environmental selection favors the occasional rare mutations that display adaptive phenotypes; environmental selection eliminates the more commonly-occurring, non-adaptive phenotypes. In still other words, the great majority of mutations that do result in an immediate phenotype display are rejected by environmental selection. On rare occasion, when a rare beneficial genotype mutation is immediately expressed as a new phenotype that does have increased environmental survival value, it will become incorporated in the genetic bias of natural selection, which allows the phenotype, and other hitch-hiking genotype mutations (non-displayed), to continue into the future conditions of life. Previously noted, only the genotype is inherited, but only the phenotype is selected for survival.

Simple life reproduces by simple cell division; any favorable mutation-based adaptive process is immediately inherited and propagated by the offspring. As previously stated, when reproduction only occurs in the form of sexual reproduction (as seen in higher forms of complex life) mutations that occur in body (somatic) cells are not passed on. In higher forms of complex life, somatic cells far outnumber germinal cells; somatic cells will experience the majority of mutations. Mutations leading to favorable phenotype display cannot be passed to future generations unless the mutation occurs in the source of reproductive (germinal) cells that survive to produce viable offspring; these offspring will contain the mutation in all cells. Another way of restating the above is to say that vegetative reproduction passes favorable environmental phenotype adaptations directly to daughter cells during cell division, and during sexual reproduction only germ cells pass genetically-inherited favorable environmental phenotype adaptations to offspring. In sexual reproduction, the recombination of parental alleles increases the various combinations of existing alleles (within a population's gene pool); meiosis itself may involve a form of mutation (crossover). Whether or not a genotype mutation is expressed as a phenotype can be influenced by the environment (conditions of life).

A discernable change in the phenotype may occur with changes in the environment, which leads to genotype expressions of more environmentally-suitable phenotypes. During frequently encountered environmental conditions, normal variation in the phenotypes displayed (from the genotype norm of reaction) understandably does not mean that a sudden mutation has just occurred. Yet, on rare occasion, a completely new phenotype change may occur with changes in the environment. And it may or may not mean that a mutation caused this change. (Supposedly, the environment can increase the rate of random genotype mutations, as in the above "hypermutation" example; however, in reality, hypermutation probably doesn't even exist.) When a new phenotype is expressed by the genotype, it can only change form or function by utilizing the expression of existing genes. The environment cannot directly control a mutation; nevertheless, under stress, the environment can cause the formation of random changes in genetic expression. So, the environment indirectly initiates normal environmental modifications (in the plastic range of expression) as well as unique changes in genetic expression under unique stress (also in the plastic range). In an isolated stressful environment, the formation of unique phenotypes from (the plastic range of expressing) the genotype's norm-of-reaction would not necessarily involve great changes in the genotype's environmentally-connected displays; however, there may be wide variation in the random phenotypes expressed. Under severe stress, there may even be a unique expression outside the norm-of-reaction, expressing a mutation that occurred long before this particular random phenotype display was formed. This unique phenotype display may even become a new adaptation (environmentally-integrated phenotype); that is, the environment that caused the generation of random genotype expressions may even preferentially stabilize the random new phenotype. And if the new phenotype has this environmental support (increased advantage in the isolated, changing, stressful environment), the new environmentally-integrated expression may become integrated into a new norm of genotype expression (adaptation).

At one time in genetic studies, any discernable change in a phenotype was called a mutation. In Douglas Futuyma's *Evolutionary Biology* text, he states that: "A gene mutation is an alteration of DNA sequence…. Before the development of molecular genetics, however, a mutation was identified by its effect on a phenotypic character… that is, a mutation was a newly arisen change in morphology, survival, behavior, or some other property that was inherited, and could be mapped (at least in principle) to a specific locus on a chromosome. In practice, many mutations are still discovered, characterized, and named by their phenotypic effects. However, not all alterations of DNA sequences have phenotypic consequences. Hence, in a molecular context, the term 'mutation' refers to a change in DNA sequence, independent of whatever phenotypic effect it may have." The old practice of equating mutation with a change in the phenotype opens the possibility for a major misunderstanding and misuse of the term. Because of the possibility for misunderstanding when mutation is used in the old context (the old practice of equating mutation with a sudden change in the phenotype), normal phenotype variations as well as any newly displayed phenotypes (reactive random expressions to a stressful environment), could be confused with the formation of a sudden mutation. It is helpful to be aware of this source of confusion.

In an effort to overcome this old terminology confusion, mutation will (again) be redefined. In *The Plausibility of Life,* authors Marc Kirschner and John Gerhart define mutation as: "A change in the sequence of A, T, G and C elements (nucleotide bases) in DNA, due to causes such as chemical or radiation damage to DNA, unrepaired errors in the replication of DNA, errors of recombinations of DNA strands, movement of virus-like DNA sequences to new sites, and insertion or deletion of pieces of DNA. These changes occur randomly in the DNA sequence." However, they also define a mutant as: "An organism that has undergone a genetic change, usually one that has observable consequences." Problems can arise when the time of a phenotype change is considered as the time of the genotype change. This problem in understanding the timing of the mutation and the timing of expressing the mutation still creates confusion; redefining an ambiguous use of mutation only redefines the confusion. It is helpful to remember that the great majority of preserved mutations are silent; i.e., they have no influence on the phenotype (like a neutral mutation) unless they are actually expressed by the mechanism that forms the phenotype. Any expressed neutral mutation, which has no discernable effect on the phenotype, effectively remains neutral (silent). When mutations produce recessive alleles (that are seldom expressed in phenotypes), they are only expressed in low numbers, and become part of a vast reservoir of hidden potential in genotype variability that can show up in future generations. Whether a mutation is dominant or recessive depends largely on the role of the affected protein in the cell. Whether or not a genotype mutation is expressed as a phenotype can also be affected by other mutations in the organism.

23

Phenotype and Environment

The phenotype is any detectable characteristic of an organism (i.e. structural, biochemical, physiological and behavioral) determined by an interaction between its genotype and the environment. Many phenotypes are determined by multiple genes and influenced by environmental factors. Thus, the identity of one or a few known alleles does not always enable prediction of the phenotype. The interaction between genotype and phenotype has often been described using a simple equation: genotype + environment → phenotype (Wikipedia—phenotype). Said otherwise, an organism's phenotype is what you see when looking at it. The genotype is inside the organism; genes are locations on strands of DNA that code for amino acids to make proteins, and are difficult to see (recall Genotype and Phenotype chapter). What you see, the phenotype, is determined by these nearly-invisible genes interacting with nutrition, temperature, social conditions and many other aspects of the environment. As an example, Siamese cats and Himalayan rabbits are both animals that have dark colored fur on their extremities. This pattern is caused by an allele that controls pigment production, which is only able to function at the lower temperatures of those extremities. The environment interacts with the genotype to determine the phenotypic pattern of genotype expression. The environment does not directly interact with the genotype; its interaction with the phenotype is indirect. Environmental conditions indirectly affect the type of phenotype that is displayed by the genotype; they indirectly determine non-random environmentally-connected genetic displays as well as the unique random displays to unusual stress, and the phenotype stability. The environmental conditions directly affect survival by environmental selection. The indirect environmental effect on the phenotype is greatest early in development. The phenotype is a heritable trait with an environmental component. **It is naïve to think that because a phenotype is inheritable, that genetics alone creates the phenotype.**

Acclimation is a (familiar phenotypic, norm-of reaction) change occurring in an (adult) individual as a result of prolonged exposure to a particular environmental condition, such as a <u>horse</u> shedding its winter coat to produce a lighter summer coat; acclimation is a tool in maintaining a being's natural <u>homeostasis</u>. In general, acclimation is a short term response to an environment (that is) expressed within one generation of an organism, as opposed to adaptation which is a long term multi-generational response to an environment (Wikipedia—acclimation). The genetic information for developing the (familiar, norm-of reaction) phenotype characteristic is already present in the organism; all it takes is the proper environment for the phenotype to be expressed. In other words, somatic change (body change) as a result of functional adjustment (acclimation) to environmental conditions cannot be inherited unless the genetic information is already present in the cell's genetic code. Functional phenotypic change (acclimation/somatic "adaptation") in adult humans is demonstrated by the stress of high altitude first increasing breathing and heart rate, and eventually increasing red blood cell counts. In another example, the stress of physically-demanding environments will necessitate greater amounts of exercise, accompanied at first by increased adrenaline and heart rate, and eventually resulting in somatic changes such as increased muscular blood supply, muscle mass, bone density, and cartilage thickness. In yet another example, stress from reduced food supply first promotes some improved efficiency in food utilization, but eventually affects the body size of the developing young; there seems to be a correlation between calorie intake and body size, with no noticeable differences in gene frequencies, in of both humans and animals. These are all norm-of-reaction responses to familiar environmental stress, not random changes from an unusual environmental stress.

Early in development, complex-life individuals with the same genotype often become different in phenotype when they encounter different environmental conditions of temperature, latitude, light, water currents, food source, predation or any other external factors. There is a range of responses from phenotypic plasticity (where the same genotype produces two distinct variants under two distinct environmental conditions), through genetic polymorphism (codominant alleles and dominant alleles), to no phenotypic variation (even with marked genotypic difference). In the case of genotype variation that demonstrates no variation of the phenotype, (environmental) selection has no variation on which to act; it is of obvious environmental value to elaborate a single head, despite variations in environmental temperature (Levinton—chapter on Patterns of Morphological Change in Fossil Lineages). Other traits, such as surface epidermal melanin, hair density, and hemoglobin concentration in the blood are not canalized (unchanging), but are phenotypically plastic (Levinton). In the same chapter on Patterns of Morphological Change in Fossil Lineages, Levinton continues: "One must be careful to distinguish between the genetic architecture of phenotypic variability involving a multi-allelic system (number of genes that add for effect) and a single genetic program that allows a single genotype to respond to different environments (variable expressions to environmental signals) by producing different phenotypes. Such phenotypic plasticity allows an individual to increase in fitness during its lifetime, thus reducing the evolutionary cost of having an inappropriate phenotype when exposed to a given microenvironment. Phenotypic plasticity is a measure of the evolved phenotypic response to the environment." In multi-alleic systems, characters (traits) are influenced by several genes; in the simplest instance, this multi-alleic (polygenic) effect is from the simple addition of the genes for the total effect, as in height of humans. In any real population, each genotype would vary in height from the polygenic addition as well as from environmental interaction; this is known as continuous variation. Further complexities occur if one gene locus (location on a chromosome— including the alleles) governs the expression of another (epistasis). A special form of epistasis occurs in threshold traits, which are characters that may occur in two (or more) discrete (absolute) states where the phenotype is controlled by multiple loci (genes), rather than a single pair of dominant and recessive alleles.

So, an organism's phenotype is determined not only by the conditions of life at the moment, but also by the whole succession of environments experienced during its lifetime. The greatest amount of environmental influence on the phenotype occurs in early development. Differences in organisms of similar heredity are referred to as environmental variations or modifications. From the population damage observed as a result of environmental stress, it is possible to determine the extent of adaptations available from the norm-of-reaction phenotype variations. The phenotype is a heritable trait with an environmental component. It is naïve to think that because a phenotype is inheritable, that genetics alone creates the phenotype.

The norm of reaction for a genotype is the variety of different phenotypes it can express in different environments. Different genotypes often differ in their norms of reaction. For example, genotypes at the *Bar* locus in *Drosophila* (fruit flies) differ in the effect of temperature on eye size. *Infra-bar* and *Ultra-bar* genotypes have very different phenotypes at high temperatures, but have the same phenotype at low temperature. Therefore, not only can a single genotype produce different phenotypes under different environmental conditions, a given phenotype can be produced by different genotypes under one environmental condition. "Because most phenotypic characteristics are influenced by both genes and environment, it is meaningful only to ask whether the differences among individuals are attributable more to genetic differences or more to environmental differences, recognizing that both may contribute to the (phenotype) variation" (Futuyma—Evolutionary Biology).

In frequently-encountered environments, the environmental determination of the phenotype (from the norm of reaction) involves indirect interaction of an organism's <u>genotype</u> and the environment. Genetic adaptation to a unique environment also involves indirect interaction of an organism's <u>genotype</u> and the environment. Both processes involve indirect interaction of an organism's <u>genotype</u> and the environment. **If the driving force of evolution is adaptation** (begins with genetic variation of the genotype, originating from mutation, eventually leading to variation in phenotypes, which must favorably interact with a changing, isolated environment), **resulting in a new genetic bias of natural selection** (adding a new phenotype into the norm-of-reaction); **this must be connected to somatic change of the phenotype from the everyday interaction of an organism's genotype and the environment.** It would make sense to think the same cellular mechanisms responsible for forming norm-of-reaction phenotypic display are involved in forming an adaptation change. An adaptation is also a heritable trait with an environmental component. It is naïve to think that because an adaptation is inheritable, that genetics alone creates the adaptation.

Adaptation will be discussed in detail in an upcoming chapter. But any discussion of the phenotype and the environment must include at least something on adaptation. It has been noted that the norm of genotype expression is the variety of different phenotypes expressed in different frequently-encountered environments. However, new phenotype adaptations are associated with difficulty, literally, in every sense of the word. Severe environmental stress that has no adaptive response may cause the formation of new random phenotypes, some even outside the norm-of-reaction in genotype expression. Environmentally-supported random phenotypes will become adaptations, and these adaptations become part of a new norm-of-reaction. Adaptations are far less common, and far more complicated than everyday acclimations, especially in more complex life where only sexual reproduction exists. In a reproductive-environmental comparison, vegetative (somatic) reproduction allows immediate transfer of an acquired adaptation to all daughter cells. Intimate contact with the environment offers advantages and disadvantages. Subtle environmental change over longer periods of time will enhance cellular adaptation possibilities to this environment. Greater environmental stress over shorter periods of time will be more damaging, both for somatic reproduction and sexual reproduction. However, there is no genetic difference in daughter cells of vegetative reproduction, so an environmental selective agent that is fatal (non-adaptive) to this genetic expression will more easily kill this entire population that cannot vary genotypes.

Sexual reproduction offers the possibility of greatly varied genotype combinations; this increase in individual biodiversity helps to somewhat compensate for the external environmental feedback isolation of germinal cells; this isolation can also offer some protection from an adverse environmental selection of the germ cells, providing the somatic cells of some individuals manage some degree of survival and reproduction. In gamete forming times, and in early stages of development, the complex life developing from fertilized germ cells may still have the option of responding to signals in the internal environment of the body that contains them. Under non-recognizable environmental stress, it is likely the response will include random genetic expressions of variable phenotypes, mostly from the genotype norm-of-reaction. In the above example of temperature effect on fruit fly eye size, it would seem that increased eye size has no adaptive advantage to an increased temperature; it is only a random change in expressed phenotype. A change that offers environmental advantage, an adaptation, can be integrated into the previous norm-of-reaction. In sexual reproduction, any adaptation continuity requires genetic possibility in the germinal cells, in order to have the adaptation potential continue in the developing offspring. Upon fertilization, the earlier stages of a developing organism are highly sensitive to any environmental stimulus, providing great possibility in genetic expression (random changes) of the phenotype; this environmental sensitivity decreases as maturation increases.

Unusual environmental stress generates random changes in the phenotypes displayed. It will not necessarily be an adaptive change; in fact, it is usually a non-adaptive change. In *The Plausibility of Life*, authors Marc Kirschner (chair of Systems Biology @ Harvard) and John Gerhart (professor in Berkeley's Department of Molecular & Cell Biology) discussed experiments (in the chapter on Physiological Adaptability and Evolution) where unusual forms of environmental stress led to random changes in the phenotypes (expressed by the genotype), some of which were outside the norm-of-reaction. The classic study was first performed by Conrad Waddington on fruit flies. During embryogenesis, he stressed fly populations with high temperature environments, as well as to ether fumes. He found that many changes in the phenotypes were expressions of phenotype variation already present in the population (norm-of-reaction); but, some of the changes were novel phenotypes. So, during an unusual environmental stress, nearly all random changes are constrained within the norm-of-reaction; but occasionally, a random change will appear that is slightly outside of the norm-of-reaction. Flies exposed to heat during development revealed a small percentage without supporting cross-veins on their wings; this was a random change to a specific environmental stress. Except for the environmental selection from heat, there was no other selection for or against the phenotype; the wing change provided no protection from the environmental stress of heat (a non-adaptive phenotype). After 20–25 generations (all exposed to heat early in development), a loss of wing cross veins was increasingly present in successive generations; i.e., cross-veined wing phenotypes that were initially common in the population became increasingly less common in the population. In genetically-inbred populations without this cross-veinless variation in the population, this effect did not occur. When the cross-veinless phenotype had offspring, the increased numbers of this type was likely due to inheritability of the new heat-induced (non-adaptive) expression. That is, increased numbers of the new heat-induced phenotype had increased numbers of offspring that looked like the parents. This loss of wing cross-veins eventually became independent of the heat treatment. So, after many generations, the heat was no longer needed to carry-on the inheritable phenotype. Waddington showed that the cross-veinless phenotype can be "fixed" (to occur in close to 100% of the population). In other stress experiments, flies exposed to ether fumes developed another pair of wings; the extra wings did not seem to protect against ether exposure (another random non-adaptive change of phenotype in response to an unidentified stress). Ether was an environmental selector for a double-wing non-adaptive phenotype, a random non-adaptive change to a specific environmental stress, and the double-wing phenotype was inheritable.

Kirschner and Gerhart elaborate on work beyond Waddington's experiments. "Lindquist and her colleagues tested not just *Drosophila* (fruit flies) but also *Arabidopsis*, a small flowering plant. They lowered (an important regulatory protein) Hsp90 activity in several ways, by heat (as described above), by mutation of the Hsp90 (gene) itself, and by exposure of the organism to a chemical agent that inhibits Hsp90 specifically. Many alterations of morphology were seen and, if desired, could be put through stabilizing selection (sorting by environmental induction) so they would continue to form even after Hsp90 was restored to full activity (similar to Waddington's fruit flies). The variety of alterations was great: different altered parts and different degrees of alteration. Lindquist refers to Hsp90 as a 'capacitor for phenotypic variation.' (Capacitors in electric circuits accumulate a reservoir of charge and release the charge when there is a change in circuit.)

Several evolutionary biologists, including Mary Jane West-Eberhard, (along with others) emphasize the possible broad applicability of an "adaptation-assimilation hypothesis" in explaining the origin of complex phenotypic variations. "In West-Eberhard's view, evolution of a novelty proceeds by four steps. In the first, called TRAIT ORIGIN, an environmental change or genetic change affects a preexisting responsive process, causing a change of phenotype (often reorganization). At this initial step, she regards environmental stimuli as being more important to evolution than genetic variation. The traits may or may not be adaptive.... In the second phase, the organism adapts or accommodates to its changed phenotype by COMPENSATING in part for the perturbed condition by using what we would say are its highly adaptive core processes. In the third phase, RECURRENCE, a subset of the population continues to express the trait, perhaps owing to the continued environmental stimulus. In the final phase, GENETIC ACCOMMODATION, selection drives gene frequency changes that increase fitness and heritability, although the phenotype change is not necessarily ever completely under genetic control. While having a heritable component, it could retain an environmental dependence (recall gene atrophy in icefish). Thus this model... has a phenotypic accommodation phase followed by a genotypic accommodation phase. Most elements of phenotypic novelty would not be new, and the role of mutations would be to provide small, heritable regulatory modulations rather than to create major innovations." This is a genetically or environmentally-determined (either-or) view of evolution. It does not require genetic expression to form the environmentally-induced traits; however it does allow this as a genetically-induced option. The environmental importance of generating phenotypes seems to be strongly supported here.

Non-adaptive change in a phenotype will not be supported or maintained by environmental selection. Adaptations are new phenotype variations, modifications of form, function, behavior, etc., that are supported by environmental selection; they are intimately involved with (often adverse) environmental selection. Only adaptive phenotypes survive environments that are lethal to non-adaptive phenotypes. Acclimations are (inheritable, common, norm-of-reaction) existing genetic variations of form, function, behavior, etc. that are already in (genetic) place in a high percentage of the population, have been encountered previously on a more or less frequent basis, and the minor environmental changes are within the range of a rapid phenotypic modification. Acclimations are environmental adaptations from the past; the continuing ability to acclimate to more common environmental stress, is inherited by the offspring.

The adaptation perspective can be illustrated by an immune system response to disease in humans, sometimes known as "hypermutation." Adaptive immunity is antigen-specific and found only in vertebrates. A new disease, like the virus H5N1 strain of bird flu, could mutate and enable it to be transmitted from human to human. The human immune system has had no chance to adapt to this new virus strain and produce an effective defensive response; the new virus could be virulent enough to cause a great many deaths. A specific viral (environmental) adaptation response must occur in time to prevent a fatal outcome. Because of previous mutations in the human genome, some infected patient immune systems will adapt in time to overcome the new threat and generate a new phenotypic immune system expression (in the form of an effective defense specific for the viral antigen) that will enable survival of the infected patient. This can only happen from having a genome that can express a saving genetic code that opens a combination padlock on (or "key" to), the environmental selection gate, allowing passage into the future. "Hypermutation" is not the best term to enhance understanding of this process; I believe "hyperadaptation" would be a better description, as it describes a new phenotypic expression of the pre-existing genotype. This example further illustrates the recently-discussed (previous chapter) confusion on phenotype change and mutation. The somatic immune system phenotype is highly antigen-specific (environmentally-specific). The immune system adaptation process provides a new and continuing active resistance for the infected individual; however, it is only effective for that particular strain of virus infection. The virus could mutate once again and bypass the newly-acquired human immune-system defense. Unfortunately, this frequently occurs in many influenza strains. Active immunity per se does not appear to be inherited in humans. The adaptation is a somatic functional adaptation of the body cells, not transferred-on in the germinal cells. However, the genetic potential to form this active immune system adaptation (new phenotype expression of genotype variation that will neutralize the environmental bird flu virus strain) will be passed on in the germ cells of survivors, and inherited in the offspring. Offspring will be supported in pregnancy and infancy by passive humoral immunity antibodies transferred across the placenta, as well as by passive humoral immunity antibodies in the mother's milk. Regarding the potential for inheritance in sexual reproduction, a co-dominant additive benefit would mean that favorable odds follow a pattern of some heterozygote advantage, but the greatest advantage lies with a matching pair of homozygote "lucky combinations" on both alleles (that represent the greatest potential for forming a specific disease resistance). The groups of genes that code for immune resistance are located on the short arm of human chromosome #6, residing on four closely associated sites. Each locus is highly polymorphic; i.e., each is represented by a great many alleles within the human gene pool. These alleles are expressed in co-dominant fashion; all alleles are active and all traits are expressed at the same time. In the case of each parent having a heterozygote allele with maximum resistance potential (all 4 sites), the offspring inherits one chromosome #6 from each parent, so the offspring have a one in four chance of having received the same maximum (all 4 sites) disease-resistant paternal code expression on chromosome #6, and maximum (all 4 sites) disease-resistant maternal code expression on chromosome #6, in order for offspring to be allele-matched (the most-favorable homozygote combination of the adaptive genetic expression code). The environmental selective process of this disease will become an agent of natural selection preservation of the immune potential. Environmental selection provides support and direction to the randomness of genetic variation; natural selection is the resulting genetic bias. Conversely, natural selection is the genetic bias in allele frequency that is primarily a result of environmental selection. Recall (from The Rest of the Story chapter) that the conditions of life (environmental conditions) are the overall determining factor in any natural selection process. In proper perspective, there is a strong environmental connection to evolution; the environment selects the situational and directional path of life's random genetic variation (for the phenotype, which is a genotype expression). The environment does not control genetic combination of alleles, but it does indirectly influence their expression and it directly affects their individual survival; so, it selects the path of evolution's direction in this way. Natural selection ("a change in gene frequencies (alleles) in a population, owing to fitness of phenotypes' reproduction or survival among the variants"—Levinton) is the result of an environmental selection process. Natural selection genetic bias is involved in the coding of any new adaptation in the offspring; the adaptation will be an environmentally-selected genetic change in phenotype.

24

Genetic Constraint

There are many constraints on life, but in any discussion of environmental biology or evolutionary biology, genetics, particularly genetic constraint, plays a significant role for any possibilities in the changes in types of life. After the most recent "snowball earth" animal life suddenly appeared in the fossil record and underwent an explosion in diversity (Cambrian explosion). It appears that all animal major body plans (phyla), almost simultaneously, were present. In the half-billion years since, animal life never again experienced such a fundamental change in major types of body plans. Any changes in body plan, from that time to the present, have been below this phyletic (major body plan) level. Changes have certainly occurred on lesser levels, but these changes have boundaries, within whatever mechanism that changes the phenotypes. In other words, some changes do seem to occur, under genetic and environmental influence, seemingly without much problem, yet there appears to be limits in the type and direction of change.

The environment cannot generate a "needed mutation;" that is, the environment cannot even initiate just any new genetic expression of the genome. There does seem to be a protocol for environmentally-influenced genetic expression. An older generation of biologists will recall Haeckel's Biogenic Law: "Ontogeny Recapitulates Phylogeny! (the development of the organism follows the pattern of its phyletic evolution)," which has since fallen into disfavor. While it was not completely accurate, the old observation does illustrate a point. In addition to a strong environmental role, not only does phenotypic expression of the genome depend upon the biased preservation of existing genetic combinations (alleles), and different combinations of the any new mutation (allele) into the variation, the phenotypic expression of the genome and the development of an organism appear to be restricted to a lineage of very similar cellular function pathways, which also appear to be functionally-linked to phyletic roots. This functional linkage should not be confused with genetic linkage, which refers to genes located in close proximity on one chromosome. The functionally-linked inherited cellular pathway patterns control phenotypic expression at different levels of genetic hierarchy. The genes involved in the environmental threshold responses and the development of body plans can only accommodate certain genetic expressions from an environmental influence during development, and this restricted expression is even more constrained after the full development of the phenotype (a developmental constraint). The bottom line is that there are genetic and environmental limits to all forms of environmental adaptation.

In the chapter on Development and Evolution under a section of Ontogeny, Phylogeny, and some evolutionary trends, Levinton reflects on Biogenic Law: "If, as an organism grows larger, a defined sequence of morphological events best suits it to function within its environment, then those stages might be added in an evolutionary sequence and preserved as adaptive solutions to an ontogenetically (developmental) ordered series of changing environments…. If terminal addition were the mechanism of for the assembly of such a series of changes, then phylogeny would assemble ontogeny by natural selection… (the) character states early in ontogeny are more general (i.e., ancestral), whereas states later in ontogeny are most likely to be more special (or derived)…. The embryos of related forms resemble each other only because of the lack of change in earlier embryos in phylogeny, and not because they represent stages achieved by adults of ancestors…. The likely modification of later stages in development is supported by the argument… for a developmental ratchet." In a section on Development, Genes, and Selection: The Evolutionary Ratchet, Levinton states: "The developmentalist point of view… leads to a model, which I call the evolutionary ratchet, generating some of the constraints in evolution. The model characterizes an organism as being part of a long evolutionary history, where the evolution of timing, rates, and localization leads to a complex developmental process that can be disrupted less and less easily as time progresses. Any spatiotemporal interactions in the developing phenotype that, when accumulated, cannot be disrupted because of integration, are accumulated by the epigenetic ratchet (comprehensive layer of gene function above the level of simple gene function), and lead to epigenetic constraints…. The genetic ratchet refers to new genes that, when incorporated, cannot be easily lost by process such as genetic drift."

Levinton continues, in the Order in form in ancestors and descendants: "Any violation of the biogenic law is due to the overall process of heterochrony, which de Beer defined as a difference in order of appearance of structures between ancestors and descendant. This must entail a phylogenetic change in the onset or rate of development of a structure, relative to another. In other words, descendant structures are accelerated or retarded in appearance in development, relative to those of an ancestor." Also discussed is the Interaction of developmental units with natural selection and the external environment: "Adaptive evolution may indeed involve a conflict between function in the external environment and the maintenance of an organism with orderly development... a developmental program can be built gradually, and the intermediate steps are likely to consist of a harmonious arrangement of the independent characters. Geographic-ecological range extension (the creation of a special new environment) seems to be important in such evolution. If the most derived forms were subject to evolution by arrest of development, it is possible that the resultant forms would be, in a sense, preadapted."

In the same chapter on Development and Evolution, Levinton notes that embryos can vary from a mosaic pattern of development, where cells have fates determined very early in development, to regulative embryos, where cell fate is determined more by the surrounding environment. He references studies indicating that fruit flies, amphibians and mammals tend to the latter form of development; echinoderms tend more to the former form of development. Levinton goes on to discuss the nature of organization, timing, genes, diffusible gradients, localization, induction, rates of cell division, extracellular signals and thresholds. He notes that "no vertebrate system shows the simple switches characteristic of the arthropods.... Segmentation in insects is also determined by a compartmentalization process.... The concept of compartmentalization in insects can be generalized to other organisms as developmental fields or regions of strongly presumed developmental interaction, relatively independent of other regions.... The special scale of the effects of morphogens suggests that major effects within a developmental field can only occur over short distances. Longer-distance effects within a developmental field might be determined by cell-cell communication systems, modulated by a transmembrane protein.... These proteins act as gateways to communication to gene action or changes in metabolism within the cell... there is strong evidence for developmental gene determination of compartmentalization and regionalization in development; opportunities for evolutionary change abound by means of allelic variation within populations, which allows responses to selection of developmental expression (Brakefield) and apparent co-opting of developmental genes to developmentally novel functions (Lowe & Wray)... the entire phenotype is not an overall developmental unit so tightly bound that it cannot be broken up by natural selection. Perhaps strong (environmental) selection tends to favor larger-scale developmentally correlated changes. If selection is relatively weak, genetic correlations may have time to be broken up... intraspecific (within species) polymorphisms of discontinuous traits (adaptations) of small magnitude is the stuff of evolutionary change." In the chapter on Development and Evolution, Levinton also states: "Most likely, currently discontinuous developmental events were built up gradually and in continuous interaction with the external environment. On completion of the developmental program, it may have been internalized, i.e., insulated from direct interaction with the external environment."

And it is in multicellular animals that the internal environment gains great significance. In *The Plausibility of Life*, authors Marc Kirschner (chair of Systems Biology @ Harvard) and John Gerhart (professor in Berkeley's Department of Molecular & Cell Biology) argue that Darwinian explanation is incomplete, and that the results of recent discoveries in cell biology and developmental biology can be used to remedy this effect. In a chapter on Physiological Adaptability and Evolution, they observe: "Developmental plasticities differ from physiological plasticities in that after a critical time in the animal's Development, they are irreversible. The alternative phenotypes can be distinguished by their morphology, physiology, or behavior. Organisms with alternative phenotypes are polyphenic. There are two types of polyphenism, sequential (mostly marine) and alternative (different expressions in form for the same genotype); both basically involve bugs. Taking sequential phenotypes first, the cases include animals with complex life cycles of two or more different developmental stages (such as larva, juvenile and adult). Most animal phyla inhabit the ocean, and pass through strikingly different developmental stages…. When stages of a life cycle differ dramatically, they are connected by a drastic metamorphosis. In the metamorphosing caterpillar, most larval tissues are destroyed and replaced by newly developed adult cells. The transition is usually dependent upon external conditions, although it must be appreciated that this dependence is evolved and selected. Two hormones control insect metamorphosis… they globally release the internal means to propel the organism into a new state…. These effects, though far-ranging, are mechanistically no different than gene regulation that occurs in all cells of our body…. Different tissues respond differently to juvenile hormone and ecdysone, various cells responding to one or the other, both, or neither. Where there has been a response to external conditions, it is the organism (interaction between genetics and the environment) that specifies the response, the readiness to respond, and the specificity for responding to a particular environmental agent. Thus the same genome can read differently at different times to drastically alter the phenotype. The timing of these events can be linked to the external environment or can be driven purely by internal (environmental) means. Sequential phenotypes have in some cases provided the phenotypic variety for the founding of new races or species…. In some ways the more interesting forms of developmental plasticity occur when the organism has alternative adult phenotypes developed in accordance with environmental social conditions, the second kind of polyphenism. The irreversibility of this polyphenism is due to an environmentally dependent branch point, controlled by a sensitive switch, at one episode of development. When this decision has been made, it cannot be repeated or undone. Social insects, such as ants, wasps, bees and termites offer dramatic examples of alternative phenotypes (female queen and worker bees).... How do they develop one way or the other? Both come from the same kind of diploid egg at the start, as shown by experiments transferring eggs between royal cells and worker cells—whatever egg is in the royal cell becomes a queen. Workers construct royal cells at a time when the queen mother produces insufficient queen substance to inhibit them from doing so, and the workers start preparing royal jelly and feeding it to larvae in these cells. Royal jelly is a nutritious substance, which in honeybees includes high concentrations of vitellogenin, a protein found in both vertebrate and invertebrate egg yolk.... Polyphenism is not limited to insects.... Plasticity and fixation may underlie much evolutionary change."

In Kirschner and Gerhart's chapter on Invisible Anatomy, in a section under A Map in the Embryo, developmental processes at the cellular level are discussed. It begins discussing cellular development as a simple egg. Starting with an (environmental) localization initiated by fertilization, genetic influence within the fertilized egg is able to organize into compartments "under conditions set by the surrounding environment. It is a bootstrapping process where a few small initial differences are acted on to make further differences. When a particular component map is completed in the embryo, it is a map shared by all members of a phylum, the largest group of organisms based upon anatomy and physiology. Though not strictly identical in each species of a phylum, the map is the most conserved (constrained) feature of the phylum's anatomy. Once organized in an advanced multicellular stage, but well before the cells have differentiated into their final cell types, the compartment plan gives each cell its address, its identity, and its location relative to cells in the rest of the body. This address will serve the cell and all its descendents into the adult. Each compartment will develop multiple tissues. Similar cell types in different compartments (with different addresses) may appear similar in structure and activity, such as bone deposition or nerve impulse conduction, but will differ from one another in other ways, such as their capacity to proliferate, to migrate in the embryo, or to adhere to other cells.... We call the compartment plan 'invisible anatomy' because the compartments are only identifiable if one can establish which genes are expressed there. At these early stages, compartments cannot be distinguished by anatomical features. The actual differentiation of the organism will depend not only on the compartments but also on the interactions of cells of one component with signals from other components. The component map is an extensible map: individual components can expand and shrink independently, while overall neighbor relations are maintained. The flexibility occurs not only in development, when certain areas grow relative to others, but also in evolution where there is disproportionate growth, for example the neck of the giraffe relative to the neck of the whale. The compartment boundaries, though curiously arbitrary with regard to final anatomy, nevertheless divide the embryo into regions. In different regions, the same target genes can be controlled differently. The result is a platform for local differentiation and for use by the genome in many different ways... the compartment map makes possible the use of different combinations of processes at different places in the body. In fact, it provides those places. The map makes possible the use of different genes in different places. The concept of (isolated environmental) compartments has been the most valuable contribution of cellular and developmental biology to understanding evolution, at least the evolution of complex multicellular animals. The basic compartment structure independently identifies the phylum and hence must have remained unchanged since the Cambrian, for more than 500 million years." "It is a compartmentalization of the organism that allows independent evolution and development, while maintaining overall coordination of activities related to the body plan of the phylum."

In the adult world, genetic constraint of the phenotype is totally entangled with environmental (selection) constraints. Why would this suddenly appear in the mature adult, unless this was always the relationship in the very beginning (even if as a more controlled internal environment)? In addition to the previously-mentioned extensive phenotypic changes during all development there are other limited phenotypic expressions after development. Genetic constraints are firmly fixed in the adult; acclimation was discussed in the preceding chapter. Life history traits are genetically constrained by the genotype norm-of-reaction, have phyletic connections, and are related to environmental influence (e.g., damage) during the entire life cycle (what doesn't kill you, nearly kills you, or eventually does kill you—the adult version of reality).

Douglas Futuyama's 3rd edition of *Evolutionary Biology* is an excellent up-to-date text that contains much valuable information on the evolution of life histories; life history phenotypic traits affect growth rates of populations. "Most all sexually reproducing organisms, if they escape predation or other extrinsic causes of death, undergo physiological changes (senescence or aging) that increase their likelihood of death as they grow older. Among animals, life spans range from about 10 days in some rotifers to more than 2 centuries in some mollusks. And individual trees may attain ages of 3200 or even 4600 years…. Sea anemones and corals are known to live for close to a century, and may have the physiological potential to live indefinitely. This may also be true of plants that propagate vegetatively by rhizomes, or other means…. The assumption that reproduction is costly—that there exist trade-offs between reproductive effort at one age and components of fitness at other ages—underlies much of life history theory, and is supported by considerable evidence…. After reproduction begins, the selective advantage of survival declines with age…. Since early reproduction is correlated with lowered subsequent reproduction, and since there is always an advantage to reproducing earlier in life, we should expect organisms to reproduce at the earliest possible age and to reproduce only once… (nevertheless, a different life history pattern of a late-blooming genotype) can have a higher rate of increase… if adults have a high enough survival rates to reproduce repeatedly and if older parents produce enough offspring to compensate for their lower value. This is especially true if the rate of population growth is low… as may be the case in a population regulated by density-dependent factors (such as competition for resources) that causes high juvenile mortality. Thus this iteroparous (this late-blooming type of life history) is likely to be advantageous if juvenile mortality is high, adult mortality is low, and population growth is low. Conversely (the early blooming pattern of)… life history is likely to evolve if the population growth is high (at least at times), juvenile survival is high, and adult survival is low (in which case the probability of surviving to reproduce a second time is low). In an iterparous (late blooming) life history, an individual of age x allocates some energy not to reproduction, but to self-maintenance and perhaps growth—and therefore to subsequent reproduction. In species with indeterminate growth, such as many plants, fecundy (successful reproduction) is often correlated with body mass. In such species, clearly, saving energy for growth is like saving energy for future reproduction. As individuals age, however, the benefit of withholding from reproduction declines, both because larger bodies require more energy for maintenance and because the intrinsic demographics of reproducing late in life become greater. Therefore we would expect, as a rule, that species with indeterminate growth should be (late blooming life history pattern) iteroparous, but the proportion of energy or other resources devoted to reproduction should increase with age…. All else being equal, a genotype with higher fecundity (reproductive success) has higher fitness than one with lower fecundity…. The very low fecundity of species such as humans, albatrosses, and the California condor… requires explanation… a parent's fitness is maximized by laying an optimal clutch size…. A similar effect often arises from a trade-off between the number and size of offspring due to the finite amount of energy and materials that a parent can allot to yolk, endosperm, or nourishment of embryos…. When, because of population density or simple scarcity of resources, competition for resources is intense, juvenile survival can be enhanced by large size, which can confer a competitive advantage or simple resistance to starvation… traits associated with low intrinsic rates of increase—delayed maturation, production of few, large offspring, a long life span—are likely to evolve in species that occupy stable, competitive, or resource-poor environments…. Reproductive effort in plant species that occupy disturbed or early successional environments is generally higher than in plants that typically grow in mature 'climax' communities." Futuyma then went into details on sexual selection and sexual reproduction aspects of life histories, and then into other aspects of behavior. All phenotypic traits are entangled expressions resulting from the marriage of heredity and environment. All are genetically constrained, as well as environmentally constrained. At one time, these all were new environmental adaptations.

In contrast to the above considerations of firm fixation of genetic constraints in the adult, stem cells (multipotent adult progenitor cells) in the human body are undifferentiated; they have unlimited potential for differentiation into any cell type, offering great promise in the future of medicine. Of particular concern today is stem cell mutational change, induced by factors in the environment, which can lead to cancer (unconstrained pathological cell growth), Cancerous cells are characterized by bizarre chromatin patterns displayed in the cell nucleus and unrestricted invasive cell division; this diseased condition eventually results in death of the organism.

25

Genetic Variation

In a unique environment created by isolation, a changing environmental condition of life known as time, allows genetic variation (from mutation) to occur within a population. If this mutation is immediately displayed, most of this genetic variation is damaging to life; environmental selection eliminates harmful phenotypes and they are lost to future generations. A displayed phenotype may be neither helpful nor harmful; this would be a neutral phenotype. Most commonly (but not always), neutral variation is carried as a recessive allele in the genotype and it is not expressed in the phenotype. This neutral genetic variation has no current value, but is not harmful. Neutral genetic variation may continue in the genotype of a population for a long time. If the genome expresses a previously silent mutation (e.g., with homozygous recessive alleles), it usually expresses a change. Neutral genetic (genotype) variation that had no survival value in the past could eventually gain importance; at a later time, this genetic variation may combine with other forms of genotype variation and express a phenotype with improved survival in a newly-created environment. In Levinton's chapter on Development and Evolution, he states that "The stuff of phenotypic evolution is small developmental units of relatively low burden on fitness. Some of these, such as doubling of structures early in development, might eventually have strong evolutionary significance." Levinton makes a case for small amounts of variation over time as being the most likely to survive. Darwin showed that some variation will be advantageous to survival (in a special isolated environment). Darwin suspected, but did not have today's evidence, that mutational change, the original source of inherited (genotype) variation is completely random. Environmental selection's support for a valuable expressed variation (phenotype) increases the number of offspring that survive under the conditions of life in that environment; environmental selection's limiting factors have the opposite effect.

In the struggle for life, within the conditions of life, the costs of living are dear. Life has to "make do" with whatever genetic potential for adaptation that is available (within its genetic variation/natural selection genetic bias). History has shown that this is often not enough. Additionally, the basic needs of water, energy, and other resources are never enough; populations grow until resources are exhausted. Recall the "struggle for existence" of Thomas Malthus (there is a strong and constantly operating check on population from the difficulty of subsistence). In other words, it takes not only the right genetics, but it takes energy and other costly (environmental) resources to make and maintain a structure (Jared Diamond, *The Third Chimpanzee*). Structure, formation and maintenance must be cost-effective in terms of survival value (natural selection). In a world of limited resources, there is not a great deal of room for neutral variation in structure (phenotypic). Environmental selection (the gatekeeper for genetic success) sorts out survivors; this environmental selection results in a natural selection preservation of favorable phenotypes (a change in gene frequencies in a population, owing to fitness of phenotypes' reproduction or survival among the variants—Levinton).

Phenotype variation is packaged as combinations of traits within one individual organism. Expanding one trait may then be at the expense of other traits (Diamond—3rd Chimpanzee chapter on Why Do We Grow Old and Die?). To be beneficial, this new combination must enhance the overall survival chances of the individual and/or offspring. The most effective expression of the phenotypes in an organism is the most optimum balance of all traits (Diamond—3rd Chimpanzee chapter on Why Do We Grow Old and Die?). One form of primitive hominid (*Australopithecus robustus*) took a vegetarian pathway towards developing an energy-intensive a large gut. Other forms of primitive man (*Homo habilis and erectus*) utilized a diet of meat and developed a more energy-intensive large brain; this followed the pattern of gracile australopithecines with a lighter build (especially skull and teeth).

What happens when (phenotypic) structures are no longer needed in a changed environment? Cave fishes have eyes in varied stages of degeneration. The complex morphological structure that once provided the function of (trichromatic) sight is now only a neutral degenerating phenotypic expression in a sightless world. **The genes that formed the structure and function of vision received no environmental indirect support. Without environmental selection, whenever mutation occurred in these genes, the mutations persisted; i.e., were not eliminated.** Structure and function was compromised, irreversibly. In addition, using limited resources that could be used to maintain the (phenotypic) function of sight is not cost-effective for this special environment of total and eternal darkness. The combined effects of environmental selection and natural selection genetic bias have favored the redirection of limited resources needed for survival elsewhere. **Nearly all phenotype variation has been influenced by the interaction between environmental selection and natural selection genetic bias. Genetic expression and genetic survival are not random events; they are both influenced by environmental conditions.** Recall again the icefish that lost the function of its globin genes, due to loss of environmental support of the globin-bearing phenotype. Specifically, environmental selection did not eliminate the mutation that caused this gene to malfunction; the non-functioning gene was passed to offspring while resources were directed elsewhere.

Genes can be added; genes can be duplicated and modified. In a process that is duplicated today by genetically-modified crop technology, bacteria may have combined genetic material to create new life variations, different and more complex than those existing before. If genes can be transmitted separately from genomes, perhaps the appropriate unit of evolutionary change could be the gene, and not the genome; yet, without a doubt, the genome does influence genetic function. Early plant and animal life may have been modified by a similar genetic combination process. Polyploidy (more than two sets of chromosomes per nucleus) is widespread in plants and is known to exist in animals (Levinton). Animal genetic variation, polymorphism, may be genotypic or phenotypic; nearly every trait has allelic variation (Levinton). Allozymes, enzymes that are one of a series of alternative gene products of a given locus, may exhibit protein polymorphisms (Levinton). Nearly every morphological trait in plants and animals has a genetic basis under polygenic (governed by the cumulative effects of many genes) control, including polymorphisms involving discrete (absolute—not additive) morphological traits (Levinton—chapter on Patterns of Morphological Change). In addition to the genetic architecture of both discrete (absolute) and continuous traits being polygenic (additive traits whose variations are measured with a scale), the genetic determination of the phenotype is a complex result of interactions among genes that perform different functions; the control comes from much of the genome and usually cannot be restricted to switch genes of major effect (Levinton—chapter on Patterns of Morphological Change). Developmental thresholds may be involved (Levinton). A correspondence between genetic correlations and functional relatedness of traits suggests that the genotype is a co-evolved unit, designed to serve the entire organism (Levinton—chapter on Developmental Evolution). Revisiting an earlier discussion of development, Levinton notes: "If, as an organism grows larger, a defined sequence of morphological changes best suits it to function within the environment, then those stages might be added… and preserved as adaptive solutions to… series of changing environments. Adaptation, improved function of a phenotype, involves the evolutionary change of the structure-function relationship (Levinton—chapter on Patterns of Morphological Change). "Sudden environmental shifts may be the basis for adaptation and subsequent stasis; speciation, when it occurs, is more the effect of natural selection than the generator of variation" (Levinton). In this last statement, substituting environmental selection (a process) for "natural selection" seems to clarify communication; recall that Levinton previously defined natural selection as a change (a result) in gene frequencies in a population, relative to fitness.

26
Environmental Adaptation

Nearly all adaptations are environmental adaptations. It is environmental adaptation that enables "the preservation of favored races in the struggle for life"—by indirectly influencing design in the favored race. I hypothesize that natural selection is the (genotype) preservation of successful phenotypic adaptations to previously-experienced environments. The adaptive process uses the genetic bias of natural selection as raw material to create a new environmentally-integrated more-fit design. Both the new adaptation to the new environmental stress, as well as the continuing ability to acclimate to more common environmental stresses, are inherited by the offspring. Indirect environmental integration begins in the adaptation process itself; this will inevitably occur early in development. Sometimes, early in development may mean as early as possible. In humans, high level brain function is maximized in the pre-frontal cortex. But mankind's most significant adaptation (high level brain function in the pre-frontal cortex) is not completed until age 20, via an environment-related use-it-or-lose-it selective process. The adaptation of high-level brain function is developmentally constrained to realize maximum potential from environmental input until the age of 20. Mankind's most important adaptation is a heritable trait with an environmental component. Said otherwise, this unique adaptation requires an environmental component, but transfer of the environmental adaptation potential to offspring is only through genetic inheritability. Learning is an acquired adaptation, and it cannot be inherited. Only the ability to learn is inherited. There are some differences between inherited and acquired adaptations; acquired adaptations will be addressed shortly.

An adaptation is a new trait that helps an organism to be more suited to its environment; it involves genotype expression of a new environmentally-integrated phenotype. In detail, an adaptation is the improved function of an organism in the environment, pertaining to an environmentally-integrated change in genotype expression, the new phenotype. The new adaptive phenotype is a specific manifestation of any detectable characteristic of an organism, whether its structural, physiological, behavioral, and/or biochemical, which involves an interacting evolutionary change between the environment and the previous structure-function relationship. **Evolution is a change in phenotypic adaptations, in response to variable environmental conditions, over time.** Recall that the driving force of evolution may well be the product of the interaction between environmental selection and genetic variation, in a changing, isolated environment. **The driving force of evolution is (genetic) adaptation (to a unique environment). The creation of a changing, isolated environment must occur to initiate this life-changing phenomenon.** There is a strong environmental connection to evolution; the conditions within the changing isolated environment indirectly influence the directional path to life's random genetic variation. Environmental selective forces in the changing isolated environment and their influence on then-present genetic patterns (existing variation from natural selection genetic bias) act in concert to initiate and stabilize a new genetic expression to environmental challenge, a potentially successful adaptation. **In the end, the environment will select for favorable adaptive changes of the phenotype.** Environmental selection will not support a non-adaptive change in phenotype. There are three scenarios for environmental selection and adaptation advantage. Only adaptive phenotypes survive extreme environments that are lethal to non-adaptive phenotypes. Only adaptive phenotypes manage in environments that are more limiting to non-adaptive phenotypes. Only adaptive phenotypes prosper (positively supported by the selective conditions of life) in environments most favorable to that adaptive phenotype. Non-adaptive phenotypes may not be necessarily limited, but in this last case there is no positive support (advantage) for the non-adaptive phenotype. **Adaptations are phenotype variations, modifications of form, function, behavior, etc., that are either supported or not eliminated by environmental selection; they are intimately involved with (often adverse) environmental selection.** The adaptive process results in an environmentally-integrated more-fit individual. Genetic variation from mutations is only randomly provided; there is no directedness or anticipation of future needs. Even though existing environmental conditions do not form mutations that are the source of inherited genetic variation, they do influence the initiation of genetic expression under stressful environmental conditions (Waddington experiments in the Phenotype and Environment chapter). So, the original genetic variation is random (chance variation from mutation), as is the genotype formation of a unique randomly-formed phenotype variation under environmental stress. Environmental selection is not randomly selective; it is directional, and this direction is determined by the existing environmental conditions. **A new environmentally-stabilized adaptation is an environmentally-integrated expression of the genotype, even if it began as the genotype's expression of a random phenotype resulting from unusual environmental stress.** Both simple and complex life's cells have many functions that may be random (exploratory); however, random exploratory functions are environment-sensitive. **Environmentally-advantageous feedback stabilizes an exploratory function and the genetic expression pathway to this function.** Random phenotypes (random expressions from the norm-of-reaction) may be more easily stabilized in the increasingly internalized developmental environment in higher forms of complex life. **Successful new phenotypes pass these advantageous adaptations to offspring. Environmental selection will then continue to stabilize, or destabilize, the developing organism.** So, adaptations are environmentally-supported new phenotypic expressions, made from newly-modified developmental pathways that utilized previous developmental pathways in the genome (natural selection genotype bias for existing norm-of-reaction phenotypes) as raw material. Indirect environmental stabilization of a random genetic expression is the mechanism of environmentally-integrated design. The function is genetically-heritable but must be environmentally-supported. In order for an organism to adapt to a new environment (increase in fitness), genetic expression must harmonize with selective forces in an isolated environment. The new adaptation is a union of heredity and environment; like any other phenotype, it's pointless to argue over which entity is more important. The adaptive process creates a new environmentally-integrated more-fit individual. The new adaptive trait joins other genetic traits (expressions) previously preserved, and all continue in time as the genetic bias of natural selection. **It is naïve to think that because an adaptation is inheritable, genetics alone creates an adaptation.** "Adaptation" does not always refer to the formation of a "brand new" phenotype in response to unusual environmental stress (which may or may not become an environmentally-integrated phenotype, preserved as the genetic component of an adaptation in the genome's natural selection genetic bias). "Adaptation" may have more than one meaning; this creates confusion, but the confusion is easily managed. **"Adaptation" to normal environments involves expressing norm-of-reaction phenotypes; that is, expressing previously-formed inheritable adaptations from natural selection genetic bias** (preserved pathways to previously-experienced environments). Functional phenotypic change (acclimation/somatic "adaptation") in adults was introduced in the chapter Phenotype and Environment. And in developing organisms, frequently-encountered environments do not require a novel adaptation process either. As with acclimation, the developing "adaptation" is already preserved in the inheritable norm-of-reaction, only waiting for the signal to express it. Adaptation components in the genome's norm-of-reaction are biochemical pathways to previously-preserved phenotypes. For instant adaptation, all a developing organism adds is the environmental components (like an instant meal—just add water). In day-to-day "adaptation," environmental feedback initiates the formation of the most appropriate phenotype available from natural selection genetic bias (the preserved phenotype with the best match to the conditions of life). An isolated, unusually stressful environment may give this ongoing process of "adaptation" a hard kick in the seat of the pants. Providing the kick isn't fatal (can be), the creation of novel phenotype expressions seems to be initiated with greatly increasing environmental stress. If pre-existing adaptations from natural selection genetic bias cannot provide environmental advantage, reality's options are—come up with something that does work—or die. There are no survivors when adaptation to change is too slow to meet demands of rapid environmental change. **Genetic isolation, genetic potential for survival, and the time to respond to a deadly environment, before extinction occurs, are contained in the recipe for adaptive evolutionary success.** Evolution is genetic success (a life-saving inheritable adaptation) in response to environmental conditions (an isolated changing environment).

In a chapter on the Constructional Aspects of Form, Levinton recognized: "Evolutionary change can be said to involve adaptation to environmental challenge... and the 'superior' genotype is favored by selection." He also noted that surviving individuals with a new adaptation are commonly associated with having survived stress-induced population damage from a stressful environment. In his chapter of' Coda: Ten Thesis, Levinton observed: "adaptive evolution in a novel habitat type will not occur if the initial drop in fitness, concomitant with new habitat change, is too great to compensate for the slow incremental increases in fitness conferred by adaptations that secure increased ability to find the organism's current preferred habitat.... Catastrophic directional selection is therefore likely to be usually ineffectual." Life seems to prefer finding home over adapting to a new changed habitat. Sometimes home is impossible to find; sometimes adaptation is the only viable alternative. Using existing naturally-selected genetic variation (previous phenotype adaptations), organisms continually interact with the environment, sometimes generating a new phenotype under unusual stress. A random phenotype may continue to be expressed under the stress as long as the stress persists; it may even continue expression after the stress is gone, but if the random phenotype has no environmental advantage, the random phenotype will likely not persist for many generations. Survival alone, in a stressful environment, may stabilize a non-adaptive phenotype through limited environmental feedback. When non-adaptive traits are stabilized and become established in a population, environmental selection eventually eliminates phenotypes that are not supported or cost-effective. Only adaptive phenotypes survive environments that are lethal to non-adaptive phenotypes. Surviving newly-created phenotypes, stabilized by environmental feedback, become functionally-integrated into genetic hierarchy to become part of a newly-modified norm-of-reaction. Therefore, any random phenotype that has an advantage in environmental compatibility can become an inheritable adaptation to the conditions of an isolated, stressful environment. The selective forces of the environment play a major role in the preservation (natural selection genetic bias) of more-fit individuals (with better adaptations). Novel adaptations are new and better environmental expressions (phenotypes), integrated into the inheritable genome.

The Sargassum fish (*Histrio pictus*) is so specifically adapted to the Sargassum kelp environment of the Sargasso Sea, it is indistinguishable from its surroundings. In *The Origin of Species*, Darwin acknowledged that natural selection does not necessarily achieve perfection; however, in this case it is difficult to imagine a better camouflage for a predator. Teleology argues that mindless processes, such as chance genetic variation and chance natural selection (survival of the fittest), could never achieve such a level of complex perfection, and that this is proof of a grand design. Chance is not such an issue for environmentally-integrated design. While chance is involved for a fundamental genetic modification to arise from randomly provided mutations (there is no directedness or anticipation of future needs), chance ends here. Initiation of new genetic expression (phenotype), stabilization of the new phenotype, and phenotype survival (preservation) are not random events; they involve environmental interaction. Previously noted, unique, severe, or long-lasting stress in the environment can initiate the formation of unique, random phenotype expressions, some even outside the norm-of-reaction. Environmental selection provides direction and support to the randomness of genetic expression variation (from natural selection genetic bias). Random exploratory cellular behavior can be indirectly stabilized by environmental feedback and an advantageous environmental design can be functionally-integrated into genetic hierarchy (more on this later). Adaptations are constructed from components of previous pathways that are developmental pathways to previously-encountered environments. This non-random indirect influence is the basis of environmental (integration into phenotype) design. Natural selection genetic bias (preservation of more fit-types) and new adaptations to new environmental conditions influence the outcome of evolution.

In the special environment of the Sargasso Sea, created by ocean currents accumulating the floating Sargassum kelp masses, the environment provided selective indirect directional influence for genetic expression (design), with very limited options for predator camouflage or stealth. From the perspective of natural selection's variation (the preservation of favorable variations and the rejection of injurious variations) and the environment's (gatekeeper for genetic success) directional influence, specific adaptations such as Sargassum fish camouflage are more easily understood as progeny of the marriage of heredity and environment. **The selective forces of the environment play a major role in the natural selection (preservation) of more-fit individuals; environmental selection determines the pathway of survival.** The adaptive process, involving new genetic expression (environmentally-integrated design), allows gateway passage into the future. The new environmentally-stabilized phenotypic adaptation is inherited, and becomes part of the genetic bias of natural selection, in a newly-modified genotype norm-of-reaction (that displays different phenotype variations). Recall that under normal conditions, organisms often have a series of genetic expressions in response to frequently-encountered environmental conditions. The norm of reaction is an inherited pattern of genetic responses to different environmental conditions that enhance an organism's survival in frequently encountered environmental conditions. I hypothesized that these inherited genetic responses are expressions of previous adaptations to previously-encountered environments, and are passed-on to offspring by natural selection's genetic bias (preservation of favorable traits). Recall that environmental conditions do not control the genetic source, but can influence genetic expression and survival of the individual and its offspring, providing indirect control of available genetic combinations, resulting in the genetic bias of natural selection.

Darwin discussed numerous examples of biological variation, noting adaptations that were closely associated with the environment. The voyage of HMS Beagle took him to the Galapagos Islands off the South American coast. These islands had fewer types of organisms; however, these organisms filled unique roles of adaptation. The island species varied from the mainland species, as well as from island-to-island. There were endemic species or sub-species with particular unique traits. Each island had a variation of tortoise; each type of long or short-necked tortoise correlated with the height of different vegetation on that particular island. Each island had only a tortoise with inheritable traits (natural selection genetic bias-phenotype adaptations) that fit the directional environmental selection from the vegetation present. Finches on the Galapagos Islands resembled a mainland finch, but there were more types here also. Finch species on the Galapagos Islands varied according to nesting site, beak size, and eating habits. The finches adapted to their habitat; the size and shape of their bills reflected their specializations. The vegetarian finch and the ground finch all have a crushing bill; the tree finch all have a grasping bill; the cactus finch, warbler finch and woodpecker finch all have a probing bill. These adaptations were designed from an environmental selection of favorable genetic variations in their ancestors. Darwin used the finches in *The Voyage of the Beagle* to quietly announce his theory of evolution: "Seeing this gradation and diversity of structure in one small, intimately related group of birds, one might really fancy that from an original paucity of birds in this archipelago, one species had been taken and modified for different ends." In other words, Darwin suggested then present-day species (different types of life) resulted from a common ancestor and changes (natural selection preservation of the most fit competitors) occurring in each isolated population.

Alfred Russel Wallace independently conceived of a natural selection preservation of favorable traits, sending Darwin a manuscript entitled *"On the Tendencies of Varieties to Depart Infinitely from the Original Type"* in 1856. Darwin then had his previously-written 1844 essay presented orally along with Wallace's manuscript at the scientific society in London, and shortly thereafter published *Origin of Species*. Although Darwin was the first, Wallace is considered as a co-author of current evolutionary theory (natural selection). Wallace greatly contributed to the understanding of geographic distribution of species; this laid the foundation for further study in the conditions of life. Wallace also wrote *"On the Law Which has Regulated the Introduction of Species,"* a paper which was published in the Annals and Magazine of Natural History in September 1855. In this paper he gathered and enumerated general observations regarding the geographic and geologic distribution of species. By definition, biogeography is the study of the distribution of biodiversity over space and time; it reveals where organisms live, at what abundance and why (Wikipedia). Wallace is credited as the father of biogeography.

Between 1913 and 1924, Richard Hesse wrote *Tiergeographie auf oekologischer Grundlage,* which was translated into English by Hesse, Allee and Schmidt as *Ecological Animal Geography* in 1937. This is the classic text on environmental considerations for animal life. It deals extensively with environmental conditions in countless numbers of habitats; it deals extensively with animal distribution and adaptations to these specific environments. "In general, adaptation to an environment enables an animal to flourish under changed conditions that are unfavorable to unadapted animals." Hesse extended Liebig's "Law if the Minimum" to apply to animal life. (In the growth of plants, the needed resource that is least available becomes the limiting factor for growth.) He stated that all life has an ecological valence, a range of minimum requirements and maximum tolerances for certain environmental conditions. Hesse clearly understood limiting factors and their effect. This is the classic text on environmental selection; it discusses the effect of environmental selection on animal distribution, worldwide. Hesse defined <u>environmental selection</u> as "<u>factors that condition animal existence being favorable in varying</u> <u>degrees at different places on the earth's surface;</u>" everything old is new again (sort of). Although Hesse recognized "germinal change" (biologic variation), he placed the emphasis of environmental selection on the importance of somatic change through use; "the most frequent and most important form of adaptation is somatic, or functional, adaptation." This is similar to the Lamarckian belief that organisms become adapted to their environment during their lifetime and pass on these adaptations to their offspring. Experiments fail to uphold Lamarck's inheritance of acquired characteristics (in adults). Molecular mechanisms of inheritance show phenotypic environmental acclimation (somatic) changes in complex life (after development) do not result in genetic changes that can be passed on; the acclimation process is dependent on genetics, not vice-versa. The single cells of simple life are constantly exposed to external environmental feedback. They are in closer contact with the outside environment than are the germ cells of large complex life. The single cells of simple life, which don't reproduce sexually, can pass-on genotype-possible phenotype adaptation (somatic) change via vegetative mitotic division. For multicellular complex life's sexual reproduction, the environment of early development is more internalized, within the organism. Adaptation potential to the outside environment is more dependent upon inherited variation from the genotypes of the parents. Complex multicellular life adaptations are heritable changes that can only be transferred to future offspring through the germ cells. Sexual reproduction's germ cells dictate genetic patterns in all cells of offspring. Previously discussed in the chapters on Mutation, and Phenotype and Environment, with sexual reproduction (includes recombination and crossing-over), the germ cell exchange of genetic material increases population variation; this helps to somewhat compensate for the great adaptation challenges of complex life.

For complex life, an acquired characteristic's advantage (improved performance through acclimation, or true somatic adaptation as seen in active immunity) cannot be completely discounted in survival. Active immunity in humans is not inherited. The potential for generating active immunity is genetic; however, the active immunity itself is an acquired (somatic) adaptation. Continued resistance through ongoing active immunity is another acquired adaptation. Said otherwise, favorable genetic patterns may lead to an established life-saving active immunity (somatic function) against a disease; prolonged active immunity to the same disease may result (also somatic function). Ten per cent of infected Caucasians may become resistant to the AIDS virus, as a result of inheriting earlier smallpox or plague epidemic resistance potential that was developed in ancestors. The genetic potential must be present, but the actual performance of establishing active immunity is of deciding survival importance.

In the case of an increasingly life-threatening stressful environment, assuming an existing genetic range of phenotypic expression (that can offer some degree of coping short of an effective new inherited adaptation) to environmental stress, the worsening situation will demonstrate increasingly severe population damage and sharply depressed reproduction, but not necessarily immediate population extinction or a complete lack of reproduction. Partial environmental acclimation potential for survival is maximized, by utilizing the full plastic range in the genotype norm-of-expression for existing phenotypes, during the early development time of offspring. Partial acclimation to moderate environmental stress can buy time to enable genetic expression of a more-adaptive new inheritable phenotype adaptive response, in the developing offspring of a declining population, which may occur during this time extension. This would be a likely scenario for the occurrence of a new adaptive process that can be inherited. The greater the biodiversity, the greater is the chance that this adaptation can occur.

Environmental conditioning is important to survival in a challenging environment. Northern Europeans, acclimated to cooler temperatures, suffer far more from (global warming) high summer temperatures than Americans (of Northern European descent) suffer from the same temperatures; the difference is primarily due to climate conditioning (throughout life). Climate is more variable and extreme in America; many Northern Europeans died in a European heat wave (e.g. 35,000 in 2003) that was similar to numerous ones that occurred in the U.S. without a similar effect (even long before air conditioning). Eskimos, conditioned to living in frigid temperatures, seem to be more cold-tolerant than their southern Central America cousins; cultural skills alone do not account for all cold-tolerant differences. They can easily survive exposures to low temperatures that kill outsiders. In matters of survival, conditioning does make a difference. Limited somatic acclimation changes, in response to stressful environmental conditions, furnish some small functional change advantage in the changing environment, providing the environmental stress acts early and slowly enough to permit this advantage from a partial acclimation. A well-conditioned free-diver can hold his breath submerged in water over 6 minutes; partial pressure of oxygen at depth helps. The average person is challenged to hold their breath for 1 minute. I knew a fit young man that could hold his breath on the surface very close to 5 minutes, but he cannot do that today. He could also swim (completely unprotected by an insulated suit) for extended periods of time in a lake that would freeze solid within a day. The average person is not able to do this either. Today this man's life is far more limited in respect to environmental challenges. In an environmental crisis, conditioning differences even between identical genetics (no genotype variation between individuals) allows advantage from improved somatic functional change. For example, identical twins have identical genetics; however, somatic functional change (improved conditioning or acclimation response) still could mean the difference between life and death. Hypothetically, if the identical twins are chased through the environment by a predator, a well-conditioned identical twin only has to run faster than a poorly-conditioned twin. It takes more than good genetics (good potential) to win an Olympic Gold Metal (in predator evasion). As another example, when hunting, a well-conditioned twin is able to run further than a poorly-conditioned twin (who cannot run the distance in order) to outlast an overheated animal running on the hot African savannah. A poorly-condition twin will not have the same success in hunting, or feeding his family. Genetic potential is inherited; but, genetic potential must turn into performance for best results. It takes both good genetics and good environmental conditioning (improved somatic change through use— Hesse) to make the fastest and/or longest-enduring runners in the race for survival (survival of the fittest).

Genetic potential differences between individuals can be like comparing cards of different value that are dispersed to different players in a game of chance. However, it can make a difference on the game that is played (the rules can change, as in an asteroid impact), and how it is played (the winning hand may not always hold the best cards—behavior, even luck, can be a factor in survival). Odds favor the best genetics, but odds also favor the best somatic response to environmental stress; i.e., survival is enhanced by both genetic potential and somatic functional change (good conditioning or better acclimation). Relocation to a more favorable environment is often the first response of acclimation change (instinctive behavior) to a stressful environment. Behavior is an interactive part of the organism that links it to the environment; it links the nervous system of complex life to the ecosystem. **Instinctive behavior is a heritable phenotypic trait** (adaptation) **that is environmentally-integrated.** This is the *ne plus ultra* for adaptive environmental feedback transfer to heritable mechanisms; **it simply wouldn't happen unless environmental feedback stabilized neural circuitry that was linked to adaptive-genetic pathways**. Organism-controlled patterns of behavior can ensure habitat constancy. In Levinton's chapter on Coda: Ten Thesis, he emphasizes: "Habitat choice is a major component in the stabilization of the phenotype.... Larval selectivity of marine invertebrates assures that a dispersal will 'find' the parental habitat." Instinctive behavior is not only an environmentally-integrated inherited phenotypic trait, hard-wired into the norm of reaction; but, it can be modified by learning. Learned behavior (change in behavior) can be even more important to survival. Learned behavior is an acquired adaptation, parallel to an active immune response; neither of these adaptations are inherited by the offspring. The ability to change behavior allows animals far greater control over their environment. Learning may be the ultimate adaptation to environmental challenge. As with the immune response, this somatic adaptation has an environmental component, but transfer of the environmental adaptation potential to offspring is only through genetic inheritability. And as with any adaptation, the inheritance potential doesn't last forever; over time, it may be changed by mutation, gene expression modification, or different forms of environmental selection. The use-it-or-lose-it environmental option ultimately protects against mutation and gene expression modification, by environmentally selecting against these changes. In humans, grammar basics must be learned by age 3 in order to develop functional language skills (a use-it-or-lose-it environmental selective process). Language is one of the greatest human (somatic) adaptations. And for a developing child, language is an environmental phenomenon. Humans just cannot implement coherent sentence structure if they are not exposed to language interaction prior to the age 3 developmental milestone. Recall that this chapter began with a discussion of learning; it is mankind's ultimate adaptation. The somatic adaptations of learned behavior have enabled humanity to overcome many environmental difficulties in the past. Mankind has used the somatic adaptations of learned behavior to even modify the environment; however, the environmental limits never go away.

Therefore, somatic functional change (acquired adaptation, acclimation +/or conditioning) can make a difference, even if it cannot be directly inherited. In times of moderate environmental stress, the most-fit individuals will have the best available genetics as well as the best somatic functional response; both will combine for maximum advantage, enabling the survival of some individuals. In the long run, a slightly better genetic potential combined with a slightly better acclimation response and/or conditioning of the parents may help some individuals buy more time for a truly effective inheritable adaptive change to occur in the developing offspring (that will manage the moderate environmental stress); or, even to manage the moderate environmental stress utilizing existing genetic potential (e.g., ability to learn), combined with somatic adaptation performance (e.g., learned behavior). In the latter case, if the existing genetic potential provides enough learning structure (e.g., large highly-functional brain), and this permits a somatic adaptation (e.g., appropriate learned behavior) that effectively manages the moderate environmental stress, a viable adaptation to changing environmental conditions is achieved. This combination may serve as well as any other inherited environmental adaptation, providing continued somatic performance is assured. In this situation, the ability to adapt to environmental stress truly depends upon somatic performance, changing behavior as a result of learning. If learning does not occur, there will be no adaptation. This is the foundation of managing mankind's future. There are adaptation issues related to mankind's present environment (in Humanity chapter). The continuing selective forces of any moderate environmental stress will continue to eliminate less-adapted individuals; they will not pass through this time into the future.

The views of Hesse and Lamarck seem directly opposite to those of Darwin. Recalling Darwin's *Origin of Species* chapter on laws of variation, he stated: "How much direct effect of climate, food and etc. produces on any being is extremely doubtful. My impression is that effect is extremely small in the case of animals, but perhaps rather more in that of plants. We may at least, safely conclude that such influences cannot have produced the many striking and complex co-adaptations of structure between one organic being and another, which we see everywhere throughout nature. Some little influence may be attributed to climate, food and etc.... such considerations such as these incline me to lay very little weight on the direct action of the conditions of life." Darwin's natural selection (the preservation of favorable variations and the rejection of injurious variations) was in contrast to Darwin's conditions of life (everything that affects an organism during its lifetime, the circumstances of biological existence, or the environments effects upon an organism). It would seem that environmental selection as the gatekeeper for genetic success is still closer to and more intimately involved with Darwin's view on natural selection than is Hesse's environmental selection based upon the greater importance of somatic change through use. And since Darwin said: "Natural selection is the preservation of favorable variations and the rejection of injurious variations," doesn't that really mean natural selection is the (genetic) preservation of favorable variations and the (environmental) rejection of injurious variations? **Natural selection genetic bias is the genome's preservation of environmental adaptations** (genotype display part of environmentally-integrated phenotypes). It is interesting to reexamine Hesse's text content, with the understanding that new adaptations result from environmental selection's (gatekeeper for genetic success) directional influence on new expressions from the genetic bias of natural selection, the (genotype) preservation of successful phenotypic adaptations to previously-experienced environments. While some of the text material is outdated, the adaptations discussed by Hesse still have relevance today. In the same way that Darwin's *The Origin of Species* will forever remain as the classic work on natural selection, Hesse's *Tiergeographie auf oekologischer Grundlage* will forever remain as the classic work on environmental selection. Any discussion of changes in types must include the inheritable genetic considerations; however, the last words in any discussion of evolutionary process involve the closing argument of environmental selection (the mechanism of a natural selection process) and its role of influence early in development, during the adaptation process. **Early developmental adaptations are environmentally-integrated, and are inherited.**

27

A Foundation of Modern Evolutionary Biology

In this text, natural selection was first discussed as a process or result of evolution that preserves favorable variations and rejects unfavorable variations; it was primarily influenced by success in a process of fatal competition for limited resources. Later in this text, natural selection was seen as a process or result of evolution, influenced primarily by environmental selection, which leads to a genetic bias (preservation) of more environmentally-fit types; said otherwise, natural selection is the (genetic) preservation of favorable variations and the (environmental) rejection of injurious variations. I have hypothesized that natural selection is the (genotype) preservation of successful phenotype adaptations to previously-experienced environments. Natural selection is undeniably the foundation of modern evolutionary biology, but the term seems to have more than one meaning. When natural selection has more than one meaning, it makes understanding of a difficult concept even more difficult for the general public to comprehend. Natural selection requires further clarification. Douglas Futuyama's 3rd edition of *Evolutionary Biology* is an excellent up-to-date text that deals with the confusion on the various definitions of natural selection. In this text, the author includes sexual selection as a form of the natural selection process; recall that Darwin treated sexual selection separately from his natural selection of traits resulting from a fatal competition for limited resources. Other than sexual selection, Futuyama uses ecological selection to refer to all other forms of natural selection. Ecological selection (including competition) should be the same as environmental selection. Sexual selection is also one of the conditions of life; the conditions of life are the conditions of the ecosystem, which includes all interactions within both the biological community and the associated physical environment. Futuyama acknowledges the process of natural selection as one of the two essential components of evolution (mutation is the other). All of this would seem to relegate ecological (environmental) consideration to a very high level of importance. Futuyma clearly recognizes the conditions of life.

It seems as if any natural process with selective evolutionary impact could be considered as part of a natural selection process. Confusion occurs when natural selection is treated as one process. It is comprised of several processes that often combine to generate a result. Futuyma defines natural selection as "any consistent difference in fitness (i.e., survival and reproduction) among phenotypically different biological entities." This is very close to Levinton's definition of natural selection (a change in gene frequencies in a population, owing to fitness of phenotypes' reproduction or survival among the variants). Both definitions seem to treat natural selection more as a result of the process of evolution, than a process of evolution; any "process" seems to be a combination of mostly environmental processes. Both definitions acknowledge that: "Selection acts on phenotypes, but may change allele and genotype frequencies, if the phenotypes differ in genotype." Both authors are prominent professors at The State University of New York at Stony Brook. Although there are strong similarities of evolutionary views in their texts, there appears to be some differences between them as well.

Natural selection and adaptation are delayed for discussion until midway through Futuyma's text (Chapter 12: Natural Selection + Adaptation, Chapter 13: The Theory of Natural Selection); he also lists several definitions of adaptation: "An adaptation is a phenotypic variant that results in the highest fitness among a specified set of variants in a given environment." (and) "For a character to be regarded as an adaptation, it must be a derived character that evolved in response to a specific selective agent." (but seems to prefer) "A feature is an adaptation for some function if it has become prevalent or is maintained in a population (or species, or clade) because of natural selection for that function." There is good reason for professor Futuyma to discuss adaptation in combination with natural selection; they are inseparable. There would be no (natural selection) preservation of traits if heredity and environment could not coordinate in the production of new traits that become adaptations to a new environmental challenge. I have hypothesized that natural selection is the genetic preservation of past environmental adaptations. Both genetics and the environment were equally involved in the original formation of past adaptations that are preserved (as genotype expressions of phenotypes), are inherited, and are expressed today as normal phenotype variations. Simply said, natural selection is partly-preserved adaptations. Recall that Darwin hypothesized that the cause of evolution, changes in the types of life, is a natural selection of more-fit individuals. Changes in the types of life (phenotypes—indirectly genotypes), occur through inheritance of favorable (environmentally-selected) genetic variations.

In common phenotype formation (determining the developing phenotype from a genotype's norm-of-reaction, in response to existing environmental conditions), Professor Futuyma allows environmental considerations significant status, comparable to the role of genetics. "Variation in most phenotypic traits includes both a genetic component and a nongenetic ('environmental') component; the proportion of the phenotypic variance that is due to genetic variation is the heritability of the trait." The environmental contribution to the formation of familiar phenotypic traits is acknowledged here; but, does he also allow the environmental contribution to the formation of novel phenotypic traits to be on an equal status with genetics for the first-time formation of any new trait (that will become an adaptation and be preserved)? In a directional selection discussion, he states: "The replacement of relative disadvantageous alleles by advantageous alleles is the fundamental basis of adaptive evolution. This occurs when a homozygote (sameness in paired alleles) for an advantageous allele has fitness equal to or greater than that of the heterozygote (different alleles paired) or of any other genotype in the population. An advantageous allele may have been initially very rare if it is in a newly arisen mutation or if it was formerly disadvantageous, but the environment has changed so that it is now favorable; or, it may have been initially very common, if under previous environmental circumstances it was selectively neutral, or was maintained by one of several forms of branching selection." Continuing with Futuyma's discussion of natural selection, he notes that: "Genotypes that are heterozygous at several or many loci appear to be fitter than more homozygous genotypes…. The all-important concepts of allele frequency and genotype frequency are central to the Hardy-Weinberg principle, which states that in the absence of perturbing factors, allele and genotype frequencies remain constant over generations. Biological evolution is change, over the course of generations, in the properties of populations of organisms, or groups of populations; thus, it consists of descent with modification, and often includes diversification from common ancestors. Evolution would be very slow if populations were genetically uniform and only occasionally mutations arose to replace preexisting genotypes; however, populations of most species contain a great deal of genetic variation—this variation includes rare alleles at many loci, which usually appear to be deleterious, but it also includes many common alleles, so that many loci (perhaps up to 1/3rd) are polymorphic. Many phenotypic traits, including morphological, physiological, and behavior features, exhibit polygenic variation; although the individual loci cannot usually be distinguished and studied, the magnitude of this variation can be estimated by breeding experiments and by artificial selection." Natural selection, "the major process" of evolution, is further discussed by Futuyma: "Natural selection is not the same as evolution; evolution is a two step process: the origin of genetic variation by mutation or recombination, followed by a change in the pattern of variation, such as replacement of some genotypes with others. Natural selection is one agent of change in the pattern of variation; genetic drift is another. Both can be responsible for the spread of traits through populations, but neither natural selection nor genetic drift accounts for the origin of variation. Natural selection is different from evolution by natural selection. Just as evolution can occur without natural selection (e.g. by genetic drift—random change in the frequency of alleles), natural selection can occur without evolution. In some instances... genotypes differ in each generation in survival and fecundity (production of offspring), yet the proportions of genotypes and alleles stay the same from one generation to another.... This happens when homozygotes' (same paired alleles) fitness are different, but both are fitter than the heterozygote (different alleles paired)... selection cannot move the population from the less fit to the more fit condition; thus a population is not necessarily driven to the most adaptive possible condition." In Futuyma's example with two types of homozygous alleles at the same locus having greater fitness, if population adaptation means were graphically represented, there would be two peaks of optimum conditions in the environmental landscape. It is far more common to find only one peak of optimum conditions for a population of like phenotypes (assumes like genotypes); this would be the case if there is only one form of phenotype-expressed homozygote advantage, or in the case of phenotype-expressed heterozygote advantage. Assuming an unchanging environment, increased "specialization" to that particular environment causes adaptive peaks to increase (# of environmentally-favored phenotype individuals) and to narrow (ecological valence), enhancing the narrow advantage. There is a trade-off; generalists with a broader adaptive peak have a broader environmental tolerance (ecological valence), but do not do as well (increase in #) in a constant, unchanging environment. (The epilogue will discuss peak shifts.)

Futuyma's more genetically-determined views on evolution strongly differ from Hesse's environmental emphasis in the evolutionary process (recall that Hesse minimized germinal change in deference to environmental adaptation). Futuyama states: "It is naïve to think that if a species' environment changes, the species must adapt or else become extinct. That is true when some factor changes beyond the tolerance limits of the majority of the population, but many environmental changes, even if they reduce the species' abundance, do not press these limits. Some changes enhance the abundance of certain species; the conversion of farms to suburbs, for example, increases the abundance of some bird species while reducing others. An environmental change that does not threaten extinction may nonetheless set up selection for change in some characteristics. Thus white fur of the polar bear may be advantageous, but not necessary for survival. Just as a changed environment need not set in motion selection for new adaptations, conversely, new adaptations may evolve in an unchanging environment. For example, new mutations may arise that are superior to any pre-existing genetic variations." In the beginning of this chapter, recall that other than sexual selection, Futuyama uses ecological selection to refer to all other forms of natural selection. So, shouldn't ecological selection (including competition) be the same as environmental selection? (Sexual selection is also one of the conditions of life; the conditions of life are the conditions of the ecosystem, which includes all interactions within both the biological community and the associated physical environment.) Futuyama appears to believe natural selection (ecological selection) is a chance phenomenon. Futyma seems to believe natural selection (ecological selection) acts primarily on existing traits (variations) and sudden trait changes from mutation. It also appears that Futuyma believes that gene change (mutation) causes adaptation, an improvement in function. In the above, it seems as if the environment doesn't play a very significant role in evolution. This genetically-determined view of evolution is incomplete. There is little wrong with genetically-determined evolution, except that it is incomplete. Extinction is primarily an environmental process; habitat destruction is more the issue than genetics. Extinction is environmentally-determined, not genetically-determined; however, just as genetically-determined evolution allows some role for environmental variation, genetics may still play some role in extinction. Evolution is not genetically-determined, any more than it is environmentally-determined. Evolution is genetically AND environmentally determined.

Although I may be misinterpreting, I believe the "conversion of farms to suburbs increases the abundance of some bird species while reducing others" example will create increasing difficulties for most already-stressed endemic species. The numbers of a well-adapted population may increase, but species diversity will fall. Population decimation (10% loss is conservative) in number of bird species, and particularly in numbers of individuals for an endemic bird species, is almost assured on a manicured lawn with free-roaming cats. Conversion of the original wildlife habitat to farmland was the beginning of the end for indigenous wildlife, approaching local species loss percentages equaling those of mass extinctions, and this loss preceded the suburbs. Unless polar bears can adapt to the upcoming environmental change, habitat loss from global warming will lead to their extinction. The habitat loss will involve a loss of adequate polar-ice based food sources to sustain a wild polar bear population (they can tolerate warmer climates in zoos). They need their white fur to tolerate the cold; it is essential for survival during continuous exposure to the harsh winter of the Arctic environment. Closely-related brown bears are not found here; unlike the winter-loving polar bear, they must seek winter shelter. The polar bear's white fiber-optic fur transfers heat from sunlight to the black skin below, while it insulates from the extreme cold and provides excellent predator camouflage for hunting on the ice. Unchanging environments aren't necessarily non-changing from a habitat change perspective; all environments change. Spatial and temporal changes are part of any environment of any size. All environments undergo constant change; any new phenotypic change would undergo immediate environmental selection. However, it is helpful to recall that stable environments are commonly associated with a relative lack of change, and trait changes are normal trait changes. Normal changes in genetic variation involve normal phenotype change, expression of previously-preserved phenotypes that are adaptations to previously-encountered environments. Significant evolutionary changes are commonly associated with significant environmental change. Surviving individuals with a novel adaptation are commonly associated with having survived stress-induced population damage from an abnormally-stressful environment. In other words, major adaptive change parallels major environmental change. Small changes in genetic variation will be more commonly associated with small environmental variations.

Adaptation and natural selection genetic preservation involve both environmental selection AND genetic selection. Natural selection's genetic bias is the result of an environmental selection process; variable environmental conditions indirectly selected variable genetic expressions. The genetic expressions that were part of successful environmental adaptations were preserved in a newly-modified norm-of-reaction, as a newly-modified natural selection genetic bias. Today the environmentally-sensitive genotype norm-of-reaction expresses environmentally-integrated phenotypes for frequently-encountered environments. Genetic variation is important. But environmental conditions, like genetic combinations, also vary. So far, *Environmental Biology* has established co-equal roles of environmental variation and genetic variation in changes in types of life over time. Natural selection doesn't act on genetic variation; environmental selection and natural selection (preserved genetic variation) interact. Natural selection is preserved genotype variation—for the phenotype. Natural selection genetic bias is the genome's selected phenotypic adaptation components, which were successful in previously-encountered environments; it is also the raw material used to form new adaptive phenotypes. **Natural selection (the preservation of genetic variation) is the result of an ongoing environmental selection process.** I have hypothesized that natural selection is the genome preservation of successful phenotypic adaptations that were formed from exposure to previously-experienced environments. In short, natural selection is the genome's preservation of adaptations to frequently-encountered environments. Shorter still, natural selection is the genetic preservation of adaptive phenotypes. Natural selection is partially-preserved adaptations (preserved genetic variation). Genetic constraint must operate above the species level to maintain the genetic preservation of an adaptation, which may be lost later to either mutation or environmental selection.

The long view of evolution over time is that random genetic variation does occur (chance variation—mutation and variable mutation expressions), but that environmental selection is not a random process; it is directional and this direction is determined by the existing environmental conditions. Nearly all adaptations are environmental adaptations; adaptation is the basis of environmentally AND genetically-integrated design. A new environmentally-stabilized adaptation is an environmentally-integrated expression of the genotype, even if it began as a random phenotype expression resulting from unusual environmental stress. The environmentally-integrated adaptive process may enhance a random phenotype's stability and survival, resulting in the preservation of more-fit individuals. This favored survival of adaptations (natural selection) continues in a population from the time (of initiation) of the environmentally-integrated adaptation advantage, up to the time of extinction.

When it comes to putting it all together, making evolution the unifying theory of biology, no one else does it better than Futuyma. I have seen no better text that clearly explains the basics of evolution. He may know more details about evolution than anyone else. And, I would fully expect him to find the more environmentally-determined evolutionary view, discussed in this text, to be both narrow and unconvincing; in fact, he already has. In a chapter on The Evolution of Genetic Systems, under the topic of sex and recombination, he dismissively mentions an environmentally-deterministic hypothesis (in 0.002% of the text). "In environmental deterministic hypothesis, selection acts not on new mutations, but on existing variation, by reorganizing it into new combinations." I seem to have difficulty following Professor Futuyma here. Is the old usage of mutation (new phenotype variation) involved? I understand that a new mutation in the genotype may or may not be expressed in a phenotype. If a new mutation immediately affects the phenotype, environmental selection of the new phenotype variation occurs immediately (new phenotype may be environmentally positive, negative, or neutral). If a new genotype mutation is silent or neutral (a phenotype variation is not expressed or it makes no change), the genotype is unaffected by environmental selection. Environmental selection may initiate a new phenotype expression from the new genotype at a later time (now with the new mutation), which may include the particular mutation's expression in a new phenotype; this new environmentally-integrated phenotype expression (variation) may be environmentally positive, negative, or neutral, and will then undergo future environmental selection over time. Futuyma continues: "The most likely circumstance in which this reorganization can provide an advantage is when a polygenic character is subject to stabilizing selection (selection against phenotypes that deviate in either direction of an optimal value in character), but the optimum fluctuates, due to a changing environment.... Let us assume that alleles A, B, C, D... additively increase a trait such as body size, and alleles, a, b, c, d... decrease it. Stabilizing selection for intermediate size reduces the variance and creates negative linkage disequilibrium, so that combinations such as AbCd and aBcD are present in excess. If selection fluctuates so that larger size is favored, combinations such as ABCD may not exist in an asexual population, but they can arise rapidly in a sexual population. This can provide not only a long-term advantage to sex (a higher rate of adaptation of the population), but a short-term advantage as well, because sexual parents are likely to leave more surviving offspring than asexual parents." This truly underscores the advantage of genetic biodiversity in individuals. In Futuyma's coverage of an environmentally-deterministic hypothesis, emphasis still seems to be on genetics, isn't this an environmentally AND genetically deterministic acknowledgement?

In a somewhat-related topic of heterogeneous habitats, Futuyma notes: "Spatial, rather than temporal variation in the environment can also provide a strong advantage to sex.... Before explaining this hypothesis, we must point out that a genetically variable asexual population, consisting of various clonal genotypes, may actually utilize a variety of habitats or resources more effectively than a sexual population. Suppose an individual animal of a certain size can effectively feed on only a narrow range of certain particle sizes. The distribution of body size in a sexual population will be more or less normal due to recombination, so that some food types such as the largest and smallest will be underutilized. However, if different asexual genotypes span the range of body sizes, each genotype can increase in frequency to the level set by the abundance of its favored resource.... Populations that include combinations of several clones are more abundant, in the aggregate, than sexual species, as we would expect if they can collectively use a wider range of resources. If a combination of asexual genotypes were to use the full spectrum of resources, it could drive an asexual population to extinction.... Heterogeneity of resources could, however, favor (sexual) recombination in two related ways. First, suppose offspring disperse at random into small patches of habitat where competition occurs, and only one or a few of the individuals best adapted to conditions in a patch survive. Williams described this situation as a lottery, in which a sexual female has many 'tickets' with different numbers (different offspring genotypes), but an asexual female has numerous copies of the same ticket. The probability is greater that the sexual female will have 'the winning ticket.' The second hypothesis assumes that even within an apparently homogenous habitat patch, genotypes can differ in their use of limiting resources; for example, plant genotypes may vary in the ratios of various nutrients they require for growth. If siblings compete, then a patch of habitat can sustain more progeny from a sexual family than from an asexual family, because asexual siblings compete more intensely." Recall that in an earlier chapter on environmental biology, Alone, that it was suggested that speciation is an isolating mechanism, and in complex animal life it is increasingly the result of sexual selection, a behavior phenotype, which may itself be further constrained by other isolating factors in the environment. A species is a sex club; it supports harmonious development and reproduction of complex life. Sex enhances variation and enables the isolation of speciation.

What is not delayed or diminished in the Futuyma text is an early-on discussion of the principal claims of the Evolutionary Synthesis, or Modern Synthesis. (It is a genetically-determined view of evolution, and it is incomplete.) "The Modern Synthesis is forged from the contributions of geneticists, systematists and paleontologists (certainly not paleontology)… , and reconciling Darwin's Theory with the facts of (Mendelian) genetics… developed a mathematical theory of population genetics, which showed that it is the conjunction of mutation and natural selection (among other things) that causes adaptive evolution; mutation is not an alternative to natural selection, but is rather its raw material (for preserved genetic bias)." Futuyma further elaborates on the "Major Tenets of the Evolutionary Synthesis: The phenotype is different from the genotype, and the phenotype differences among individual organisms can be partly due to genetic differences and partly to direct effects of the environment. Environmental effects on an individual's phenotype do not affect genes passed on to its offspring (acquired characteristics in adults are not inherited); however, the environment may affect the expression of an organism's genes (and this change in expression is inheritable). Hereditary variations are based on particles (genes) that retain their identity as they pass through the generations (genes do not blend with other genes); this is true… of those genes that have discrete effects on the phenotype (e.g. brown vs. blue eyes) , but also of those that contribute to continually varying traits (e.g. body size, intensity of pigmentation); variation in continuously varying traits is largely based on several or many discrete genes, each of which affects the trait slightly (polygenic inheritance). Genes mutate, usually at a fairly low rate, to alternative forms (alleles); the phenotypic effects of such mutations can range all the way from undetectable to very great; the variation that arises by mutation is amplified by recombination among alleles at different loci. Environmental factors (e.g. chemicals, radiation) may affect the rate of mutation, but they do not preferentially direct the production of mutations that would be favorable in the organism's specific environment. Evolutionary change is a population process; it entails... a change in the relative abundances (proportions) of individual organisms with different genotypes (= often different phenotypes) within a population; over the course of generations the proportion of one genotype may increase and gradually replace the other type—this process may occur within only certain populations or in all populations that make up the species. The rate of mutation is too low for mutation by itself to shift an entire population from one genotype to another; instead, the change in genotype proportions within a population can occur by either of two principal processes: random fluctuations in proportions (random genetic drift) or nonrandom changes due to superior survival and/or reproduction of some genotypes compared to others (natural selection), both can operate simultaneously. Even a slight intensity of natural selection can (under certain circumstances) bring about substantial evolutionary change in a relatively short time; very slight differences between organisms can confer slight differences in survival or reproduction—hence natural selection can account for slight differences among species and for the earliest stages of evolution of new traits. Selection can alter populations beyond their original range of variation (by recombination with other genes that affect the same trait) by increasing the proportion of alleles that give rise to new phenotypes. Natural populations are genetically variable: the individuals within populations differ genetically and include natural genetic variants of the same kind that arise by mutation in laboratory stock. Populations of a species in different geographic regions differ in characteristics that have a genetic basis; the genetic differences among populations are often of the same kinds that distinguish individuals within populations; a genotype that is rare in one population may be predominant in another. Experimental crosses between species, and between different populations of the same species, show that most differences between them have a genetic basis; the difference in each trait is often based on differences in several or many genes (i.e. it is polygenic), each of which has a small phenotype effect... the differences between species evolve by small steps rather than single mutations with large phenotypic effects. Natural selection in natural populations occurs at the present time, often with considerable intensity. Differences among geographic populations of a species are often adaptive (hence, a consequence of natural selection) because they are frequently correlated with relevant environmental factors. Organisms are not necessarily different species just because they differ in one or more phenotypic characteristics; phenotypically different genotypes often are members of the same interbreeding population... different species represent distinct gene pools which are groups of interbreeding or potentially interbreeding individuals that do not exchange genes with other groups; the reproductive isolation of species is based on certain genetically determined differences between them (I wouldn't believe this genetically-determined view.)... even a mutation that causes substantial change in some phenotypic feature does not necessarily represent the origin of a new species. Nevertheless, there is a continuum of degree of differentiation of populations, with respect to both phenotypic difference and degree of reproductive isolation, from barely differentiated populations to fully distinct species... an ancestral species differentiates into two or more different species by the gradual accumulation of small differences rather than by a single mutational step. Speciation, the origin of two or more species from a single common ancestor, usually occurs through the genetic differentiation of geographically segregated populations; geographic separation is required so that interbreeding does not prevent incipient genetic differences from developing. Among living organisms, there are many gradations in phenotypic characteristics among species assigned to the same genus, to different genera, and to different families or other higher taxa... higher taxa arise through the prolonged, sequential accumulation of small differences, rather than through the sudden mutational origin of drastically new types. The fossil record (which does not support the last statement) includes many gaps among quite different kinds of organisms, as well as gaps between possible ancestors and descendants; such gaps can be explained by incompleteness of the fossil record... the fossil record also includes examples of graduations from apparently ancestral organisms to quite different descendants... the evolution of large differences proceeds by many small steps (such as those that lead to the differentiation of geographic populations and closely related species)... we can extrapolate from the genesis of small differences to the evolution of large differences among higher taxa, and can explain the latter by the same principles that explain the evolution of populations and species. Consequently, all observations of the fossil record are consistent with the foregoing principles of evolutionary change." (Don't believe this one either; changes in fossils are far more consistent with environmental variations.)

Futuyma endorses the above claims as "the foundations of modern evolutionary biology." Both Futuyama and Levinton seem to hold the long-standing geneotype-determined view of evolution; this view is incomplete. The incompleteness of genetically-determined underlies much of the confusion (discussed in this text) associated with the usage of "mutation," "adaptation," and "natural selection." Still, it does seem as if Levinton recognizes a greater environmental role in evolution. Levinton's depth of paleontology expertise and ecological insight is most exemplary. Levinton supports the hierarchy of the individual organism's survival as the level of evolutionary selection, whereas Futuyma discusses "Examples of... forms of selection have been described, but we do not know which are most important for explaining the abundant genetic variation found in most populations, nor whether selection, rather than genetic drift, is the most determinant of evolution at the molecular level." Futuyma accepts natural selection at many levels of hierarchy. Does evolutionary selection hierarchy occur at the level of the individual or at the level of a molecular change? Yet, at other times, as in his discussion of traits in life histories, Futuyma will discuss the importance of fitness in the individual organism; individual traits, such as behavior, can benefit individuals within a population. Levinton discusses the genetic ramifications of natural selection in great detail; however, there seems to be an even greater focus on paleontology and on ecological differences as related to different populations. For example, he discusses adaptive peaks undergoing major shifts with sudden landscape change; these are associated with a pattern of "constancy, sudden change, and then constancy; morphological change is to be expected." This strongly supports the primary role of environment in any natural selection process. In his opening chapter on Macroevolution, The Problem and The Field, Levinton notes that in the last half-billion years, evolutionary change seems to be largely at the family level; "families appear and disappear continuously throughout the Phanaerozoic." Levinton references studies supporting that families seem to correspond to ecologically adaptive zones better than species: "family-level divisions may represent minor evolutionary changes that came and went in response to minor changes in earth and biotic history." In his chapter on Genealogy, Systematics, and Macroevolution, Levinton references research that indicates "taxon longevity depends upon biogeographic range." Levinton further states that "adaptations of individuals influence the susceptibility of species and larger taxa.... Species become extinct because of character traits they bear." In a chapter on Genetics, Speciation and Transspecific Evolution he further notes that: "The origin of a highly divergent group is often associated with the movement into a major new habitat and lifestyle... the accumulation of genetic differences might be quite slow... ecologically significant differentiation might be much more rapid." Regarding genetic considerations in fruit flies, he references: "Some viability differences among individual organisms bearing different inversions (genetic mutation with repair causing inverted gene order along chromosome) may relate to the effects of temperature and crowding. In some cases, strong regional differences in seasonal variation are related to climatic differences.... This may indicate that superiority is conferred by favored gene combinations and not by any innate superiority of chromosomal heterozygosity (this inversion phenomenon seems to be restricted to fruit flies)."

In all fairness, Futuyma's text, in a chapter on A History of Life on Earth, does contain an excellent chronology of environmental change and biologic change. Although Futuyma recognizes that natural selection does have an ecological component, he seemingly does not want to give the environmental conditions a primary role in the evolutionary process. Futuyma illustrates (the co-primary) role of genetic (genotype) determination extremely well. Futuyma does note that while the environmental stress does not directly influence a mutation, it can influence mutation rates (Futuyma observed that: "There is abundant evidence that the rate of crossing over, a mutation associated with sexual reproduction, is genetically controlled and can evolve rapidly—Brooks study.)" and that organisms can vary genetic expression at different times and in different areas (no substantial environmental tie-in, but recognizing ability to vary expression is important). Futuyma indirectly supports the co-primary environmental role in natural selection when he recognizes that: "The history of increase in marine animal diversity through the Phanerozoic (Paleozoic, Mesozoic and Cenozoic) can be explained, in part, by increases in the occupancy of ecological space.... Evolutionary radiation, rather than sustained, progressive evolutionary trends, is probably the most common pattern in long-term evolution." Surely, nothing else causes an evolutionary radiation quite like exploiting a new environment following a significant environmental destruction and the associated extinctions. Both texts provide a far better overall perspective of *Evolutionary Biology* and *Genetics, Paleontology and Macroevolution* than the narrow environmental aspect of evolutionary theory contained within this *Environmental Biology* text.

28
Difficulties on Theory

Different usages of "natural selection" can create confusion. So far, it has been demonstrated that <u>natural selection is not one process, but a combination of several processes that lead to the preservation of favorable variations and the rejection of injurious variations</u> (result). **If natural selection is ever discussed as a process, chances are that it is really environmental processes that combine for the effect (result), and that result is natural selection genetic bias.** The case has been made for the co-primary role of the environment in forming the genetic bias of <u>natural selection (a change in gene frequencies in a population, owing to fitness of phenotypes' reproduction or survival among the variants— Levinton).</u> The conditions of life play a major selective role in the natural selection (preservation) of more-fit individuals. Environmental selection determines the pathway of survival; survivors often show a change in the genotype, compared to the genotype of the former population. Therefore, **the environment indirectly changes the genotype.**

The environment indirectly changes the phenotype. The genotype's expression of the phenotype is environment-sensitive; the organism uses natural selection genetic bias in displaying the phenotype that is expressed under normal conditions. Environmental signals influence natural selection's genetic bias (comprising the genotype's norm-of-reaction of phenotype variations) to display the most appropriate phenotype, and, under severe environmental stress, alternative phenotypes. **Nearly all adaptations are environmental adaptations; adaptation** (genetically and environmentally-integrated design) **is the basis of the phenotype variation preserved in the genotype as natural selection genetic bias.** <u>Natural selection is the preservation of adaptations; it is the genetic inheritance for environmentally-proven phenotypes.</u> **Natural selection genetic bias is the genome's selected phenotypic-adaptation-components, which were successful in previously-encountered environments; it is also the raw material used to form new adaptive phenotypes.**

The environment has an indirect influence on changing phenotype anatomy and/or physiology; during development this influence is time-sensitive. Recall environmentally-integrated design will be involved in the formation of any new adaptations (new inheritable environmentally-selected phenotypes). The environment is the gatekeeper for success or failure in phenotype variation. Adaptation is improved function of an organism in the environment; adaptation to an environment is a change in fitness. So, in isolated, changing environments, natural selection genetic bias may or may not allow an organism to cope with changing environmental conditions; it often does, but sometimes it does not. If the existing norm of genotype reactions to the environment cannot permit survival, a randomly-formed new phenotype expressed as a reaction to the uncommon environment may allow life to continue. The environment interacts with stress-induced phenotypes. Positive environmental interactions are adaptive; negative environmental interaction is non-adaptive. Only adaptive phenotypes survive environments lethal to non-adaptive phenotypes. Stress-induced phenotypes may persist for several generations; they are not necessarily adaptive. The stress-induced phenotypes can be stabilized by remote environmental feedback, and the genotype expression of the random phenotype can be functionally-integrated (supported) by the genetic hierarchy. This has been verified by icefish hemoglobin gene atrophy and the inheritance of instinctive behavior (in respective The Rest of the Story and Adaptation chapters). The stress-induced phenotypes are constructed from components of previous pathways that are stabilized genetic expressions of phenotypes in previously-encountered environments. A new environmentally-selected adaptation represents a new phenotype expression from an individual organism's genotype. Over time, the new adaptation's environmental advantage increases the numbers of the better-adapted phenotype in a population. This population has a <u>new natural selection genetic bias</u>; it now includes <u>a newly-preserved adaptation pathway that is better-designed for enhanced survival in the newly-changed environment</u>. The adaptive process enables the creation of more-fit individuals; i.e., environmental adaptation allows "the preservation of favored races in the struggle for life." This favored survival occurs in a population from the time (of initiation) of the genetically and environmentally co-designed adaptation advantage, and continues as a phenotype expression of the genotype in the bias of natural selection. Environmentally and genetically integrated design is everywhere. Environmentally-integrated indirect design was also created in past, isolated, changing environments from environmental interaction with natural selection genetic bias (preservation of past adaptations).

In *The Origin of Species*, in his chapter of Difficulties on Theory, Darwin expressed concern for natural selection's ability to form organs of extreme perfection and complication. "To suppose that the eye, with all its inimitable contrivances for adjusting the focus to different distances, for admitting different amounts of light, and for the correction of spherical and chromatic aberration, could have been formed by natural selection seems, I freely confess, absurd in the highest possible degree. Yet reason tells me, that if numerous gradations from a perfect and complex eye to one very imperfect and simple, each grade being useful to its possessor, can be shown to exist; if further, the eye does vary ever so slightly, and the variations be inherited, which is certainly the case; and if any variations or modification of the organ be useful to an animal under changing conditions of life, then the difficulty of believing that a perfect and complex eye could be formed by natural selection, though insuperable by our imagination, can hardly be considered real."

It is now known that each (genotype's) successive elaboration of an environmental adaptation (phenotype) increases the efficiency of an already serviceable organ. Many such steps collectively produce a combination of characters (adaptations) that could have never existed in an ancestor. There are many examples of intermediate states that can be observed in living forms. In the knowledgeable scientific community, Darwin's interpretation has successfully withstood the test of time. In Futuyma's text, in the chapter on Pattern and Process in Macroevolution, he states: "Since Darwin's time, a great deal of information on the photoreceptive organs of various animals has been amassed.... These organs are exceedingly diverse in structure and function, ranging from small groups of merely light-sensitive cells to the complex structures, capable of registering precise images, found in many arthropods, some mollusks, and vertebrates. Many protists (single-celled organisms), such as dinoflagellates, have an eyespot consisting of an aggregation of visual pigment associated with the chloroplast or base of the flagellum. These organisms can move in response to changes in light intensity. This photosensitive structure resembles that of ciliated photosensitive cells that are widely distributed among the animal phyla. In a few instances, the phylogenetic sequence of photoreceptor evolution has been clarified; in its place, we can recognize grades of complexity among unrelated animals that show the adaptive feasibility at each stage.... The simplest grade is a mere aggregation of a few or many photosensitive cells, found in some flatworms, rotifers, annelid worms, vertebrates (lamprey larvae), and others. The next grade is a simple ectodermal cup lined with photic cells; this structure, which can provide some information on the direction of a light source through the differential illumination of different parts of the cup, has evolved independently in numerous lineages of flatworms, cnidarians, mollusks, polycheates, cephalochordates, and others. From this grade, there are numerous transitions to pinhole eyes and thence to closed eyes in which translucent cells or cell secretions (vitreous mass) act as a rudimentary lens. Closed eyes, usually with some kind of lenslike structure have evolved independently in cnidarians, snails, bivalves, polycheate worms, arthropods, and vertebrates. A closed eye with a lens allows an organism to more accurately determine the direction of incident light and to orient by it, to detect movement of objects, and by principles of the pinhole camera, to form at least elementary images. Image formation reaches its apogee in insects, in which each element (ommatidium) of the compound eye subtends a small angle of the field of view, enabling the many elements together to provide a detailed mosaic image; and in cephalopods and vertebrates, in which muscles move the lens or alter its shape in order to focus.... Insect and vertebrate eyes as such are truly not homologous (the same in common ancestry), since they differ so greatly in structure. It is likely, instead, that... gene primitively had the more general function of governing the differentiation of photosensitive cells, which are more widely distributed among animals that they doubtless predate the divergence of the animal phyla (of complex life forms from single-celled organisms). In various phyla, other genes that organize the eye structures came under control of this master gene. Salvani-Plawen and Mayer estimated that at least 15 lineages have independently evolved eyes with a distinct lens. The evolution of eyes is apparently not so improbable! Each of the many grades of photoreceptors, from the simplest to the most complex, serves an adaptive function. Simple epidermal photoreceptors are most common in slowly moving or burrowing animals; highly evolved structures are more typical of more mobile animals. The mystery of how a simple eye could be adaptive is no great mystery at all."

Just as reason told Darwin that there must be a gradual change of serviceable phenotypes to arrive at present phenotype complexity, reason tells me that phenotype expression must be stabilized by environmental signal feedback, and that the environmentally-stabilized design must be functionally-integrated into genetic hierarchy. Consider the previously-discussed Waddington experiments on stress-induced phenotypes in this text's chapter on Phenotype and Environment. **There is no new mutation every time there is a newly-expressed random phenotype. Almost always, at the time of new random phenotype expression, the genes** (would) **remain unchanged.** Mutations that suddenly change the phenotype do occur, but they do not occur frequently. New genes come from old genes that were present in ancestral types. Recall the timeline in the icefish's hemoglobin gene atrophy (in Rest of the Story chapter); genes change, but the changes are slow, over time. A phenotype's long-term survival depends upon environmental selection support from the stressful environment (culling the mutation-induced change). Gene expression not supported by environmental selection leads to gene atrophy due to mutation. Somewhat related to gene atropy (see upcoming enhancer discussion), Sean Carroll, in an article entitled *Regulating Evolution* notes that "The loss of features is very common, though much less appreciated.... Loss of features may or may not be beneficial for survival or reproductive success, but some losses are adaptive because they facilitate some changes in lifestyle. Hind limbs, for example, have been lost many times in vertebrates—by snakes, lizards, whales and manatees—and those losses are associated with adaptation to different habitats and means of locomotion.

The single cell of complex life has random behaviors that are stabilized through environmental feedback. Marc Kirschner (chair of Systems Biology @ Harvard) and John Gerhart (professor in Berkeley's Department of Molecular and Cell Biology) argue that Darwinian explanation is incomplete and the results of recent discoveries in cell biology and developmental biology can be used to remedy this. In their chapter on Exploratory Behavior, in *The Plausibility of Life* text, they state: "The architecture of cells is achieved without an architect. No central regulation is discernable. Cells are in fact capable of many structures; many are chameleons that change their structure in response to circumstance. The free-living *Amoeba proteus* was aptly named for the Greek sea god Proteus, who could transform himself into any shape. This capacity relates to a huge capacity of somatic adaptations for cells, which becomes a substrate for evolutionary change, much of it based on exploratory principles. The proteins used to generate cell structure are like all other proteins encoded in the cell's DNA. Although DNA sequences control the time and circumstances of expression, DNA provides no instructions on where to place proteins in the cell. There is no genetic information for large-scale cellular organization. Furthermore, cells having the same DNA and inhabiting the same environment can have very different shapes. Cell shape responds to developmental and environmental cues independently of genetic control. The capacity to change shape underlies significant processes, such as directed migration of cells into the margin of a wound for repair, the extension of nerve axons to different targets in the development of the nervous system, the contortions of white blood cells when it engulfs a large particle or when it infiltrates the lining of a blood vessel to hunt down an infection, and the extensive remodeling that occurs in each cell during cell division. Part of the process of achieving cell organization relies in trial and error, a form of physiological variation and selection at the level of protein assembly." "Cells have an internal skeleton called the cytoskeleton, made up of three different families of long, thin filaments. These filaments criss-cross the cell interior in arrays reflecting its own kind of globular protein unit; in each filament a hundred thousand or more identical globular units may be linearly assembled. The mitotic spindle, whose fibers were perceived by early microscopists to connect the chromosomes to the poles at cell division, is made of microtubules, which are also widely spaced in the cytoplasm of non-dividing cells. The two major filament types, actin and intermediate filaments, along with microtubules, play structural roles in the cell—a different role for each. The cytoskeleton is both rigid, giving the cell specific cell stability against mechanical deformation, and versatile, capable of being reassembled and used over and over again to support different shapes."

Kirschner and Gerhart continue: "The key to adaptability of microtubules is their dynamics. In a typical nondividing cell, hundreds of microtubules radiate out toward the cell membrane from a central nucleating structure. In this configuration, microtubules, like spokes of a wheel, seem to give rigidity to the roughly polygonal cell. Yet this initial characterization of microtubules, as rigid rods giving a cell its shape, was misleading. It was an impression gleaned from the early fixed histological preparation of cells. (In a similar vein, a single snapshot of a football game, as opposed to a movie, would also give a misleading impression of the event.) When specialized methods allowed movies of microtubules to be made from living cells, the microtubules proved to be dynamic. They continually grow, disintegrate, and regrow, each individual microtubule persisting for only about five minutes. When a single microtubule grows out for a period, then spontaneously shrinks back to its point of origin, it is replaced by a new one—which grows in a different direction. Over time, the number and nearly random distribution remains about the same, although individual microtubules change. The entire process of microtubule growth requires energy. This requirement was initially puzzling because much more complicated structures such as viruses assemble spontaneously. In actuality, energy is not involved in assembling the microtubules but instead causes them to disassemble and keeps the turnover dynamic. The purpose of turnover was initially unclear. This process, now called dynamic instability, seemed to amount to nothing more than a futile cycle of growth and shrinkage of individual filaments, without changing their overall distribution. We had glimpsed physiological variation without selection; but, when (environmental) selection was included, it revealed a new and powerful mechanism of somatic adaptation. The function of dynamic instability lies not in the individual assembly of microtubules, but in the capacity to organize them in different arrays. Microtubules extend randomly from their tips and depolymerize back to the nucleation center by loss of their tips…. They continue doing so until they encounter a stabilizing activity in the cell periphery (environmental), which blocks depolymerization at the tip, far from the site of nucleation. Microtubules that randomly enter the region of stabilization persist, while rapid turnover eventually eliminates the others. A particular polarized or asymmetric array is achieved by local stabilization. In the end, most microtubules extend from the center to the stabilizing region at the periphery. When the cell structure is finally achieved, the dynamics of the microtubules may be reduced, and the arrangement will be more permanently stabilized. Thus rapid turnover (random variation) and local stabilization (environmental) can transform an irregular dynamic array into a polarized stable one.

The adaptability of this process is such that stabilizing signals can come from any direction; the microtubule array and the cell will respond appropriately. The mechanism is avowedly selective (indirect), rather than instructive. There is no evidence of instruction from external signals directly causing the microtubules to polymerize in specific directions." **Environmental feedback stabilizes random cell behavior.** The stabilizing influence for the random growth of microtubules is the environment of the cell periphery. **The indirect environmental stabilization of random exploratory cell function, applicable to changing environments, solves a great mystery of adaptive strategy. And, indirectly, environmental feedback must also integrate with genetics.** Microtubules are present in all complex life, but this doesn't mean they are all exactly the same. The introductory chapter of Sean Carroll's *The Making of the Fittest: DNA and the Ultimate Forensic Record of Evolution* notes that "In (warm blooded) mammals, microtubules are unstable below 50 degrees F. If this were the case in Antarctic fish, they would certainly be dead. Quite the contrary, microtubules of Antarctic fish assemble and are very stable at temperatures below freezing. This remarkable change in microtubule properties is due to a series of changes in the genes that encode components of the microtubules, changes that are unique to cold-adapted fish, both icefish and their red-blooded Antarctic cousins." **Without environmental feedback, indirect genetic integration with the environment would not occur.** The two-way connection between genetics and random cellular activity is widely accepted. Genetic function is surely involved in protein production and its initiation and/or inhibition; i.e., the use of variable control switches to regulate the protein production mechanism. "What the genome codes is the means to explore, not the outcome of the exploration.... Exploratory systems, based upon simple rules of interaction,... provide a way to generate complex signals and respond to them with simple functions (K&G)." Genetics is not involved in every aspect of phenotype formation; random exploratory cellular activity plays a significant role. Environmental feedback is also involved in the random cellular activity of phenotype formation; it changes random exploratory cellular activity into a physiologic response.

Complex life modifies the environments within it. Epithelial tissue folds on its basement membrane (collagen), creating an environment within; this environment can be better maintained (homeostasis) or modified. Turner, in the chapter on embryonic origami (*The Tinkerer's Accomplice*), credits epithelial folding and the value-added physiology that goes with it, as the most important event in the history of complex life on earth. The activity in these enclosed environments is mostly random cellular behavior, modified by environmental and genetic feedback. "A sea-urchin embryo, for example, is really a very close-knit family, an assembly of specialized descendents of a single fertilized cell, the zygote (fertilized egg). The embryo is not simply a collection of cells, though, it is the embodiment of a ritual family dance, a highly disciplined and stylized series of steps and maneuvers of the cellular dancers. The dance begins when the zygote's descendants first organize themselves into a spherical hora, a hollow ball of cells called a blastula. The embryo's cells then perform a maneuver called gastrulation, in which the blastula folds in on itself, forming a new partly enclosed space called the archenteron, which will eventually form the animal's digestive tract. The archenteron initially opens to the outside through a small opening called the blastopore, but on its closed end on the opposite side, the dancers part to form a new opening, which will become the mouth, leaving the blastopore to become the anus. (Another group of animals switch this and have the blastopore become the mouth; this caused the entire nervous system to form-upside down.) Soon after, new circles of dancers spin off each side of the archenteron, forming new spaces inside the embryo called coeloms." (The space of the coelom provides an environment for a sheet of cells to fold and expand into a third dimension.) "There is a... metaphor: development as a form of embryonic origami. This... offers a powerful way of thinking about evolution of different body forms. Just as an origami master conjures the most wonderfully complex shapes from just a few simple folding maneuvers, assembled in different patterns and sequences of folds, so do embryonic cells create wonderfully diverse types of animals." "The dance finishes when the archenteron pinches in on itself to form interconnected compartments for the oral cavity, stomach, and intestine. At this point, the embryo can swim, find food and eat."

In the multicellular forms of complex life, the cellular environment of early development is increasingly internalized. Natural selection genetic bias is increasingly varied through sexual reproduction genetic recombination. In the common norm-of-reaction phenotype expression, environmental integration during early development generates the appropriate norm-of-reaction phenotype display to common variations in the environment, using a repertoire of preserved pathways to existing phenotypes as construction materials. Environmentally-integrated indirect design is deeply engrained in existing norm-of-reaction genetic bias; this preserved phenotype-forming genotype bias (of natural selection) is the raw material used to form any new adaptation. In a highly stressful environment, a successful adaptation must be expressed in a timely manner, or an environmentally-integrated response to the stressful environment cannot be genetically transferred to germ cells and offspring (see Environmental Adaptation chapter—Inherited and Somatic Adaptations). Previously, in this text's chapter of Genetic Constraints, recall Kirschner and Gerhart's *Plausibility of Life* chapter on Invisible Anatomy, in a section under A Map in the Embryo, where developmental processes at the cellular level was discussed. (**Environmental creation is achieved through local differences along chemical gradients produced by developmental +/or regulatory genes**.) It followed cellular development from a simple egg. Starting with a localization initiated by fertilization, genetic influence within the fertilized egg is organized into compartments "under conditions set by the surrounding environment… a few small initial differences are acted on to make further differences. When a particular component map is completed in the embryo, it is a map shared by all members of a phylum, the largest group of organisms based upon anatomy and physiology. Though not strictly identical in each species of a phylum, the map is the most conserved (constrained) feature of the phylum's anatomy. Once organized in an advanced multicellular stage, but well before the cells have differentiated into their final cell types, the compartment plan gives each cell its address, its identity, and its location relative to cells in the rest of the body. This address will serve the cell and all its descendents into the adult. Each compartment will develop multiple tissues. Similar cell types in different compartments (with different addresses) may appear similar in structure and activity, such as bone deposition or nerve impulse conduction, but will differ from one another in other ways, such as their capacity to proliferate, to migrate in the embryo, or to adhere to other cells…. At these early stages, compartments cannot be distinguished by anatomical features. (Developmental +/or regulatory genes create isolated environments through local differences in chemical gradients.) The actual differentiation of the organism will depend not only on the compartments but also on the interactions of cells of one component with signals from other components. (Compartments must interact with other compartments and maintain the body plan; this is accomplished by signals at the compartment margins.) The component map is an extensible map: individual components can expand and shrink independently, while overall neighbor relations are maintained. The flexibility occurs not only in development, when certain areas grow relative to others, but also in evolution where there is disproportionate growth, for example, the neck of the giraffe relative to the neck of the whale. The compartment boundaries, though curiously arbitrary with regard to final anatomy, nevertheless divide the embryo into regions. In different regions, the same target genes can be controlled differently. The result is a platform for local differentiation and for use by the genome in many different ways… the compartment map makes possible the use of different combinations of processes at different places in the body. In fact, it provides those places. The map makes possible the use of different genes in different places. The concept of compartments has been the most valuable contribution of cellular and developmental biology to understanding evolution, at least the evolution of complex multicellular animals. The basic compartment structure independently identifies the phylum and hence must have remained unchanged since the Cambrian, for more than 500 million years." Compartments in the embryo allow a modular approach to organism development, simplifying and freeing regional development. "It is a compartmentalization of the organism that allows independent evolution and development, while maintaining overall coordination of activities related to the body plan of the phylum." **Compartments are isolated, changing environments.** Kirschner and Gerhart further expanded compartmentalization into the individual cell itself, and further reading of this reference provides even greater understanding. Their text provides far more detail than the few brief examples referenced herein.

Limb formation in the embryo can be used to illustrate regional development. The environment of the developing embryo is carefully controlled. The environment within the compartments exerts significant influence. Bone formation is consequential in the developing embryo; it will be an ongoing focus in this text. Under regional genetic and environmental control made possible by compartment individuality, it begins with a single fibroblast, which joins other fibroblasts in secretion of the collagen intercellular matrix, becoming fibrocytes, chondroblasts, chondrocytes, and later differentiating into osteoblasts, osteocytes, and osteolytic osteoclasts. In a fully developed adult, these cells form bone tissue that is forever in a state of dynamic turnover (e.g. 10%), whether this is in the marrow cavity or in cortical bone. (The latter turnover is through tunneling populations of osteoclasts, followed by osteoblasts forming tunnel-shrinking circular lamellar layers of new bone inwards and transforming into osteocytes.) The bone will increase in size (diameter) under the periostium, with sequential depositions of cortical circumferential layers of lamellae. At some point, new tunnel systems may also form in the bone-circumference areas of outer circumferential lamellae, following the above turnover process. A similar process of surface resorption and deposition occurs in the less-dense framework of the marrow cavities. Bone will be discussed in even greater detail, shortly. **In the developing embryo, cartilage and bone formation initiated by regional gene activity and unique environmental compartments begin a process of further differentiation.** Bone modifying proteins play a significant regulatory role.

Kirschner and Gerhart continue shedding light on what occurs next: "Muscle precursor cells are formed in the trunk in clusters close to the nerve cord (via embryonic induction). From this site, they migrate outward and follow an exploratory path into the neighboring appendage. There they associate with the bones and cartilage in whatever arrangement they find. The muscle precursors proliferate and differentiate in response to local cues.... Experimental studies... demonstrate that muscles find their way to associate and proliferate properly relative to the bone." This is an environmentally-stabilized feedback of exploratory random cellular behavior, occurring in a larger form of complex life. It involves bone environment proximity feedback signaling, which modifies the random cellular exploratory behavior.

Kirschner and Gerhart continue with limb development in a discussion of the nervous system (inducted by neural crest): "We have learned that the central nervous system, both the brain and spinal cord, produces far more neurons than are ultimately needed for the nerve connections to the peripheral targets. These cells extend their long, thin axons somewhat randomly into the periphery of the body, like foraging ants. If an axon tip by chance enters the anatomically appropriate organ, it receives (a) survival factor produced by tissues, and it persists. If it enters the wrong region, it receives no survival factor and commits suicide. Since there is only a limited amount of survival factor even in the appropriate region, competition and selection occur among neurons." This is another example of environmentally-stabilized local feedback of exploratory random cellular behavior, occurring in a larger form of complex life.

Kirschner and Gerhart continue with limb development in a discussion of the circulatory system: "In normal development, blood vessels are formed in two ways: from specialized embryonic stem cells called angioblasts, and from existing blood vessels. Early in development, angioblasts coalesce (around a matrix of collagen fibers, formed and marked by fibrocytes) and form hollow tubes, initiated by a hormone-like signal, vascular endothelial growth factor, which is secreted by the surrounding tissue (fibrocytes). This step of blood vessel development is highly deterministic (from the compartment map). Large vessels, such as the aorta and the major embryonic veins are formed in this manner. Later the small vessels form by a different process, called angiogenesis. Vessels sprout from existing vessels, almost a type of vegetative growth like the sprouting of shoots of a plant. In response to a local vascular endothelial growth factor signal (as well as other signals), the vessel swells and becomes leaky (large crews of fibrocytes relax their tension on ropes of collagen). Individual capillary cells dissociate from one another (allowing angioblasts to begin their random exploratory behavior), migrate towards the signal, and proliferate in response to it. Ultimately the capillaries reseal (fibrocyte crews reestablish tension on ropes of collagen, making them taunt once again) and rejoin a continuous (closed) network. As more blood flows through the vessels, they increase in size (fibrocytes deposit more collagen ropes and readjust tension).... In this model of a totally responsive vascular network, ready to sprout new vessels in response to local signals, the position of each capillary must be determined by the prelocalization of billions of sites of vascular endothelial growth factor release. The 'design problem' becomes one of placing those local signals. In fact, the growth factor is not localized. All tissues have the capacity to signal (these signals are cellular environmental signals). What is uniquely adaptive about the system is that signal by tissues is directly related to their need for oxygen. Therefore, most of the vasculature is generated by a functional feedback process: the local need for oxygen drives a local cellular response, which leads to a local signal (vascular endothelial growth factor production), which, in turn, leads to increased capillary growth, increased delivery of oxygen to the previously oxygen-starved tissue, and finally termination of the signal and the process."

Environmental feedback is deeply involved with random exploratory cellular activity in phenotype formation; it changes the random activity into a physiologic response. In development, it is the physiology that forms the anatomy. Signals from the external and internal environment initiate flexible, indirectly regulated, modifications in the phenotypes of complex animal life. Random exploratory cell behavior changes in the phenotype are indirectly stabilized by environmental feedback. In the previously discussed example, both genetically-directed change (genetic variation—causing compartment gradient and changing the shape of a bone), and environmental change (environmental variation—environmental feedback signals) induced corresponding changes in the placement of blood vessels, nerves, etc., without requiring additional genetic changes to manage infinite details. These fundamentals of limb development illustrate the random cellular exploratory processes involved in embryological development; random cellular exploratory behavior forms the phenotype. And, genetics is not involved in every aspect of phenotype formation; random cellular exploratory processes are involved in most of the details of phenotype formation. Geneticists may argue that ultimately, these random cellular exploratory processes are still under genetic control, even if it is indirect. The significant point is that the genetic control of phenotype formation is indirect; i.e., it is not a precise control of every detail. The other significant point is that the environmental influence on these random cellular exploratory processes that form the phenotype is also indirect. Physiologic cell function (modified random exploratory cellular behavior) depends upon environmental feedback; the environment is indirectly involved in many of the details in phenotype formation. **This means that the formation of the phenotype is under the indirect control of the environment as well as it is under the indirect control of heredity.**

The fundamentals of limb development illustrate how different environments (both internal and external) **normally influence the genotype to produce the formation of different phenotypes, within the norm of phenotypic variation.** Genetics must be integrated with the environmental feedback regulating random exploratory cellular processes at some level, at least indirectly, and **it seems to integrate at the time of expression. Genetic integration is required for phenotype trait inheritability.** Earlier referenced experiments (Waddington and Lindquist experiments in this text's Phenotype and Environment chapter), where environmental stress resulted in different phenotypes that became inheritable phenotypes after many generations (sometimes independent of the stress), seem to support a short-term genetic integration of environmentally-inducted phenotypic change. These experiments also support a belief that **even though genetic expression isn't directly modified by the environment, the indirect modification of genetic expression is likely a remote, doubly-indirect modification.** Adaptation, an environmentally-integrated change in the norm of expression (assembled from past phenotype adaptation components of natural selection bias), occurs only rarely, primarily during unusual environmental stress. **Randomly variable genetic expressions, which are increased in frequency by unusual environmental stress, result in randomly variable phenotype expressions that may be remotely stabilized by positive environmental feedback.** The indirect stabilization of random cellular exploratory functions by the environment must be indirectly linked to genetic expression for inheritability to occur.

Kirschner and Gerhart greatly minimize the importance of the environment, and even minimize "natural selection" to some degree. Kirschner and Gerhart primarily focus upon an organism's variability in phenotype expression, particularly on the unchanging, pan-phyletic level of random exploratory cellular processes that express phenotypic changes and this widespread connection to evolutionary change. Theirs is more of an orthogenetic approach, "a view that organisms evolve according to internally-directed rather than externally-selected paths." In the chapter on The Source of Variation, Kirschner and Gerhart discuss their (orthogenic) theory of how genetic variation is used in the generation of heritable phenotypic variation. "It is a theory of how inherited genetic material, along with the (note this) environment, constructs the individual organism in each generation, from the egg to the adult and on to the next generation. The organism's anatomy, physiology, and behavior are only remotely connected to the DNA sequence through all the complex processes of growth, development, and metabolism, though they depend on it. A change in the DNA sequence is therefore only indirectly connected with a change in the anatomy and physiology of the organism." It is easy to support the indirect genetic control of phenotype formation; but, in their view there is a lack of support for granting random cellular exploratory control (physiologic control) of the phenotype formation to indirect environmental feedback. Physiology is environment-responsive. In Kirschner and Gerhart's text, the environment is noted, then disregarded. It is nevertheless encouraging to find mutual agreement on the indirect integration of random cellular exploratory processes and genotype expression. The orthogenetic view (of Kirschner and Gerhart), inter-relating cellular (physiologic) function to genetic function, partially confirms the environment-cell physiology-genome doubly-indirect linkage involved in expressing the phenotype. The indirect influence of the environment on random cellular explorative processes (cell physiology) is well known. This influence was introduced in the preceding discussion of limb formation, and will continue with an environmentally-integrated discussion of bone anatomy and physiology.

The feedback from the environment, to physiologic cellular function and dynamic cellular anatomy, and then to genetic function, continues into late developmental stages and throughout adult life. Many functional and dynamic structural accommodations result from this indirect interaction between the environment, physiologic cellular function and the genetic norm-of-reaction. (Acquired adaptations are not inheritable.) The responses to environmental selection influence involve physiologic and structural changes, sometimes these changes are even pathologic. These responses are part of life's daily activity and its acclimation to change; random exploratory cell behavior is the underlying mechanism of change and it is influenced by the environment as well as genetic variation. Turner, in the Knowledgeable Bones chapter in *The Tinkerer's Accomplice* provides a fine anatomical and physiological example. "Bone has a dynamic architecture that represents a balance between patterns and rates of construction and demolition. In bone, the principal actors include three types of cells. One type, known as osteoblasts (bone formers) builds up bone by depositing a mineral precipitate of calcium phosphate, called hydroxyapatite, into the interstices of collagen felts or blocks of cartilage. Osteoblasts descend from the same cellular tribe as fibroblasts. A second group, the prosaically named osteocytes (bone cells) are essentially mature osteoblasts (more about them momentarily). Finally, there are osteoclasts (bone eaters) which dissolve the bony material laid down previously by the bone formers (osteoblasts). Osteoclasts are large monsters with large nuclei, and hail from a different cellular tribe that includes the macrophages and other killer cells of the immune system.

A bone begins as a foundational structure, such as a block of cartilage, that appears early in an embryo's life. Once this foundation is plumbed into the circulation, the blood carries in hordes of colonizing osteoblasts. Close on their heels, though, are osteoclasts, sweeping in like the Golden Horde. Just as European peasants did in the thirteenth century, the osteoblasts defend their homes by closing themselves into fortified chambers, called lacunae, which open to the world only through tiny tunnels called canaliculi. Once ensconced in their little rooms, the osteoblasts mature into osteocytes, leaving the osteoclasts to sulk like itennerant bandits, waiting for opportunities to sweep in and pillage. Because these cell's invasion routes are blood vessels, the osteocytes lacunae are laid out on a cylindrical plan surrounding the vessel. Once one layer (of bone) is complete another is laid down atop it, then another and another, typically up to a dozen layers or so. What ultimately grows around the vessel is a city of hermits, like some ancient Coptic hermitary. This tiny mineralized cylinder, just a tenth of a millimeter wide, is called a Haversian cylinder. Bone, or properly (cortical-outside) Haversian bone is made of woven networks of millions of these Haversian cylinders that snake through the bone along the blood vessels. (As a bone gets larger, layers of lamellar bone and possibly more Haversian systems are laid down under the periosteum—the outer bone cover-sheet; sometimes yearly cycles produce a tree-ring pattern.) The Haversian cylinder is the bone's principal load-bearing structure. Like a cotton thread, a Haversian cylinder does best when it is supporting tensile loads, those that pull on either end. A bone is strong because its innumerable Haversian cylinders line up with the commonly imposed loads. Thus a bone that is mostly loaded in one direction along its axis, as a gibbon's arm bones might be, will be built from numerous Haversian cylinders running parallel to the bones axis, like cotton threads bundled into a rope. Where a bone must bear complex loads that come from many directions, as in the head of the femur (leg-hip joint) its Haversian cylinders follow complex arcs, building structures that approach the sublime elegance (arch support) of a Gothic cathedral (**form follows function).**

Bones do not start out elegantly built. That comes about when the bone is adaptively remodeled, which draws on the rivalry between besieged osteocytes and the pillaging osteoclasts. Normally, the osteoclasts are held at bay by a protective layer of protein that envelopes the bones mineral layers. That changes if events—irritation, injury, or something else—strips this protein layer away, exposing the mineral. Then osteoclasts swarm in, like Mongols into a breached fortification, plowing up the mineralized matrix the supporting foundation of collagen, cartilage, and blood vessels, ultimately displacing the osteocytes. Usually, this is only a limited skirmish, occupying an area a few tenths of a millimeter across.

Adaptive remodeling begins once osteoclasts finish their grim work and move on. Fibroblasts then move back in, laying down new felts of collagen (along lines of tensile strain), followed by angioblasts and blood vessels, followed in turn by a new generation of osteoblasts and the new Haversian cylinders they build. Fracture of a Haversian cylinder is one of the most common ways that bone mineral is exposed and brought to the osteoclast's attention. This need not be the dramatic fracture that involves the entire bone: microfractures of individual Haversian cylinders are always occurring within bone, and this means that bone is continually being plowed up, resettled, and rebuilt (turnover). Adaptive remodeling occurs because the most common cause of a microfracture is loading a Haversian system perpendicular to its axis. This occurs, of course, when a Haversian cylinder is misaligned with the load it is… to bear. Things begin to be set right once the misaligned cylinder breaks and is cleared away by osteoclasts. As the next generation of fibroblasts move in, the collagen webs they weave will tend to be orientated more closely to the prevailing area of strain." **In development, and throughout life, it is environmentally-sensitive physiology that forms the anatomy.**

Kirschner and Gerhart argue that if random cellular exploratory behavior is largely responsible for the formation of the phenotype, these basic properties of cells and their interactions during development of the organism, have profound consequences for the properties of the variability available for use by selection. Because random cellular exploratory behavior is largely responsible for the formation of the phenotype, there is indirect control of the random cellular exploratory behavior coming from both heredity and from the environment. Developmental adaptation involves random cellular exploratory behavior stabilized by an environment-response feedback system. **Complex life's response to the environment is increasingly internalized into genetic hierarchy; but at the same time, phenotype formation is simplified when random cell functions are stabilized by environmental signals.** Changing the activity of developmental and/or regulatory genes changes the environment of structural/functional (body-building) genes. This may or may not involve changes outside the developmental norm-of-reaction. Adaptation to environmental conditions is simultaneously restricted and supported in the following discussion on adaptive cellular behavior considerations. Kirschner and Gerhart state that during development, these cellular properties and interactions both constrain (limit) the possible types of alteration to the organism's structures, as well as offer opportunities for the rapid evolution of novel structures. The exploratory processes are highly conserved; it is the stabilizing signals that change. Kirschner and Gerhart call the latter "facilitated variation" (of the phenotype, not the genotype) which they define as: "an explanation of the organism's generation of complex phenotypic changes from a small number of random changes of the genotype (mutations). We posit that the conserved components greatly facilitate evolutionary change by reducing the amount of genetic change required to generate phenotypic novelty, principally through their reuse in new combinations and in different parts of their adaptive ranges of performance…. Instead of a brittle system, where every genetic change is either lethal or produces a rare improvement in fitness, we have a system where many genetic changes (mutations) are tolerated with small phenotypic consequences, and where others have selective advantages, but are also tolerated because physiological adaptability suppresses lethality." In this way, a new genetic expression for a new somatic change that occurs early in the developing organism can more easily produce an inheritable change that can be passed to future offspring. In light of the above facilitated-variation concept and the previously discussed material in this *Environmental Biology* text, **an adaptation is a new environmentally-integrated inheritable phenotype, indirectly stabilized by environmental feedback of random cellular processes, which indirectly stabilizes random changes in the expression of the genotype**. Existing cellular mechanisms that normally form the phenotype (in the normal range of phenotypes that coincide with environmental conditions present during development) actually support the adaptive process by enabling the creation and survival of a totally new phenotype, through modifying what already exists, in order to accelerate its formation, and physiologically supporting new phenotype expression, while acting within existing genotype parameters.

J. Scott Turner, in The Joy of Socks chapter of *The Tinkerer's Accomplice*, describes an environmental selection problem of early development. "Fish embryos often thrash about within their eggs, scarcely at first, but more often, and more intensely as the embryo grows. These movements have often been ascribed to a need to stir things up, to circulate fluids within the egg to promote exchanges of oxygen or wastes between the egg and the environment. They may have another purpose: providing a 'test pattern' of self-imposed tensions on the developing mechanical systems of muscles, tendon, and spine. These form a template that guides the embryo's fibroblast… machines that (functionally + anatomically) build the embryo. This test pattern can be disrupted in various ways; genes are one way, but there are others. For example, acid rain is a serious environmental concern in many of the watersheds in northeastern North America. Among the concerns is high fish mortality in lakes and streams, which are traceable to developmental malformations of the tail and myotomes (swimming muscles). The malformations do not appear to be a direct effect of acidity on somite development. Rather, they appear to rise indirectly through the effect of acid conditions on the egg's water balance. Fish embryos are contained within an egg membrane, which is normally kept inflated by the regulate transport of water and salts between the egg's interior and the surrounding water. Inflating the egg gives the embryo a fairly capacious space in which it can thrash about. Acid conditions interfere with the embryo's ability to keep its egg inflated. The egg then deflates, confining the embryo in a straightjacket of shriveled egg membranes, preventing it from thrashing about as embryos normally would. This disrupts the embryo's self-imposed mechanical test patterns, which deranges the architecture of the tail and muscles. Disrupting the force environment with paralytic agents produces similar developmental malformations, even if the embryo had ample space within its egg."

In (Embryonic Origins), he continues: "Consider what may be the oldest physiological function: homeostasis (constancy) of cell volume. A cell is a created environment delineated by a membrane. The basic problem faced by a cell is this: water is commonly drawn into the cell by a physical force, osmosis, which if left unchecked would cause the cell to swell and burst. Cells maintain volume by managing the fluxes of water and salts across the membrane. If a cell takes on too much water, it can correct the situation by either limiting the rate of the water's influx, or actively bailing water and salts out of the cell. This ability enables cells to exist in a variety of environments from which they would be otherwise excluded. Moving into a dilute environment accelerates the osmotic flood of water. If the cell is unable to compensate by an increased bailing rate, it will soon burst and die. The cell's water-pumping machinery comprises a suite of proteins embedded in the cell's membrane. Water pores, called aquaporins, which mediate the flow of water; other proteins, collectively called gates, allow salts to flow passively across the membrane; embedded conveyors for the active movement of salts; gauges that measure how much the cell is swelling, and, of course, the genes that specify them. A single cell's physiological versatility (ecological valence—Hesse)—how wide an array of environments the cell can tolerate- depends directly upon the genes that specify the component proteins. If a cell cannot compensate physiologically, moving from a salty environment to a dilute environment will have to wait for the emergence of, aquaporin genes that are better suited to dilute waters. There is an alternative, though, and that is to have other cells construct an equitable environment in which to live. This is what cells living in bodies do. Human red blood cells, for example, have a fully functional system for volume homeostasis. It is not a particularly robust system, but then it doesn't really need to be, because the red cell lives in an environment that is regulated for it: the blood. That state of affairs arises from a clever trick of embryonic origami. Other cells are organized into sheets, called epithelia, which line highly folded surfaces within the kidneys, intestines, and skin. These transport water and salts across the sheets, just as cells on their own can do. But now they can impose homeostasis on the environment within their own cell membranes. Within the body, this regulated environment includes the liquid plasma inhabited by red cells. Remarkably, cells in epithelia can pull off this trick with very little in the way of genetic innovation. Mostly, it involves rearranging the same regulatory components that let free-standing cells regulate their volumes. For example, aquaporins and the other components in a free-standing cell may be distributed evenly throughout the cell membrane. When the cell is organized into an epithelium, the aquaporins may now be confined to one side of the sheet, gates to another, pumps at one face or another: same genes, but very different architecture. Upon this foundation, a diverse suite of new functions can be built, what we might call a value-added physiology. These new functions need not await the evolution of new genes for the components—simple rearranging of existing components will do."

A hypothetical adaptive scenario is helpful with any discussion of difficulties on theory. With a better understanding of water maintenance for cells and tissues, return now to the overly-confined fish embryo "waiting" for an inheritable adaptation to prevent egg shrinkage from water loss due to acid rain. This is an example of an unusual ecological stress that is widespread and has been ongoing for some time. The limiting factor, acid rain, environmentally selects against the developing embryo by interfering with the maintenance of an adequate water volume inside the egg. An effective adaptation of improved function in the acid-rain water environment could be accomplished by doing nothing at all; that is, stop bailing for awhile—but just for awhile. A simple on-off switch regulating the activity of the aquaporin gates would allow more water inside; perhaps variably control some of the switches. This should not be difficult to accomplish. Yet, there is no adaptation to manage this problem. It should be a simple matter to break-off normal membrane function for short periods and allow water from the more dilute surrounding environment to fill the collapsing void inside the egg. But it would seem as if presenting previously-preserved phenotypes from the genotype's norm-of-reaction cannot manage the problem of the acidic environmental stress. Expressions from the genome's norm-of-reaction are poised plastic phenotype responses, former adaptations incorporated into natural selection genetic bias. But if the norm-of-reaction plastic range of phenotype possibilities (natural selection genetic bias) was capable of solving the problem, it should have happened already. It appears as if it may take more than a common expression change (e.g. heterochrony-change in timing during development) from the norm-of–reaction to fix this problem. Utilizing a more significant change, from a mutation, in the display of a new adaptive phenotype would seem to offer the most likely chance of survival in this highly stressful acid-rain environment scenario. Recall Sean Carroll's earlier remark: "The important messages from (environmentally-integrated organisms)... are that mutations can be 'creative' and that the main limit to evolution is not so much what is mutationally possible, but what is ecologically necessary." Mutation and its immediate expression, or the new expression of a previously neutral (silent) mutation would seem to be necessary to generate an effective adaptation to this particular environmental problem. The mutation's expression would necessarily be sensitive to an acid environment and lead to a change in expression of the aquaporin gene (say an "enhancer" off-switch) that can variably interrupt function. Function would be preserved; an environment-sensitive regulatory change could solve this problem. This is a temporary loss of function scenario. The acid rain problem could also be managed by another temporary loss of function scenario. It would also vary the internal environment of genes in compartments. In this case a mutation (in close or remote regulatory area of a developmental or selector gene) allows the formation of an environmentally-sensitive variable switch, which affects the protein production that regulates (varies) the volume control of regional aquaporin-gate genes (overriding existing genetic constraint); function of the aquaporin gene would remain unchanged.

About 375mya, a fish-like creature left the sea and walked on land; this fossil was both a fish (lungfish with fish-scales) and a tetrapod (lobe-fins could do push-ups). *Tiktaalik roseae* was a predator 1.2–2.7m (up to 9') long with sharp teeth and a head shaped like a crocodile. *Tiktaalik roseae* had several remarkable anatomical features that show it was not only capable of wading in shallow water, but it was also capable of supporting itself on land, outside of the water, in the manner of four-limbed animals or tetrapods. Unlike other lungfish, it had a defined head, flexible neck, shoulders and a strong overlapping ribcage that would have assisted standing and walking about on land. *Tiktaalik roseae* had pectoral fins with bones that correspond to a shoulder, elbow and a wrist joint, which enabled it to crawl on the ground; the joints can be assembled to understand how the fin worked and how the animal moved. Between the most tetrapod-like fish and the first actual tetrapod there had previously been a gap in the fossil record; the gap has been filled. This was the first fossil ever found showing a structure intermediate between fins and a hand. *Tiktaalik roseae* has allowed a freeze-frame process of adaptation to land that took tens of millions of years. Plant-clogged waterways made weight-bearing fins, and eventually limbs, useful for getting around. The characteristics which the Devonian tetrapods had in common with certain lobe-fin fish were those which suggest that they were designed for a similar kind of life, predators lurking in the shallows. Along with Harvard University zoologist Farish

Jenkens, Neil Shubin, a professor at the University of Chicago and Ted Daeschler, chair of Vertebrate Zoology at the National Academy of Science (and others) greatly expanded evolutionary understanding with their remarkable discovery and the preceding commentaries. Daeschler stated: "The animal is developing features which will eventually allow animals to exploit land." This is the first real evidence of limbs in evolution. Jenkins noted that: "No other fish in the world, either living or dead, has overlapping ribs." Overlapping ribs buttress the animal's weight better against gravity, enhancing lung function. Shubin believes that "much of today's vertebrate diversity was defined by ecological and evolutionary shifts that happened during two critical intervals in the history of the Earth: the Devonian and the Triassic. The ~ 400 million year history of terrestrial animals reveals surprising patterns of anatomical stasis and parallel evolution: similar designs crop up in different species living in different environments." He believes that "the tetrapod limb development in the Devonian made possible the subsequent development of all the mammals, birds, reptiles and amphibians that have existed since that time." This perspective illustrates the essence of changes in types of life; the terrestrial environment created new opportunities for life. These creatures with their unique morphologies, which were at one time new phenotype environmental adaptations, used pre-existing phenotypes for construction materials. In the marriage of heredity and environment, ontogeny expresses phylogeny, at least to some degree.

Levinton devotes an entire chapter to A Cambrian Explosion with extensive reference to Cloud, Walcott, and other researchers that have studied this most remarkable time of variable environment and extinctions. "The notion of a Cambrian Explosion implies that a period of evolutionary eruption might have produced a peak of morphological differences among taxa, as a presumably ecologically driven radiation expanded taxa into many new ecological roles." The Cambrian is the time of origin for all major body plans of complex animal life (phyla) that exist today. It began following the last snowball earth period (worldwide glaciations = 580–750mya); multicellular animals and plants suddenly appeared. Approximately 544mya (million years ago), grazers were newly-found in the microbial mat (algae); O_2 atmospheric levels surpassed levels of today; then, predation arose; next, $CaCO_3$ (limestone) skeletons appeared; animals soon became larger to escape predation; the arms race had begun. There were four large extinctions during the course of the Cambrian. The first of these large extinctions resulted in the disappearance of the archaeocyathids (sponges) and a major group of trilobites (marine bugs). The later extinctions limited the diversity of conodonts (early vertebrate), brachiopods (looks like a clam—but not), and other triolobite groups. The Early Cambrian witnessed a mass extinction that may have been even more severe than the end-Permian extinction, in terms of absolute percentages of species lost. Some 83% of genera of hard-shelled or hard-bodied animals did not survive into the Middle Cambrian. This was double the Cambrian 40% background loss; even this background rate loss was very high compared to later geologic times. The Cambrian ended in a now-familiar ice age/sea level change pattern; this was a mass extinction with a 50% family loss. Since that time, phylogeny genetically constrains ontogeny for any new phenotypes formed in the marriage of heredity and environment.

Kirschner and Gerhart focused on the rise of the eukaryotic cell as the key breakthrough for all complex life. Following the melting of the first snowball earth (worldwide glaciation), this occurred at some time after the change of the ocean's color from (oil-of-vitriol) green to blue, and the change of the atmosphere's color from mars-orange to blue. Free oxygen arrived; this was a huge environmental change, perhaps the most significant environmental change of all. Twice a day, ocean tides, hundreds of meters high, swept across much of the small continents that were barely above sea level; the day at that time (including night) was around 12 hours long. Two billion years ago, there was no ontogeny or phylogeny, but there was a changing environment, there was changing heredity, and they were interacting, producing phenotypes that would be preserved as natural selection genetic bias.

It has now been established that random cell exploratory behavior is responsive to signals in the environment, as well as it is to genetics. The interaction of eukaryote random cellular exploratory behavior and signals (genetic and/or environmental) leading to a controlled physiologic formation of the phenotype is highly conserved. Kirschner and Gerhart discuss conserved components and processes: "The conserved components and processes... are integrated functioning pathways and circuits, the core processes of the organism buy which the phenotype is generated from the genotype. They are the essentials of synthetic and energy-generating metabolism, of the development by which the anatomy is generated, and of the organism's physiology.... Based on extensive comparisons of DNA sequences over the past 20 years, a deep conservation of coding sequences, those encoding the amino acid sequence is incontrovertible. For example, many enzymes of metabolism that are components of some of the core processes shared by bacteria and humans are conserved, though these are separated by three billion years." "The cell has hundreds of behaviors or activities that involve conserved core processes.... There is only a limited, though large, set of core cell behaviors, which change in limited and understandable ways. Novelty usually comes about by the deployment of existing (physiologic) cell behaviors in new combinations and to new extents, rather than in their drastic modification or the invention of completely new ones." "Genetic variation or mutation does not have to be creative; it only needs to trigger the creativity built into the conserved mechanisms." "Our (K&G) point is that the conserved core processes respond to genetic variation or environmental variation (note that both types of variation are recognized here) by producing their special type of phenotypic change." "All the conserved core processes possess adaptability, which they use to varying conditions.... Genes encoding the RNA's and proteins of these (core) processes are highly conserved across diverse animals, from jellyfish to humans.... Complexity must arise through multiple use of a relatively few conserved elements. Complexity arises when different parts of the adaptive range (environment) are selected." (K&G) "In our (K&G) view, the capacity for facilitating variation has itself evolved, as the core process of organisms have accumulated more adaptive and robust behaviors... there is really no alternative but to think that new core processes, such as those that first arose in eukaryotic cells, were cobbled together from the existing processes in prokaryotic cells.... Core processes may have emerged as a suite, for we (K&G) know no organism today that lacks any part of the suite.... Thus in extant (present) eukaryotes, we meet the whole span of eukaryotic processes whenever we meet a eukaryotic cell." "Facilitated variation is seen to have taken a giant step after the core processes of multicellularity were introduced, as it did at previous steps of core process evolution." Kirschner and Gerhart point out that development in multicellular animals is increasingly controlled by internalized signal-response systems (e.g. tissue oxygenation), in which many of the individual components are highly conserved (constrained) over much of the metazoan (multicellular animal life) history. "For the organism to generate different cell types and different cell behaviors, it must produce and respond to diverse signals, retrieve diverse information from the genome, and generate diverse combinations of cell behaviors. The ability to process all this information with a limited set of components underlies the somatic adaptability of the core processes. In evolution, these preexisting combinations of cell behaviors and expressed genes must be altered to give new combinations of behaviors and genes. Some kinds of somatic adaptability arise by the way conserved core processes are linked to one another. By linkage, or, better (indirect) regulatory linkage, we usually mean how information is passed from one component to another. Since all information transfer occurs on the molecular level, we mean specifically how one molecule passes information to another. Signals may come from outside the cell or the organism. They must be passed through a chain of command until a response is made. That response could then affect the environment or the organism... we use the term *weak linkage*, first coined by Michael Conrad in 1990, to mean an indirect, undemanding, low information kind of regulatory connection, one that can be easily broken or redirected for other purposes." Clearly, environmental variation is recognized here (but disregarded in other instances). "Every new gene in evolution must somehow be linked to a transcriptional regulatory program, and old genes continue to undergo changes of regulation. Every time an innovation occurs, these processes invariably change.... Presumably under strong selection, (environmental) the developmental function can be maintained but the regulatory sequences can change.... The same pathways are used over and over again within the organism for different purposes. Thus, they must be modified slightly to interact with a variety of processes and to work in different environments and cell types."

Genes (selector genes) can selectively control (regulate) other genes in environmental compartments through the production of signaling molecules. The same signaling molecules are often reused in different contexts; this "weak (indirect) linkage" (not genetic linkage, which involves different gene loci on the same chromosome) between signal and response, permits conserved (constrained) components to be combined in different contexts, allowing a unique outcome of development (adaptation) to be produced, without the invention of new individual components (to create detail unique to each structure). This all fits very well with Kirschner and Gerhart's orthogenetic approach to evolution, where the organism alone is primarily responsible for the creation of the phenotype. At the same time, environmentally and genetically integrated adaptation fits even better; the internal mechanisms of the organism indirectly respond to an internal or external environmental signal; the signal itself plays no direct role in formation of the phenotype. The weakness of the (indirect) linkage in internal environmental signals enables flexibility in response; i.e., internal environmental signals are easily modified. Body plans are phenotypic expressions of the genotype and indirect environmental signals. In the Weak Regulatory Linkage chapter they note: "Eukaryotic gene regulation is perhaps the most powerful conserved core process, responsible for much of the phenotypic variation on which selection acts. In eukaryotes, various regulatory proteins (factors) bind to specific regions of DNA sequence near the gene to be controlled and directly activate transcription (1st step in formation of proteins) in response to extracellular signals.... The typical mammalian gene is regulated by dozens of such factors binding in the vicinity of the gene at particular DNA sequences (switches) they recognize. Each carries a message from some different signaling system, saving time, space, amount, cell type, and other information. These signals are not on-off switches but partial-on and partial-off switches. The multiple (internal environment) factors result in a certain level of RNA synthesis from a specific gene (1st step in formation of proteins).... The (indirect) regulation of gene expression (through internal environmental factors produced by genes of a higher hierarchy) is one of the core processes most critical for generating phenotypic variation." "The time and place of expression of particular genes involved in embryonic development are often the same in the different species (which only subtly differ in anatomy), but the sequence of DNA in the regulatory regions of these genes has changed a great deal. Presumably under strong (environmental) selection, the developmental function can be maintained, but the regulatory sequences can change." In other words, these changes in the regulatory sequences are mutational changes of regulatory genes that regulate other genes; it involves a sudden mutation and an immediate change in phenotype. The authors emphasize the importance of this (internal environmental signal) indirect-linkage flexibility, particularly in developmental systems. "Eukaryotes have the most indirect linkage; modification of internal regulation for phenotype variability is more regional in respect to gene proximity in Eukaryotes (complex life) than in Prokaryotes (simple life), which allows more indirect flexibility than a direct-fit regulation."

Futuyma's *Evolutionary Biology* text discusses developmental genes that regulate other genes. Futuyma mentions that "Hox (homeobox) genes have been found in all phyla examined, whether they are segmented or not. All phyla, from Cnidaria (coelenterates-like jellyfish) to Chordata, have multiple hox genes, with very similar homeobox sequences, implying that a gene family, stemming from repeated duplications of an ancestral gene, has been inherited from a common ancestor of all metozoans (animals), which lived more than 550mya. Moreover, genes with homeobox sequences are carried by all eukaryotes, including fungi and plants.... Hox genes do not encode structures. Rather, they... (encode) transcription-regulating factors (signals) with spatial and temporal patterns of expression that provide positional information (create special isolated environments for local structural/functional genes)."

In chapter Is Life Plausible, Kirschner and Gerhart state: "On the deepest molecular level of selector genes and signaling pathways in their development (developmental genes indirectly regulate other genes along signaling pathways), the limbs of all vertebrates are strikingly similar. They form at exactly the same position relative to the segmented muscle blocks. The pectoral fins of fish use the same selector gene as the forelimbs of a mouse, whereas the pelvic fins use a different factor (internal environmental signal), the same as the (mouse) hindlimbs. The detailed patterning of front-to-back selector genes is the same in all limbs. These points are further illustrated with many other examples drawn from cell and developmental biology; these subjects are far better understood today than when the Modern Synthesis of Evolutionary Biology was developed during the 1930's through the 1950's. Perhaps the most striking discovery is the conservation of the basic genetic circuits underlying the body plans of animals as distant as vertebrates and arthropods, a great advance in the knowledge of the history of life." Neil Shubin, in the Handy Genes chapter of *Your Inner Fish*, discusses transcription-regulating factors (signals) that relate specifically to spatial and temporal patterns of expression for limbs, particularly to the formation of bony anatomy of limbs in fish and tetrapods (recall bone formation's prominent role in this chapter's earlier discussion of limb formation). "A strip of tissue at the extreme end of a limb bud is essential for all limb development. Remove it and development stops (delayed removal allows increased development at the distal end).... This patch of tissue was named the zone of polarizing activity (ZPA).... Experiments proved that digits formed according to a concentration gradient relative to ZPA proximity." A developmental gene known as sonic hedgehog made a regulatory protein that played a highly significant role in limb formation in all vertebrates. Variation along a concentration gradient caused variations in the bony structures formed. "Experiment after experiment on creatures as different as mice, sharks and flies show that the lessons of sonic hedgehog are very general. All appendages, whether they are fins or limbs are built by similar kinds of genes.... The great evolutionary transformation of fish fins into limbs did not involve the origin of new DNA; much of the shift likely involved using ancient genes, such as those involved in shark fin development, in new ways to make limbs with fingers and toes." Young and Badyev, in *Evolution of ontogeny: linking epigenetic remodeling and adaptation in skeletal structures*, referenced a study suggesting that "mutations in regulatory, promoting, and processing of genetic pathways of bone forming regions can generate changes in gene expression (without disrupting cohesiveness of developmental networks).... In a broad examination of molecular evolution in morphogenic genes among cichlids (fishes with high jaw variation), Terai found allelic variation in the production of BMP-4 (a bone morphogenic protein) consistent with high levels of morphological variation; this variation was related to protein folding, and thus modified downstream effects without disrupting the general function of the gene."

In their chapter, Is Life Plausible, Kirschner and Gerhart also note that change can occur without a change in DNA: "Of course, if the (environmental) selective conditions changed, (expressing) a previous mutational change in the regulatory sequences could presumably generate a new time and place of expression of the (regulatory or regional-functional/structural) gene, a new phenotype." Clearly, recognizing the expression of an earlier mutation as a result of environmental change is acknowledged here (but, as with the recognition of environmental variation, disregarded in other instances). The result in both cases is the same; both the expression of a sudden mutation and an environmentally-induced expression of a previous mutation are effectively expressing a "new" mutation. The former is mostly genetically-induced and the latter is mostly environmentally-induced; however, "mostly" does not mean "exclusively," as all phenotypes are formed by influence from both heredity and environment. And in this phenotype-producing partnership, any dominant influence should only be seen as the more dominant influence from the marriage of heredity and environment. If recognizing that the expression of an earlier mutation as a result of environmental change is acknowledged as a viable cellular function, this is different from how the cell normally varies the phenotype under frequently-encountered conditions; but, only because it requires the formation of a completely new phenotype that must be environmentally-acceptable. Other than the oppressive signals of unusual environmental stress that seem to be necessary for random phenotype expression, far less intense environmental signals from frequently-encountered environments would use the same core mechanisms to develop norm-of-reaction phenotypes. **Norm-of-reaction phenotypes** (previously-preserved environmental adaptations) **are already inheritable.** During frequently-encountered environmental change, layers of biochemical reactions operate variable switches to express (norm-of-reaction) previously-formed pathways for the genes to produce their proteins. This is the plastic range of expression for the norm-of-reaction. The inherited plastic range of gene expression represents the most common level of gene function, one that controls gene expression (not mutation) at the chromosome level. Said otherwise, plastic expression of normal variation may involve change in gene expression, but not change of any gene itself. In this case, the entire genome may function as a single unit; this plastic form of gene expression (epigenetic) will be discussed in further detail in the upcoming text.

Kirschner and Gerhart pose an interesting question: "What if evolutionary biologists were wrong to think of phenotypic variation as random and unconstrained, even though genetic variation (genotype) was random and unconstrained?" The authors note that: "the nature of the processes that increases the amount of genetic variation is still not identified." Although this is a shift away from the old genetically-determined evolution, this statement still reveals a major inconsistency if the relevant variation of innovation is only genetic variation (genetically-determined evolution); the need to increase the genetic variation by some abstract process reveals the shortcomings of genetically-determined innovation. Genetically-determined innovation minimizes the obvious role of the environment as the other source of variation, as well as it minimizes the environmental influence on the formation of the phenotype (recall chapter on Phenotype and Environment). The expression of an earlier mutation as a result of environmental change utilizes the same core mechanisms that were discussed in the preceding paragraph. Under unusual environmental stress, random phenotypes (primarily norm-of-reaction variations) are formed from genetic and environmental components. The role of genetic constraint in phenotype formation was previously discussed in the chapter by that name. **On occasion, a change in gene expression** (that may express a previous mutation) **affects random exploratory cell behavior and generates a new variation, slightly outside the norm-of-reaction.** Random exploratory cell behavior is only indirectly connected to genetics and the environment. Here, both genetics and environment join to indirectly influence the new phenotype. Signals from the external and internal environment initiate flexible, but regulated, modifications in genetic expressions that may or may not be environmentally-stabilized to become environmentally-integrated adaptive genetic expressions. Sexual reproduction's rapid recombination of natural selection's phenotypic variation helps the odds of inheriting complex life's favorable genetic potential in the developing offspring. The new genetic potential may then be realized in the creation of a new adaptation.

And while acquired adaptations are not inherited, acquired adaptations offer insight into the role of the environment in forming all adaptations. In the case of immunity and learning, recall that these new environmentally-integrated phenotypes are a somatic adaptation formed in the adult. The most important function of the human immune system occurs at the cellular level of the blood and tissues (Immune system— Wikipedia follows). The lymphatic and blood circulation systems are paths for specialized white blood cells to travel around the body. Lymphatic cells include B cells, T-cells, natural killer cells, macrophages, and dendritic cells. Each has a different responsibility, but all function together with the primary objective of recognizing, attacking and destroying bacteria, viruses, cancer cells, and all other pathogens. Without this coordinated effort, a person would not be able to survive more than a few days before succumbing to an overwhelming infection. When a pathogen has entered the body, it sets off a chain reaction that starts with the activation of macrophages and natural killer cells that reach the site of infection and destroy as much of the pathogen as possible. While this is happening, it is the job of the dendritic cells to take "snap-shots" of the battle-ground to take to the lymph nodes in order to activate T-cells, which then activate B-cells to produce antibodies against the pathogen. The immune response is highly specific to the physical chemistry configuration of an antigen, an internal environmental signal. The body can develop a specific immunity to particular pathogens by producing T-cells specifically designed to target particular pathogens; T-lymphocytes mediate immune defenses against infectious diseases through aiding B-lymphocytes to manufacture a class of proteins called antibodies. Cellular immunity mediated by T-lymphocytes can recognize infected body cells, cancer cells, and cells of a foreign transplant. The control of cellular immune reactions is provided by a linked group of genes, known as the major histocompatibility complex (MHC). These genes code for the major histocompatibility antigens, which are found on the surface of almost all nucleated somatic cells. The major histocompatibility complex class I and class II antigen-processing pathways play an essential role in the activation of pathogen-specific T-lymphocytes. This dendritic cell recognition response takes days to develop, and so is not effective at preventing an initial invasion, but it will normally prevent any subsequent infection, and also aids in clearing up longer-lasting infections. Dendritic cells constantly sample the surroundings for pathogens such as viruses and bacteria. This is done through pattern recognition receptors, which recognize specific chemical signatures found on the pathogens. Immature dendritic cells phagocytose pathogens and degrade their proteins into small pieces and, upon maturation, present those fragments at their cell surface (present peptide fragments derived from pieces of pathogen-encoded proteins on their surface using major histocompatability molecules). Once they have come into contact with such a pathogen, they become activated into mature dendetric cells. Simultaneously, they up-regulate cell-surface receptors that act as co-receptors in T-cell activation, greatly enhancing their ability to activate T-cells (+ B-cells). They also up-regulate another receptor that induces the dendritic cell to travel through the bloodstream and lymphatic system. Here they act as antigen-presenting cells; they activate T-cells and natural killer (T-cells) as well as B-cells by presenting them with the antigens derived from the pathogen, alongside with non-antigen specific co-stimulatory signals. (Wikipedia)

The major histocompatibility antigens were first discovered on the leukocytes (white blood cells) and are, therefore, usually referred to as the HLA (human leukocyte group A) antigens (Encyclopedia Britannica). The groups of genes that code for foreign protein recognition are located on the short arm of human chromosome #6, residing on four closely associated sites (designated HLA-A, HLA-B, HLA-C, and HLA-D). Each locus is highly polymorphic; i.e., each is represented by a great many alleles within the human gene pool. (There are more than 25 forms of HLA-A, more than 50 forms of HLA-B, and more than 15 forms of HLA-D; these different forms are caused by slight differences in the amino acids that make up the proteins.) Differences in alleles result in differences in immune response capability to manage infections. For example, Europeans and most Asians as a group demonstrate double the amount of HLA diversity (bacteria recognition capacity) in their populations compared to Siberian or Native American populations, providing a potential overall greater natural resistance to bacterial infectious diseases. These alleles are expressed in co-dominant fashion; all alleles are active and all traits are expressed at the same time. Selection appears to favor heterozygotes and low-frequency alleles, leading to high levels of allelic diversity at the MHC locus (Vallender). Since an offspring inherits one chromosome #6 from each parent, offspring of parents heterozygous for a single maximum protection allele have a one in four chance of having received the same maximum disease-resistant paternal code expression on chromosome #6, and maximum disease-resistant maternal code expression on chromosome #6, in order for offspring to be HLA allele-matched (the most-favorable homozygote combination of the most adaptive genetic expression code). The environmental selective process of disease will become an agent of natural selection preservation of immune potential. Previously noted, the potential for generating active immunity is genetic; however, the active immunity itself is an acquired adaptation. Continued resistance through ongoing active immunity is another acquired adaptation. That is, favorable genetic patterns may lead to an established life-saving active immunity (somatic function) against a disease; prolonged active immunity to the same disease may result (also somatic function). Up to 10% of Caucasians may become resistant to the AIDS virus, as a result of inheriting earlier smallpox or plague epidemic resistance potential that was developed in ancestors. Potential must be present, but the actual performance of establishing active immunity is of deciding survival importance.

Recall that adaptive immunity is antigen-specific and found only in vertebrates. A new disease, like the virus H_5N_1 strain of bird flu, could mutate (like H_1N_1 Spanish flu) and enable it to be transmitted from human to human. The human immune system has had no chance to acclimate to this new strain and produce an effective defensive response; the new virus could be virulent enough to cause a great many deaths. Survival depends upon a timely acquired adaptation response. Because of previous mutations in the human genome, some infected patient's immune systems will adapt to the new threat and generate a new phenotypic expression (genetic code expression of a new phenotype) in the form of an effective immune system defense, enabling the infected patient's survival. This can only happen from having a genome that can express a saving genetic code that opens a combination padlock on, (or "key" to) the environmental selection gate, allowing passage into the future. "Hypermutation" has been used to describe this process (old school). This is an acquired adaptation; "hyperadaptation" is a far better description, as it describes a new phenotypic expression of an unchanged genotype. The immune system somatic adaptation process provides a new and continuing active resistance for the infected individual; however, it is only effective for that particular strain of virus infection. The somatic immune system phenotype is highly antigen-specific; the disease resistant phenotype requires a highly antigen-specific environmental feedback mechanism. The virus could mutate once again and bypass the newly-acquired human immune system defense; unfortunately, this does occur in many influenza strains. Recall that active immunity per se does not appear to be inherited in humans. The acquired adaptation is a somatic functional adaptation of the body cells, not transferred on in the germinal cells. However, the genetic potential for this active immune system adaptation (new phenotype expression of genotype variation), to the environmental bird flu virus strain, will be passed on in the germ cells of survivors and inherited in the offspring. It will be further supported in offspring infancy by passive humoral immunity antibodies transferred across the placenta as well as by passive humoral immunity antibodies in the mother's milk. Regarding the potential for inheritance in sexual reproduction, a co-dominant additive benefit would mean that favorable odds follow a pattern of some heterozygote advantage, but the greatest advantage lies with a matching pair of homozygote "lucky combinations" on both chromosomes, providing they are an appropriate match for that particular disease.

Similar to the disease resistance potential inheritance in the immune system, the mental potential to learn is inheritable. Learning is a change in behavior, in response to environmental stress. Learning itself is not inherited; it is also an acquired adaptation. Both the ability to learn as well as the different responses to a learning situation may be inherited. Futuyma's *Evolutionary Biology* text references learning studies that revealed that: "The results of the Colorado study to date suggest that the heritability of IQ score is about 0.5 (correlation coefficient), and is much the same at ages 1 through 9.... The investigators found that the genetic component of specific cognitive abilities such as verbal comprehension, spatial visualization, perceptual speed, and accuracy was correlated to a considerable degree with overall IQ score, but that some uncorrelated genetic variation in these abilities emerged at age 7, implying that overall intelligence includes several components that are partly independent of each other. Perhaps the most interesting result, in this and other studies, was evidence of a substantial genetic and environment interaction. For instance, the researchers observed features of the homes and mother's treatment of both adopted and non-adopted pairs of siblings when they were the same age, and also scored the mother's descriptions of the family environment at those times. For both methods, the environmental score of biological siblings were more highly correlated than those of adoptive siblings. The researchers hypothesized that genetically different traits in children may elicit different responses from parents and peers, and that children's different genotypes influence their perceptions of and responses to the same objective experience. In other words, the individual may create his or her own environment, a process in which genetic differences in cognitive, personality, or physical traits may play a role. The environmental differences thus engendered may, in turn, affect the development of cognitive abilities and personality traits. This would mean that 'nature' and 'nurture' covary and interact, and cannot be meaningfully distinguished." Learning is a phenotypic adaptation (somatic adaptation) to specific environmental signals. Learning also does not always occur, and it occurs at different rates in different individuals; external environmental feedback is also crucial to the process. Learning enables a close examination of the acquired adaptation process.

One isolated mutation could lead to a suite of adaptations (those adaptations in the suite related to a bigger brain for better learning). H. Stedman, in *Nature* magazine, noted that a mutation 2.4mya could have caused man's ancestor to be unable to produce one of the main proteins (myosin-MYH16) in primate jaw muscles (sounds like a loss of function change—like the regional "enhancer switch" example in the acid rain adaptation scenario). When the timing of the autosomal dominant mutation is correlated with brain size increase, the two changes match. With loss of a *robustus*-style cranial crest following the loss of major muscle mass (bone growth physiologically responds to tensile forces), the brain was free to increase in size. Adverse environmental selection likely allowed only the survival of individuals (with compromised jaw muscles) that were inclined to really cooperate with others to survive (unselfishly— chimps won't do this). In *Regulating Evolution*, Sean Carroll notes: "Wray (and others) have identified other aspects of human biology that have evolved through mutations in enhancers in different human genes. One of the most intriguing associations revealed thus far involves divergence (pre-dates 2.4my change above) in the great ape and human regulatory sequences controlling the Prodynorphin gene, which encodes a set of small opioid proteins produced in the brain, involved in perception, behavior and memory. The human gene is more highly expressed in response to stimuli than is the chimpanzee version and strong evidence suggests that the human regulatory sequence evolved under natural (environmental) selection—that is, it was retained because it was advantageous (environmentally)." Another significant mutation, forming the Fox P2 gene, occurred in humans 200,000ya. Without Fox P2 (regulatory) gene function (relates to cognitive development), speech is greatly impaired (Enard). Recall the use-it-or-lose-it acquired adaptations (grammer usage by age 3 and highest-level brain function doesn't develop until age 20) from the Adaptation chapter. "The gene (microcephalin) MCPH1 regulates brain size during development and has experienced positive selection in the lineage leading to Homo sapiens. Within modern humans, a group of closely related haplotypes at this locus, known as haplogroup D, rose from a single copy approximately 37,000ya and swept to exceptionally high frequency (approximately 70% worldwide today) because of positive selection" (Evans, Mekel-Bobrov, Vallender and others). The brain seems to be still increasing in size. The gene ASPM (abnormal spindle-like microcephaly) also is a specific regulator of brain size during development, and its evolution in the lineage leading to homo sapiens was driven by strong positive selection only 5800ya; a genetic variant of ASPM appeared at this time and swept to high frequency under strong positive selection (Mekel-Bobrov). Thus novelty arose from enhancer switch mutations causing loss or gain of compartment gene function, mutation of a regulatory/selector gene allowing gain in function, duplication of a gene, and mutation of a regional compartment gene itself, changing function.

Kirschner and Gerhart substantially add to the understanding of the causal processes of evolution; "facilitated variation" exists. They ask: "What capacity to evolve would a hypothetical organism have if it did not have (modified random cellular exploratory behavior) facilitated variation?" (Genetically-determined evolution assumes genetic mutation is the only source of variation.) Kirschner and Gerhart greatly expand evolvability with facilitated variation of the phenotype in the developing organism. Although it is not recognized by Kirschner and Gerhart, facilitated variation also involves environmental variation; the adaptive process is not determined by genetic variation alone. Random exploratory cell behavior facilitates forming the phenotype from both mutational and environmental change. In the chapter on Constructional and Functional Aspects of Form, Levinton observes: "We must conclude that some evolutionary process has built in a capacity to change the rules as new environments are encountered." Adaptation to an isolated, changing environment involves an early developmental environmental integration into heritable variation. Its somatic change after germ cell formation that is not heritable. Kirschner and Gerhart believe that: "The main accomplishment of the theory of facilitated variation is to see the organism as playing a central part in determining the nature and degree of variation.... It is the capacity of the (environmentally-sensitive) core processes to support variation that we see as the main factor in generating phenotypic variation and in minimizing the lethality of phenotypic variation. It is the nature of these processes, which are poised to generate physiological variation within the organism, that allow genetic variation to be so effective in generating phenotypic variation on which selection acts." Without question, random cell exploratory function facilitates phenotype adaptation from mutational (genetic) variability. But also without question, random cell exploratory function facilitates adaptation from (environmental) feedback variability. Recall that **environmental adaptations do not form without the creation of an isolated, changing**

environment. (Organisms prefer to seek favored environments over risking major change that could lower overall fitness.) Recall also that signals from the external and internal environment may initiate flexible, but regulated, modifications in genetic expression from the norm-of-reaction. Layers of biochemical reactions operate variable switches to express programs stored in the norm-of-reaction; they may also express something new, as a sudden mutation or even an older gene mutation (epigenetic change). Either new expression can just affect a single gene on a local level, up to causing a change in the regulation of other genes. The preceding strongly supports evolutionary change at the level of the individual organism and/or its offspring, and is consistent with Darwin's view of variation: "Owing to the struggle for life, any variation, however slight and from whatever cause proceeding, if it be in any degree profitable to an individual of any species, in its infinitely complex relations to other complex beings and to external nature, will tend to the presence of that individual, and will generally be inherited by its offspring." But it should be obvious by this time that speciation is irrelevant, except as a vehicle to transport an adaptation. If the driving force of evolution is genetic adaptation to a unique environment, this is similar to the norm-of-reaction determination of the phenotype from the interaction of an organism's genotype and the environment. In both cases, genetic expression must be connected to environmental conditions, even if the genotype is not influenced directly by the environmental conditions. In the normal expression of the phenotype, or in the expression of random phenotypes under strong environmental stress (that may become the rare adaptation to the strong environmental stress), signals from the external and internal environment initiate flexible, but regulated, modifications in epigenome expression. And while a previously-programmed phenotype expression to previously-encountered environments is not exactly the same as a randomly-formed completely new phenotype expression to a new environment, if the same core mechanisms were used to generate a new alternative phenotype expression of the genotype, their similarity likely applies to the adaptation process. When it comes to determining common phenotype expression (in the genotype norm-of-reaction) to environmental exposure, recall the inability to separate an influentially greater role of genetics from an influentially greater role of environment. It would seem logical to think that when generating a new adaptation, **there is no single genetic influence, or single environmental influence in adaptation** (marriage of heredity and environment), **but a range of interacting hereditary and conditions-of-life involvements simultaneously and indirectly acting with random exploratory cellular processes.**

What this really means is that if the genotype can respond to environmental change by eliciting a programmed (norm of reaction) phenotype response, and if facilitated variation of the phenotype exists, facilitated variation of the phenotype uses the same mechanism to aid in producing a new environmental adaptation. Environmental and genetic interaction (in the individual organism) must occur during the time of adaptation. For complex life, it must occur early during a time in development so that adequate adaptive phenotype environmental interaction/integration can become incorporated into germ cells, which are often in an isolated controlled environment, which is itself still in an environment of some kind, not a vacuum. Ultimate understanding of adaptation cannot be achieved unless we seek both genetic expression change and conditions of life perspectives, simultaneously. Although chance variation from mutation and a sudden mutation expression may be random, environmental conditions and random phenotype generation are not random events; they are situational. And indirect environmental feedback stabilization, indirect cellular activity genetic integration and survival of the phenotype are also determined by the situation (existing environmental conditions). Contrary to the views of Kirschner and Gerhart, environmental influence cannot be separated from any process of facilitated phenotype variation; it must be integral to the process. **Facilitated variation of the phenotype includes many environmentally-sensitive physiologic cellular functions** (environmentally-modified random cellular exploratory behavior). Both here and in earlier *Environmental Biology* text discussions, some insight into adaptation has been developed by illuminating the selective role of the constantly-changing environment, discussing adaptation to environmental stress, and understanding the resultant natural selection genetic bias of survivors. The actual mechanisms of adaptation have been partially discussed. Further discussion in greater detail follows; this detailed discussion will require some understanding in the biochemistry of genetic mechanisms, and will be addressed in an upcoming chapter. This discussion will also include a discussion of epigenetic changes in gene expression, changes that do not include any change in the DNA. **Epigenetics invalidates the mindset of genetically-determined evolution.**

29

Environmentally-Determined Evolution?

Robert Carroll, in the opening chapter of *Vertebrate Paleontology and Evolution*, states: "Although fossil vertebrates were very incompletely known in the nineteenth century, remains of dinosaurs, giant marine reptiles, and mammals without modern descendents provided Darwin and other biologists with irrefutable evidence of extinction and, less directly, of the process of evolution. Evolution might have been accepted without fossil evidence, but fossils now seem inexplicable without evolution. Yet, the foremost vertebrate paleontologists of the nineteenth century—Cuvier, Owen, and Agassiz—did not accept evolution as put forth by Darwin, but argued for a succession of creations and extinctions. (This text re-introduces that argument.) Until the 1940's, vertebrate paleontologists remained outside the mainstream of evolutionary thought.... Perhaps we should not be surprised that vertebrate paleontologists did not support the prevailing view of slow, progressive evolution but tended to elaborate theories involving saltation (jumps), orthogenesis, or other vitalistic hypotheses. Most of the evidence provided by the fossil record does not support a strictly gradualistic interpretation.... Few contemporary paleontologists would deny that natural selection (genotype bias of the phenotype) controls the direction of evolution, but many would seek additional factors (like variation in environmental conditions), to account for the rapid evolution that characterized the early diversification, and radiation of groups, and the early stages in the elaboration of new structures. The great longevity of many groups and the minor evolutionary changes they exhibited pose another problem." Incorporating variable environmental conditions, along with considerations of environmental selection, extinction, creation, and adaptation, fundamentally solves all of these significant problems. In the last chapter, R. Carroll states: "The search for special evolutionary factors

must be centered on the conditions that facilitate the subsequent radiation of a group (environmental conditions- involving environmental selection, extinction, creation, and adaptation), not on the initial appearance of a species. The most spectacular evolutionary changes involve the emergence of new structures and ways of life (like adaptation to a new environment)....
The most dramatic evolutionary events are those that combine a major change in habitus (newly created environmental conditions- involves selection) with the appearance of a new structural-functional complex (adaptation). Examples include the origin of amphibians, birds, pterosaurs, and ichthyosaurs. The appearance of these groups is made all the more striking by the rarity of intermediate forms in the fossil record." And while evolutionary leaps to reptile pterosaurs and reptile ichthyosaurs may await specific intermediates, neither requires a great leap for the imagination to contemplate. There are lizards that fly (*Draco volan*), snakes that fly (*Chrysopelia ornate, C. paradisi, and C. pelias*), and mammals that fly (bats— 1100 species); there are freshwater and marine crocodilians (23 species), marine iguanas (*Amblyrhynchus cristatus*) and marine mammals (123 species—whales, dolphins, manatees and seals); and, all are present today. As for the spectacular evolutionary change from fish to amphibians, there is no longer any great leap to go from a modern lungfish like *Protopterus dolloi* and/or a lobe-finned fish like the modern coelacanth *Latimeria chalumnae,* to a tetrapod-transition like the previous chapter's fossil discovery of *Tiktaalik roseae.* And as for a spectacular evolutionary change from dinosaurs to birds, the leap is even smaller yet to go from an animal like the fossils of the feathered dinosaur *Protarchaeopteryx robusta*, or the closest known dinosaur relative of birds, *Caudipteryx zoui,* to an animal like the bird that is closest relative to dinosaurs, *Archaeopteryx macura*. The rarity of intermediates has recently become less rare. At the present time, Michael Benton, in the opening chapter of *When Life Nearly Died*, notes that "Evolution works to hone the fine details of the adaptations of organisms against commonly encountered problems, such as droughts, floods, predators and diseases, but rare events that happen perhaps once every few million years just cannot be accommodated. The phenomenon is... 'bad luck, not bad genes." So there was good reason that so many believed for so long that something other than the Neo-Darwinian view (Darwin's Theory of Natural Selection integrated with Mendelian genetics) was involved. The perfect Neo-Darwinian Hypothesis was just ruined by the facts. As long as the geneticists kept their heads in the sand, everything made sense, one way or another. And as long as the paleontologists kept their heads in the sand, their views of catastrophic extinctions and inexplicable creations, which were followed by radiations and baffling randomly variable periods of stability, worked for them. Both were a little bit wrong; yet, both were a little bit right. It was simply just not an "either-or" situation. While

environmental extinction events certainly make a case for environmentally-determined evolution, the environmental influence in evolution really cannot stand alone beyond this. **There is really no "environmentally-determined evolution," nor is there a "genetically-determined evolution."** Evolution is environmentally AND genetically-determined. Overemphasizing environmental influence in environmental adaptation is no better than overemphasizing the role of genetics and the latter error has recently been in excess. The geneotype-determined view is incomplete. The environmentally-determined view has always been incomplete. Evolutionary novelty does not arise from an "either-or" situation. It originates from a "both" situation. Neither an environmentally-determined view (alone) nor a geneotype-determined view (alone) can really explain the changes in types of life over time. That being said, the environment may not directly control genetic mutation, but it does influence the initiation of random display of phenotypes in response to stress, phenotype stabilization of genetic expression, and the survival of the individual. The role of the variable environment in extinction and adaptation has been well-documented so far, and there is more to come.

Kirschner and Gerhart believe that facilitated (genetic) variation enables the expression of a mutation. The mutation does not express the trait by itself; it only needs to change something, like a threshold, and this change may or may not make an observable change. Mutations in the genome are usually silent or neutral. If the genome expresses a previously silent mutation (e.g., with homozygous recessive alleles, it usually expresses a change; however, the observed change itself may even be neutral (no environmental selection advantage or disadvantage). And if it does not produce an environmentally-sensitive change in phenotype, the newly expressed neutral mutation effectively remains silent. But a mutation expression that is not neutral undergoes environmental selection, which promotes or diminishes any inheritable bias from that individual: "The main accomplishment of the theory of facilitated variation is to see the organism as playing a central part in determining the nature and degree of variation.... It is the capacity of the core processes to support variation that we see as the main factor in generating phenotypic variation and in minimizing the lethality of phenotypic variation. It is the nature of these processes, which are poised to generate physiological variation within the organism, that allow genetic variation to be so effective in generating phenotypic variation on which (environmental) selection acts." This is a genetically-determined view, with an indirect link to early developmental processes. Like any genetically-determined view is incomplete, it is still incomplete (but it is less incomplete).

Is there an environmentally-determined view with an indirect link to early developmental processes? And would it also be incomplete? Mary Jane West-Eberhard (K&G chapter—Phenotype and Environment) emphasized the broad applicability of an adaptation-accommodation hypothesis in explaining the origin of complex phenotypic variations. "In West-Eberhard's view, evolution of a novelty proceeds by four steps. In the 1st, called trait origin, an environmental change or genetic change affects a preexisting responsive process, causing a change of phenotype (often reorganization). At this initial step, she regards environmental stimuli as being more important to evolution than genetic variation. These traits may or may not be adaptive.... In the 2nd phase, the organism adapts or accommodates to its changed phenotype by compensating in part for the perturbed condition by using what we would say are its highly adaptive core processes. In the 3rd phase, recurrence, a subset of the population continues to express the trait, perhaps owing to the continued environmental stimulus. In the 4th phase, genetic accommodation, selection drives gene frequency changes that increase fitness and heritability, although the phenotype change is not necessarily ever completely under genetic control. While having a heritable component, it could retain an environmental dependence. Thus... a phenotypic accommodation phase (is) followed by a genotypic accommodation phase." (Over time, repeatability and predictability of an external signal should favor the developmental incorporation of the environmentally-induced developmental pathway by favoring genotypes that reliably develop a consistent phenotype across generations.)

In an article entitled: *Phenotypic Accommodation: Adaptive Innovation Due to Developmental Plasticity*, West-Eberhard discusses a general model for the origin of adaptive phenotypic novelties: "The following model is intended to describe the evolutionary origin of all kinds of adaptive traits— morphological, physiological, and behavioral, whether induced by a mutation or an environmental factor—at all levels of organization. This is a brief summary of concepts presented in more detail and with more complete supporting evidence elsewhere (*Developmental Plasticity and Evolution* text).

(a) A novel input occurs which affects one (if a mutation) or possibly more (if environmental) individuals. Individuals may experience novel inputs due to evolution in another context (e.g., which moves them into a new environment, or has novel pleiotropic, i.e., many different, effects on the phenotype via other pathways).

(b) Phenotypic accommodation (without any change in the geneotype): Individuals developmentally responsive to the novel input immediately express a novel phenotype, for example, because the new input causes quantitative shifts in one or more continuously variable traits, or due to the switching off or on of one or more input sensitive traits (causing a reorganization of the phenotype). Adaptive phenotype adjustments to potentially disruptive effects of the novel input exaggerate and accommodate the phenotypic change without genetic (genotype) change.

(c) Initial spread: The novel phenotype may increase in frequency rapidly, within a single generation, if it is due to an environmental effect that happens to be common.... Alternatively, if it is due to a positively selected mutation, or is a side effect of a trait under positive selection, the increase in frequency of the trait may require generations.

(d) Genetic accommodation (change in genotype, i.e., gene frequencies, under selection): Given genetic variation in the phenotypic response of different individuals, the initial spread produces a population that is variable in its sensitivity to the new input, and in the form of its response. If the phenotypic variation is associated with variation in reproductive success, natural selection results; and to the degree that the variants acted upon by selection are genetically variable, selection will produce genetic accommodation, or adaptive evolutionary adjustment of the regulation and form of the novel trait.

In an article entitled: *Developmental Plasticity and the Origin of Species Differences*, she further elaborates on environmental determination: "A large body of evidence indicates that regardless of selective context (genetic or environmental variation) the origin of species differences (new traits) under natural selection occurs as follows:

1. The origin of a new direction of adaptive evolution starts with a population of variably responsive, developmentally plastic organisms. That is, before the advent of a novel trait, there is a population of individuals that are already variable, and differentially responsive, or capable of producing phenotypic variants under the influence of new inputs from the genome and the environment. Variability in responsiveness is due partly to genetic variation and partly to variations of the developmental plasticity of phenotypic structure, physiology and behavior that arise during development and may be influenced by environmental factors, including maternal effects that reflect genetic and environmental variation present in previous generations. Genetic variation and developmental plasticity are fundamental processes of all living things: all individual organisms with the exception of mutation-free clones, have distinct genomes, and all of them have phenotypes that respond to genetic and environmental inputs. By 'responsiveness' and 'developmental plasticity,' I do not mean just phenotypic plasticity in the way the term is usually used, to mean only responsiveness to the external environment. Rather, I include

responsiveness to the action of genes, which may modify the internal environment of other genes and phenotypic elements within cells, with effects that extend outward to higher levels of organization and responsiveness. Any new input, whether it comes from the genome, like a mutation, or from the external environment, like a temperature change, a pathogen, or a parental opinion, has a developmental effect only if the preexisting phenotype is responsive to it. Without developmental plasticity, the bare genes and the impositions of the environment would have no effect and no importance for evolution (good point).

2. Developmental recombination occurs in a population of individuals because of a new, or newly recurrent, input. A new input from the genome, such as a positively selected mutation, or from the environment of the affected individuals, causes a reorganization of the phenotype, or 'developmental recombination.' Given the variable developmental plasticity of different individuals, this process produces a population of novel phenotypes, providing material for selection.

3. Genetic accommodation may follow. If the resultant phenotypic variation has a fitness effect, that is, it correlates with the survival or reproductive success of the affected individuals, then selection (differential reproduction of individuals or other reproducing entities with different phenotypes) occurs. If the phenotypic variation has a genetic component, selection leads to 'genetic accommodation;' that is, adaptive evolution that involves gene-frequency change. Genetic accommodation of regulation adjusts the frequency, timing, and circumstances of the novel response (e.g. by adjusting the threshold for its expression), and genetic accommodation of form refines the characteristics and efficiency of the newly expressed trait."

Unlike Kirschner and Gerhart in the previous chapter of this text, West-Eberhard believes that core processes support an environmentally-initiated, norm-of-reaction based, physiologically-supported, plastic phenotypic change of the developing phenotype; genetic (genotype) accommodation then occurs at a later time. In the same article entitled: *Developmental Plasticity and the Origin of Species Differences*, she states: "I argue that the origin of species differences, and of novel phenotypes in general, involves the reorganization of ancestral phenotypes (developmental recombination), followed by the genetic accommodation of change. Because selection acts on phenotypes, not directly on genotypes or genes, novel traits can originate by environmental induction as well as mutation, then undergo selection and genetic accommodation fueled by standing genetic variation or subsequent mutation and genetic accommodation. Insofar as phenotypic novelties arise by adaptive developmental plasticity, they are not 'random' variants because their initial form reflects adaptive responses with an evolutionary (norm-of-reaction constrained) history, even though they are initiated by mutations or novel environmental factors that are random with respect to (future) adaptation. Changes in gene frequency involve genetic accommodation of the threshold or liability for expression of a novel trait, a process that follows, rather than directs, phenotypic change. Contrary to common belief, environmentally initiated novelties may have greater evolutionary potential than mutationally induced ones. Thus, genes are probably more often followers than leaders in evolutionary change."

The environmental importance for both generating and maintaining the integrity of genetic expressions is strongly supported in West-Eberhard's environmentally-integrated developmental view; and, it is complete because it also allows genetic integration. In the Adaptive Evolution chapter of *Developmental Plasticity and Evolution*, West-Eberhard emphasizes that "selection for a trait does not (necessarily) mean selection for the spread of an allele that specifies the trait; it means selection for a change in regulation or behavior (including habitat or even diet selection) such that the trait is more readily, and therefore more commonly, produced." This would be a norm-of-reaction-limited genetic expression; here, there would be no newly-expressed mutation. Little genetic change would be needed for an inheritable change of expression (e.g. regulatory variation). However, the genotype change would occur afterwards, due to environmental selection favoring the genotype that reliably develops a consistent phenotype across generations. In West-Eberhard's Gradualism chapter in *Developmental Plasticity and Evolution*, she reaffirms that there is no primacy of selection over variation as the architect of design (i.e., all selectable variation is not just genetic, or mutational in origin): "New selectable variation can originate due either to mutation or environmental change. Variation originating in both ways can have evolutionary consequences and can lead to adaptive evolution, due to genetic accommodation under selection. This means that both variation and selection always contribute to the evolution of adaptive design. Neither can be assigned a dominant role because development is the source of all selectable variation, and selection determines which variations among these produced by development spread and persist." In the above, environmental and genetic variation and their contribution to adaptive design are clearly recognized, as is the role of a selective process (which I also maintain is very much an environmental selection process-resulting in preserved natural selection genetic bias). I strongly support the importance of her focus on development, but am not sure I totally agree with the statement "development is the source of all selectable variation."

Development can be a process that incorporates selectable variations (traits), but the variation is sourced from genetic variation and environmental variation. Development can also be the result of selectable variation, but the variation source is genetic variation and environmental variation. Development is variable, but it only exhibits variations; it is not the source of variation itself. Rather, development is the source of the adult phenotype, which also exhibits variations.

The environmentally-genetically-integrated views have much in common. We both agree that there is no primacy of genetic influence, or primacy of environmental influence in forming an adaptation, but a range of environmental and genetic interactions, utilizing plastic cellular processes to form the new phenotype. Yet, West-Eberhard's view may be somewhat more of an "either-or" view for environmental or heredity roles for initial formation of the phenotype. If the environment causes the formation of a new phenotype from the genotype norm-of-reaction, choices are limited to the norm-of-reaction plastic range of expression, which includes any ancestor-possible recombination (novelty may arise from physiologic change, heterochrony-timing, or regulatory gene functional variation). And, it is followed by a deemphasized genetic accommodation. And while her view is not that different (genetics and environment always contribute), I prefer the view that relies more heavily on genetic expression to form any phenotype, and requires little genetic accommodation after-the-fact. In this, as phenotype formation occurs, development incorporates indirect feedback from both genetic and environmental sources at the time of formation. This view of adaptation not only includes the preceding but also expands the options into greater epigenetic change, where an old mutation could be newly-expressed from the existing genotype variation, without any new mutational change in the genotype. Genetic expression (not mutations, but the expressions of mutations) can be modified by the environment. A mutationally-modified norm-of-reaction response could allow even greater evolutionary change, but would require enough time to allow formation of a random phenotype that is favored by adverse or supportive environmental selection. West-Eberhard does allow for a sudden mutation to cause simultaneous phenotype change, but at times she seems to be giving the environment a more independent role from genetics in adaptation (seemingly greater— but not necessarily a dominant role). Both are more interrelated, function in concert (a simultaneous, double-indirect interaction generating developmental novelty at the time of phenotype formation), and are not an "either-or" option. And, even though genetics is equally integrated, genetics simply cannot continue having exclusive domain as the only variation available for selection in evolution (as generally held today). It is long past the time to dismiss the deeply-conditioned religious belief of most biologists, holding that all new selective variation is due to random mutation, and that (genotype-determined) selection is completely responsible for the origin of adaptive design and direction of evolution. Normal expressions from the genome's norm-of-reaction are poised plastic phenotype responses, inheritable environmental adaptation "seeds" incorporated into natural selection genetic bias. The environment can change expression of the phenotype, causing it to vary in new ways, without a change due to a sudden mutation. Adaptation can come from existing variation. Unusual, long-lasting, strongly adverse; or, supportive environmental selective activity in the environment seems to be necessary for some adaptations to form. Mutation and its immediate expression, or the plastic expression of a previously silent mutation may be necessary to generate adaptation to an environmental problem that cannot be managed by presenting previously-preserved phenotypes from the genotype's norm-of-reaction. Recall the important messages from environmentally-integrated organisms are that mutations can be "creative" and that the main limit to evolution is not so much what is mutationally possible, but what is ecologically necessary (S. Carroll). Perhaps the best mutation candidate for a new trait is a mutational change in the regulatory area near a gene, either for a developmental gene that regulates other genes or in an enhancer that affects a compartment of body-building genes. In either case, normal gene function is preserved. While simultaneously generating novelty, heredity and environment view have interdependent developmental connections to each other. In addition, both heredity and environment play significant roles just before and just after the novelty appears. And, they both play significant roles long before and long after a novelty appears. So even though the majority of genetic modification may have occurred with expression of a random phenotype, it only makes sense that there could be some genetic accommodation from environmental feedback after phenotype formation. Both environmental biology and evolutionary biology require an integrated environmentally-determined AND genetically-determined view of life, with indirect links to developmental processes.

Unusual environmental stress generates unusual changes in the phenotypes displayed. It will not necessarily be an adaptive change; in fact, it is usually a non-adaptive change. The classic study (previously noted in the chapter on Phenotype and Environment) was first performed by Conrad Waddington on fruit flies. During embryogenesis, he stressed fly populations with high temperature environments, as well as to ether fumes. He found that many changes in the phenotypes were expressions of phenotype variation already present in the population (norm-of-reaction); but, some of the changes were novel phenotypes. So, during an unusual environmental stress, nearly all changes are constrained within the norm-of-reaction; but occasionally, a unique change will appear that is slightly outside of the norm-of-reaction. Flies exposed to heat during development revealed a small percentage increase of a type without supporting cross-veins on their wings; this was a unique change to a specific environmental stress. Except for the environmental selection from heat, there was no other selection for or against the phenotype; and, there was no protection from the environmental stress of heat (a non-adaptive phenotype). After 20–25 generations (all exposed to heat early in development), a loss of wing cross veins was increasingly present in successive generations (i.e., cross-veined wing phenotypes that were initially common in the population became increasingly less common in the population). In genetically-inbred populations without cross-veinless variation in the population, this effect did not occur. When the cross-veinless phenotype had offspring, the increase in numbers of this type was likely due to inheritability of the new heat-induced (non-adaptive) expression. That is, increased numbers of the new heat-induced phenotype had increased numbers of offspring that looked like the parents. This loss of wing cross-veins eventually became independent of the heat treatment. So, after many generations, the heat was no longer needed to carry-on the inheritable phenotype. Waddington showed that the cross-veinless phenotype can be "fixed" (to occur in close to 100% of the population). In other stress experiments, flies exposed to ether fumes developed another pair of wings; the extra wings did not seem to protect against ether exposure (a unique non-adaptive change of phenotype in response to an uncommon stress). Ether was an environmental selector for a double-wing non-adaptive phenotype, a random non-adaptive change to a specific environmental stress; the phenotype was inheritable. The preceding is a demonstration of unusual environmental stress leading to immediate formation of normal random genetic expressions (from feedback), including increased expression of an uncommon phenotype, as well as a unique phenotype (response), and a corresponding increasingly inheritable shift in the environmentally-generated patterns of genetic expression towards the unusual genetic expression patterns (which may or may not be adaptive). Both the new genetic expression and environmental feedback form the new phenotype. Signals from the external and internal environment initiate flexible, but regulated, modifications in genetic expressions that may be environmentally-stabilized to become environmentally-integrated adaptations. Inheritable changes in genetic expression occur without a (genotype) change of allele frequency.

Contrast the above to a genetic assimilation view (i.e., geneotype-determined only) that makes some sense, but seems to get lost somewhere. "Genetic assimilation is the evolutionary process by which a phenotype produced specifically in response to some environmental stimulus, such as a stressor, becomes stably expressed independently of the evoking environmental effect. First, in the absence of an environmental stimulus, a particular threshold trait (environmentally-common) is stably expressed, and phenotypic deviants remain cryptic because the environmental threshold for their expression is too high. Second, in the presence of an environmental stimulus, previously cryptic variation for the threshold trait is uncovered and the threshold for the expression of deviant phenotypes not seen under normal conditions is lowered. Third, selection in the presence of the environmental factor enriches the previously cryptic alleles determining the trait. Eventually, these alleles become so frequent that the expression of the trait overcomes the higher threshold in the absence of the environmental stimulus. (There is no increased or decreased survivability selection for or against a non-adaptive organism, but somehow there is a form of mysterious allele frequency increase within an individual organism?) Thus genetic assimilation (magically) transforms an environmentally induced trait into a phenotype which is stably expressed without eliciting the environmental stimulus; the genetically-assimilated phenotype is no longer plastic, but exhibits a genetically-fixed response independent of the environmental conditions (this needs fixing), a phenomenon called canalization." This assimilation view is just another vague and complicated expansion of the deeply-conditioned (religious) belief of most biologists: (mutation and genetic) "selection is completely responsible for the direction of evolution and the origin of adaptive design." It is far less confusing and complicated if there is no sudden mysterious allele frequency change in the absence of selection. Instead, an environmentally-sensitive regulatory gene switch or the epigenome effects the formation of a discrete trait from a single locus, or expresses a variable number of mutated multi-allelic genes for additive effect, or effects a threshold response to form a discrete trait. And the change is simply inheritable. Parsimony prevails.

Recall that Susan Lindquist "extended Waddington's experiments to discern how "heat unmasks 'cryptic phenotypic and genotypic variation.' Excessive heat, like other stress conditions, causes most proteins to fold and to lose activity. The organism produces several kinds of heat shock proteins (Hsp), or chaperone proteins as they are called, that guide the refolding of unfolded proteins back into their active form, thereby mediating recovery from heat. Hsp90 is one of these proteins. As it turns out, even without heat, the Hsp90 protein is continuously important for folding new proteins correctly, especially large proteins of signaling pathways. When the organism is heated, the Hsp90 is recruited to refold damaged proteins, but there isn't enough Hsp90 for folding the new proteins most in need of chaperone assistance. Aberrant phenotypes emerge—not just the cross-veinless kind produced by Waddington, but a wide range of others. Any (of the others) could be selected by the researcher and after some cycles of heating developing generations (environmental selection), that expression would become stabilized in the population. The spectrum of altered phenotypes differs in various stocks of flies, showing that genotypic diversity exists relative to which proteins are most dependent on chaperones." (These chaperone proteins seem to be influencing the generation of different phenotype displays, by assisting the refolding of other proteins back into their active form, thus recovering from heat damage.) Recall that "Lindquist and her colleagues tested not just *Drosophila* (fruit flies) but also *Arabidopsis*, a small flowering plant. They lowered Hsp90 activity in several ways, by heat (as described above), by mutation of the Hsp90 (gene) itself, and by exposure of the organism to a chemical agent that inhibits Hsp90 specifically. Many alterations of morphology were seen and, if desired, could be put through (environmental-sorting) selection so they would continue to form (be expressed) even after Hsp90 was restored to full activity. The variety of alterations was great: different altered parts and different degrees of alteration. Lindquist refers to Hsp90 as a 'capacitor for phenotypic variation.'" It is protein folding that implements both genetic and epigenetic change.

Young and Badyev, published an online article *Evolution of ontogeny: linking epigenetic remodeling and adaptation in skeletal structures*, in the Society for Integrative and Comparative Biology—online, May 22, 2007 (Oxford Journal), referencing a study suggesting that "mutations in regulatory, promoting, and processing of genetic pathways of bone forming regions can generate changes in gene expression (without disrupting cohesiveness of developmental networks).... In a broad examination of molecular evolution in morphogenic genes among cichlids (fish), Terai found allelic variation in the production of BMP-4 (bone morphogenic protein) consistent with high levels of morphological variation; this variation was related to protein folding, and thus modified downstream effects without disrupting the general function of the gene." And "showed that variation in timing of ossification can result in similar phenotypic patterns through epigenetically-induced changes in gene expression... such heterochronic shifts can not only buffer development under fluctuating environments (while maintaining epigenetic sensitivity crucial for normal skeletal formation), but also enable epigenetically-induced gene expression to generate specialized morphological adaptations."

30

Macroevolution and Microevolution

Kirschner and Gerhart emphasize that: "Evolution... is framed by two features: conservation on a cellular level and diversity on an anatomical and physiological level." It may be said that life's changing genetic patterns over time comprise its evolution, but there is more than genetics and time involved. Every process of evolution is dependent upon survival of the organism. Organism survival is dependent upon environmental interaction; the environment is the gatekeeper for genetic success. The environment is the ecosystem, which includes existing physical conditions as well as all life's interrelationships in the community. The primary cause of extinction today, as well as in the past, is an environmental extinction event, due to habitat degradation. **Macroevolution is the big picture on historical types of life. It is totally integrated with the small-scale processes of microevolution (**such as mutation and recombination genetic change at the molecular level).

Microevolution begins when mutation first changes the genotype; this change may or may not be expressed as a new phenotype. The adaptive response requires a small amount of genetic variation in a population, originally the result of an earlier mutation that is usually unexpressed (silent). A high frequency of cell division increases the possibility of mutation. Environmental stress increases the rate of mutation. Sexual reproduction increases allele diversity (and enhances mutation rates). Both mutation genome change and sexual reproduction's diversity of the genotype enhance the possibilities for an adaptive process to form a new phenotype. Conditions of environmental stress in an isolated environment could generate a genome-related survival response in early development that is environmentally-beneficial to the organism and could be passed to offspring. That is, at times of environmental stress, unexpressed silent genome mutations could be expressed as a new random phenotype (character trait), which could offer survival advantage to the stressful conditions in a newly-created environment. Most developing phenotypes generated as a response to unusual environmental stress are not adaptive, but may continue to be expressed in survivors as long as the environmental stress is present. Some non-adaptive phenotypes may persist for a while even after the stress is removed. In a somewhat stressful environment, survival itself may provide enough indirect environmental feedback to stabilize some non-adaptive phenotypes, at least for a limited amount of time. Non-adaptive phenotypes only continue for a short term in a stressful environment, particularly when stress levels are near-lethal; non-adaptive phenotypes are generally not cost-effective.

Evolution is genetic success (adaptation) in response to environmental conditions; evolution is a change in phenotypic adaptations, in response to variable environmental conditions, over time. In order for a developing organism to adapt to an isolated, changing environment (increase in fitness), and to pass adaptive changes to offspring, inheritable genetic change must join in concert with the selective forces in the environment to form a new inheritable change of phenotype and a modified genotype norm-of-reaction. An adaptation is a new environmentally-integrated inheritable phenotype. Adaptation to an adverse environment seldom occurs, but when it does, adaptation begins with unusual environmental stress initiating genotype formation of norm-of-reaction phenotypes, possibly including some unusual phenotype not usually represented under normal environmental conditions. Some phenotypes may even involve expressing a genotype mutation outside the norm-of-reaction, or even discontinue expressing a gene that was previously expressed. Potential adaptations, unique random phenotypes, are first indirectly stabilized by environmental feedback, and this stabilization then indirectly stabilizes the pathway for that random phenotype change in genotype expression. With this, these unique random phenotype expressions would be inheritable in offspring. The environment will select for favorable adaptive changes of the phenotype. There are three scenarios for favorable environmental selection in early development. Only adaptive phenotypes survive extreme environments that are lethal to non-adaptive phenotypes. Only adaptive phenotypes manage better in environments that are more limiting to non-adaptive phenotypes. Only adaptive phenotypes prosper (positively supported by the selective conditions of life) in environments most favorable to that adaptive phenotype; in this last case, non-adaptive phenotypes may not necessarily be limited, but there is no positive support (advantage) for the non-adaptive phenotype. The three scenarios will phase-out non-adaptive phenotypes. If non-adaptive phenotypes were to continue in an environment where selection does not eliminate them, mutation of genes expressed in the phenotype would lead to gene atrophy and further change in the phenotype, which would experience further environmental selection. Natural selection is the genetic bias of survivors (perhaps confined to fringe areas) that had the ability to adapt to the stressful environmental conditions of an isolated population. Natural selection is the preservation of past phenotype adaptations to previously-encountered environments. **The generation of shared character states is a genetic-related response to environmental challenge. Shared character states form environmentally-selected patterns over time; these patterns comprise the study of macroevolution.**

Gene pool isolating mechanisms, particularly sexual selection in animals, results in the origin of species, which genetically supports preferable adaptations. Some include speciation as an aspect of microevolution. Speciation itself is an isolating mechanism, a biologic isolating mechanism. It acts to create a special genetically-isolated environment, much as spatial isolation crates a special isolated environment for a breeding population. Noted in the previous chapter, change comes from adaptation, not speciation. No environment is constant; over time, environmental change for the isolated population will variably stress that population. On occasion, a genetic adaptive response, made possible by a genome response from variation, enhances some organism's survival and/or the survival of offspring. This new adaptive response is often a new function or morphology, and increasingly observed in the population until it is permanently established (fixed) as a new phenotype. This new phenotype (character state) is maintained by environmental selection, which notably includes selective reproduction within the species. Neither macroevolution nor microevolution is irrelevant (see next chapter).

"Evolution can be seen as a change in developmental programs that elaborate the phenotype."—Jeffrey Levinton (2001).

31

Philosophy of Genetic Mechanisms

Michael Behe, in *Darwin's Black Box—The Biochemical Challenge to Evolution,* insists that natural selection cannot account for the purposeful arrangement of parts that is seen at the molecular level. Behe uses a "black box" as a whimsical term for a device that does something, but whose inner workings are mysterious—sometimes because the workings can't be seen, and sometimes because they just aren't comprehensible. He suggests that Darwin just didn't understand what was in the black box of biochemistry, the intricate workings of life itself. He maintains that the basic workings of life essentially lie in the details of its complex molecular biochemical machinery. He believes Darwinian evolution, chance variation resulting in a competition advantage in the struggle for life, falls short in applications to the fine details of biochemistry. He firmly believes that the real work of life does not happen at the level of the whole animal, and that "natural selection working on random variation" breaks down at this most important molecular level. Still further black box analogy is made between mutation and evolution, indicating a lack of evidence of interaction between both. In a discussion of the eye and vision, he insists that the biochemistry of vision does not support natural selection; for carefully tailored machines, the leaps are just too great to be realistic. He further states that at this most important molecular level, anatomy and the fossil record are irrelevant.

Behe defines evolution "in the sense Darwin gave the word, (It) means the process whereby life arose from nonliving matter and subsequently developed entirely by natural means." Is this correct? It is helpful to return to the original work to understand what Darwin really said. Recall Darwin suggested that then present-day species (different types of life) resulted from a common ancestor and changes (anatomical adaptations via natural selection) occurring in each isolated population. Darwin's theory of natural selection (the preservation of favorable variations and the rejection of injurious variations), as the explanation for the origin of species (different types of life), seems to be less focused on the origin of life than on the evolution of changes from that time on. Today, studies on the origin of life focus not on evolution, but on exobiology, nannobacteria (alive or not?) and viruses ("poisonous slime" molecules of protein-coated DNA or RNA that inhabit the twilight zone between life and non-life: e.g. *Mimivirus*).

When natural selection was first introduced in Darwin's Struggle for Existence (*Origin of Species*) chapter, Darwin stated: "Owing to the struggle for life, any variation, however slight and from whatever cause proceeding, if it be in any degree profitable to an individual of any species, in its infinitely complex relations to other complex beings and to external nature, will tend to the presence of that individual, and will generally be inherited by its offspring." The occurrence of natural selection's survival/non-survival at the level of the individual organism is quite clear; it establishes evolutionary hierarchy at the level of the individual organism. Recall that variation is packaged as combinations of traits within one individual organism. Expanding one trait may then be at the expense of other traits. To be beneficial, this new combination must enhance the overall survival chances of the individual and/or offspring. The most effective combination for the genetic variance of natural selection then becomes the most optimum balance of all traits. This places the selective process for complex life at a different (higher) level than molecular chemistry, even though the whole is made from the sum of the parts.

Behe could maintain that the origin of life and some simple forms of life, for a period of around four billion years, function primarily at the molecular level. The origin of life is still unknown; the biochemical argument is inconclusive here (A discussion of biochemistry is just ahead). Simple forms of life, like a virus or bacteria do mutate at the molecular level, do evolve, and do seem to follow a pattern of natural selection (survival of the fittest). A virus undergoes mutation (genetic change), which enables it to invade new environments. Nevertheless, the selection outcome is still at the level of organism survival, not at the level of a chemical reaction. Bacteria develop biochemically-generated structural resistance to (an environment of) antibiotics, and pass this survival advantage on to their offspring. Antibiotic resistance in bacteria is usually conferred by genes carried by plasmids (circular double-stranded DNA molecules separate from the chromosomal DNA and capable of autonomous replication; they usually occur in bacteria, sometimes in eukaryotic organisms—Wikipedia). Again, selection relevance is at the organism level; survival is not really at the level of the biochemical reaction itself. Isn't there molecular evolution involved in the biochemistry of the mutation allowing plants to develop resistance to an environment of herbicides, and pass this survival advantage on to their offspring? It is not the chemical reaction alone that survives; it only survives due to the environmental selection influence on natural selection genetic bias, which occurs at the level of the organism possessing the adaptation advantage. Similar molecular processes occur when insects develop resistance to an environment of insecticides, and pass this survival advantage on to their offspring. Futuyma notes that: "Resistance to dieldrin (insecticide) in different populations of *Drosophilia melogaster* (fruit fly) is based on repeated occurrences of the same mutation, which, moreover, is thought (because of its map position) to represent the same mutation that confers resistance in flies." Molecular-level change in the biochemistry of genetic mutation clearly produces natural selection advantage in organism-level survival. Aren't these clear interactions where random genetic variation is environmentally-selected for a natural-selection genetic bias? Hasn't this molecular process enhanced evolutionary survival of individual organisms in a newly-created environment (that is unfriendly to most other existing life forms)?

There is a huge amount of evidence for the interaction between mutation and evolution; to say that it doesn't exist indicates a lack of something else. Understanding biochemistry does not necessarily impart an understanding of biology, just as understanding biology does not necessarily impart an understanding of biochemistry. Both subjects are enormously broad and can be extremely complicated. Failure to comprehend something does not necessarily make something invalid. Nevertheless, Behe maintains that life is not accountable to natural selection at the pure molecular level. But then what of natural selection accountability at an even more detailed, but even less appropriate, atomic or sub-atomic levels? How can anyone claiming to be an authority on evolution totally ignore a previously-published Futuyma discussion on eye evolution, which did address Darwin's own concerns of difficulties on the theory? Concurrently, Richard Dawkins presented a great deal of evidence on eye evolution in *Climbing Mount Improbable*; at the time, this information was available to "authorities on evolution."

Behe argues that many biological systems, such as the eye, are "irreducibly complex," which means they are too complicated to evolve by changing the components; i.e., in order to evolve, multiple parts would have to arise simultaneously; it assumes a creation event. It must have been created by a far more intelligent being than us; it could never evolve to this degree of complexity. This is similar to the anatomist Owens's objection to Darwin's theory nearly 150 years ago; it did not stand against scientific evidence. It does not stand against the material discussed in this text. The argument of irreducible complexity is also a rehash of the famously flawed watchmaker argument advanced by Paley at the start of the 19th century. The argument for "irreducibly complexity" lacks a credible example; an "irreducibly complex" model may not even exist. If there is such a thing, and previous components of evolution cannot be used in Behe's fabrication of an "irreducibly complex system" (e.g. eye), weren't the chemical element components in this "system" (e.g. eye) formed in the degradation of super-massive suns, long ago? Aren't these elements incorporated into molecules used in the biochemistry of all examples throughout the text? Aren't these atoms in the molecules recycled in different chemical reactions? How is it so extremely implausible that components used for other "purposes" are fortuitously adapted (recycled) to new roles in a "complex system? " If a man makes a mousetrap, logic does allow that he does have a purpose in mind for it. Yet man did not make the chemical elements that form the mousetrap, so the "purpose" of the chemical elements, as related to the mousetrap, is not demonstrable. Not everything has to have a purpose. This idea of purpose is the belief system of a man, and any belief of man is subject to error. Biochemical reactions in living systems need not have a "purpose." Evolution has no demonstrable purpose. It is what it is. Pope John Paul ll freely accepted evolution as a wonder of nature, claiming that the evidence for evolution was irrefutable. While evolution may be inconsistent with someone else's belief system, it is only incompatible with a religious belief if someone chooses to make it so. Perhaps they know more about evolution and religion than the Pope did.

How could anatomical changes and the fossil record be irrelevant to evolution? It is the best record of evolution! It would seem to me that anatomical changes and the fossil record would be essential to any comprehensive understanding of evolution; this comprehensive understanding should also include the biochemistry of anatomical structure and function. Propose hypothetically, that evolution is an "irreducibly complex" system; how could Behe ignore anatomy, the fossil record, the environment, the environmental effects on evolution, and everything else except for molecular biochemistry in the process of evolution? This seems to be an inconsistency of logic.

Previously, a case has been made for environmentally-integrated indirect design in the forgoing text. Behe claims to have established a case for design in living systems, which he calls intelligent design. There is no objection to Behe maintaining that something is causing design in nature that isn't recognized by most of the scientific community. The co-evolution of flowering plants and insects asks far too much of genetically-determined chance occurrence. Developmental structure and function, as well as acclimation to environmental challenges (includes stress-induced physiologic restructuring) all involve environmentally-integrated indirect modifications of random exploratory cellular behavior. Yet there is no scientific evidence of intelligence behind the (environmentally-integrated indirect) design of living systems. Behe only assumes that there must be an intelligent designer, behind what appears, to him, to be a created design. This is a leap away from the realm of science into the realm of a belief system. Behe, like Aristotle, seems to have even considered natural selection as a cause for the creation of different types of life and evolution; but, like Aristotle, dismissed it in favor of teleology. Behe believes that: "All of these things were designed because of the ordering of independent components to achieve some end." Teleology, a black box in itself, is still alive and well.

Religion is important to society; it fills a need. When it comes to a belief system, it is nearly impossible to challenge a predominantly cultural belief that is already in place; it would be a mistake to even try to do this. Evolution studies cannot really prove or disprove a belief system. People will believe what they want to believe. The ultimate religious argument is predictable. God should be able to do anything; however, it is logical to think that God's universe functions by natural means. Now, there's an intelligent belief! Who could argue with that?

32

Biochemistry of Genetic Mechanisms

"Periods of character stasis, strong variations in the rate of character evolution and the appearance of new morphs controlled by easily attributable and trivial genetic mechanisms are the mainstay of species-level taxonomy in the fossil record" (Levinton). The actual mechanism of adaptation, random exploratory cellular behavior modification into environmentally-sensitive independent physiological processes, phenotype-forming gene regulation via variable switches sensitive to feedback from environmental signals, and epigenome expression modification through environmental signals have been introduced; the last two require some understanding of the biochemistry of genetic mechanisms. Almost all living organisms use the same basic biochemical molecules (e.g., DNA, RNA, the same twenty amino acids and ATP) and many identical, or nearly identical, enzymes. DNA is a long double-stranded coil of phosphate-linked sugars, with attached nucleotide bases. Sugar (deoxy-ribose) and phosphate can be bonded (glued) together; this makes the assembly of long strands possible. The long strands of sugar and phosphate repeat and repeat along a line. There are four nucleotide bases (thymine, adenine, cytosine and guanine) in DNA (deoxyribonucleic acid). Nucleotide bases attach to the (deoxyribose) sugar molecules in different sequences along the sugar-phosphate line. DNA nucleotides of bacteria will number several million. DNA nucleotides of some flowering plants will number several hundred billion. The nucleotide bases of the DNA molecule form complementary pairs, zipping the double stranded coil together along its length. A nucleotide base on one of the double strands in the coil (hydrogen) bonds to the complementary nucleotide base in the other strand of the DNA forming a very stable double-stranded coiled complementary chain. This (hydrogen) bonding is specific; adenine always bonds to thymine (and vice versa) and guanine always bonds to cytosine (and vice versa). This hydrogen bonding along the molecule is always in the complementary pattern, and that holds the double-stranded molecule in a coiled form within the chromosome. DNA replicates itself for cell division and for germ cell reproduction. DNA replication occurs at a number of sites along a chromosome; the double helix is unwound in short segments, pairs of bases are separated, and DNA polymerase enzymes bring complementary bases into position along each of the two strands to create a new duplicated DNA complementary double stranded coil (resulting in 2 identical double-stranded DNA coils). Under increased environmental stress, polymerase errors may increase, increasing the rate of mutations. In the normal working state, the DNA double-strand coil is uncoiled and not tightly clumped together. During cell division, the DNA coil is condensed into a visible chromosome, and paired with a related chromosome (alleles—like a pair of shoes); these chromosomes will match each other, but not exactly. Different portions of the DNA on a chromosome are active at different times during development and adult life. New life starts small and is guided through the stages of growth and development by selective gene expression.

There is more to life than DNA; environmental interaction is all-important. There may only be small DNA differences between very different organisms; something other than DNA makes this difference. There is also a big difference between the various types of cells within one organism; yet, all of these very different cells have the same DNA genetic coding (genome). This text has introduced the concept of an environmental feedback pathway, which can indirectly stabilize the random exploratory cell behavior that forms an indirectly-expressed heritable phenotype. Layers of biochemical reactions operate variable switches to express phenotype-forming programs stored in the preserved (natural selection genetic bias) norm-of-reaction. Epigenesis, the environmentally-modified expressions of genes, doesn't involve genome (DNA) change per se, but variation in the genes expressed for phenotype change (epigenome change). The epigenome expresses the norm-of-reaction; the epigenome expresses the plastic range of natural selection genetic bias. Epigenetic moderation of gene expression endows cells with the ability to assume different properties (phenotypes) and thus function in different roles within the organism. Epigenetic changes involve environmental modifications of gene expression; they have a profound effect in development. In development, the indirect influence of the environment may influence a change in expression of the epigenome; this is the interface for environmental influence when forming the norm-of–reaction phenotype. It is an environmentally-connected gatekeeper for normal gene expression. Epigenetic changes involve normal environments and normal variation in gene expressions from the norm-of-reaction. Under environmental stress, epigenetic changes may also involve a change of gene expression outside the norm-of-reaction. In development, a change outside the norm-of-reaction pattern could even involve the expression of a mutation. The change could express something new, as with a sudden mutation, or the change could newly-express a gene mutation from an earlier time. Either type of change could cause a noticeable change in the phenotype, affecting a single gene or a group of genes in a compartment, or cause a change in the regulation of other genes. And the change could produce a new phenotype that may become an adaptation. In the adult, the epigenetic changes are generally, but not always, reversible.

Epigenesis, the overall genome response to environmental interaction, defined in detail (by Microsoft) is the development of an individual... as a result from interaction between an individual's genes, external environment and internal environment. While genome information is uniform in the different cells of complex organisms, the epigenome controls the differential expression of genes in specific cells. Programming of gene expression profiles is dependent on the epigenome. The epigenome is a layer, or layers, of biochemical reactions that operate variable switches to previously-formed patterns of genetic expression; the epigenome plays a big part in heredity. Epigenetics (Wikipedia) is the study of epigenetic inheritance, a set of reversible (in adult) heritable changes in gene function or other cell phenotypes that occur without a change in DNA sequence (genotype). These changes may be induced spontaneously, in response to environmental factors, or in response to the presence of a particular allele, even if it is absent from subsequent generations. Cellular differentiation processes crucial for embryonic development rely almost entirely on epigenetic rather than genetic inheritance from one cell generation to the next. In the past two decades, especially the last couple of years, studies have shown linkage of the epigenome to disease and development. Studies show that the epigenome changes in response to the environment, and that the changes can be passed from parents to children.

Ethan Waters, in Discovery Magazine (Nov. 2006), has an excellent discussion on epigenetics in a *DNA Is Not Destiny* article. "The epigenetic changes wrought by one's diet, behavior, or surrounding work can echo far into the future.... In 1999 biologist Emma Whitelaw demonstrated that epigenetic marks could be passed from one generation of mammals to the next. (The phenomenon had already been demonstrated in plants and yeast.).... Whitelaw focused on the agouti gene in mice, but the implications of her experiment span the animal kingdom. 'It changes the way we think about information transfer across generations,' Whitelaw says. 'The mind-set at the moment is that the information we inherit from our parents is in the form of DNA. Our experiment demonstrates that it's more than just DNA you inherit... what we inherit from our parents are chromosomes, and chromosomes are only 50% DNA. The other 50% is made of protein molecules, and these proteins carry the epigenetic marks (e.g. CH_4+) and information.'" Proteins have both structural and functional roles; in the epigenome, the latter is likely to be a regulatory form of gene expression (modulation of gene expression- epigenome change). It will respond to environmental feedback.

In 2003, Jirtle and Waterland investigated a genetic disease in fat yellow mice, caused by the above-mentioned agouti gene. In addition to making the mice yellow and ravenous, the gene makes them more susceptible to cancer and diabetes. Ethan Waters continued: "Typically, when agouti mice breed, most of the offspring are identical to the parents: just as yellow, fat as pincushions and susceptible to life-shortening disease. The particular mice in Jirtle and Waterland's experiment, however, produced a majority of offspring that looked altogether different. These young mice were slender and mousey brown. Moreover, they did not display their parent's susceptibility to cancer and diabetes, and lived to a spry old age. The effects of the agouti gene had been virtually erased. Remarkably, the researchers effected this transformation without altering a single letter of the mouse's DNA. Their approach was radically straightforward—they changed the mom's diet. Starting just before conception, Jirtle and Waterland fed a test group of mother mice a diet rich in methyl donors (CH_4)+, small chemical clusters that can attach to a gene and turn it on and off. These molecules are common in the environment and are found in many foods, including onions, garlic, beets and food supplements often given to pregnant women. After being consumed by the mothers, the methyl donors worked their way into the developing embryo's chromosomes and into the critical agouti gene. The mothers passed on the agouti gene to their children intact, but thanks to their methyl-rich pregnant diet, they had added to the gene to a chemical switch that dimmed the gene's deleterious effects. Our DNA—specifically the 25,000 genes identified by the human genome project—is now widely regarded as the instruction book for the human body. But genes themselves need instruction on what to do, and where and when to do it. A human liver cell contains the same DNA as a brain cell, yet somehow it knows to code only those proteins needed for functioning of the liver. Those instructions are found not in the letters of the DNA itself but on it, in an array of chemical markers and switches, known collectively as the epigenome, that lie along the length of the double helix. These epigenetic switches and markers, in turn, help switch on or off the expression of particular genes. In recent years, epigenetic researchers have made great strides in understanding the many molecular sequences and patterns that determine which genes can be turned on and off. Their work has made it increasingly clear that for all of the popular attention devoted to genome-sequencing projects, the epigenome is just as crucial as DNA to the healthy development of organisms, humans included. Jirtle and Waterland's experiment was a benchmark demonstration that the epigenome is sensitive to clues from the environment. More and more, researchers are finding that an extra bit of vitamin, a brief exposure to a toxin, even an added dose of mothering can tweak our genes—in ways that affect an individual's body and brain for life.

The even greater surprise is the recent discovery that epigenetic signals from the environment can be passed on from one generation to the next, sometimes several generations, without changing a single gene sequence. (Recall Waddington's environmental stress studies where heat-induced or ether-induced non-adaptive random phenotype changes could be initiated in offspring.).... The implications of the epigenetic revolution are even more pronounced in light of recent evidence that epigenetic changes made in the parent generation can turn up not just one but several generations down the line, long after the original trigger for change has been removed. In 2004 Michael Skinner, a geneticist at Washington State University, accidentally discovered an epigenetic effect in rats that lasts at least four generations. Skinner was studying how a commonly used agricultural fungicide, when introduced to pregnant mother rats, affected the development of the testes of fetal rats. He was not surprised to discover that male rats exposed to high doses of the chemical while in utero had lower sperm counts later in life. The surprise came when he tested the male rats in subsequent generations—the grandsons of the exposed mothers. Although the pesticide had not changed one letter of their DNA, these second-generation offspring also had low sperm counts. The same was true of the next generation (the great grandsons) and the next. Such results hint at an anti-Darwinian aspect of heredity (not necessarily). Through epigenetic alterations, our genomes retain something like a memory of the environmental signals received during the lifetimes of our parents, grandparents, great-grandparents and perhaps even more distant ancestors. So far, the definitive studies have involved only rodents. But researchers are turning up evidence suggesting that epigenetic inheritance may be at work in humans as well.

In November 2005, Marcus Pembrey, a clinical geneticist at The Institute of Child Health in London, attended a conference at Duke University to present intriguing data drawn from two centuries of records on crop yields and food prices in an isolated town in northern Sweden. Pembrey and Swedish researcher Olov Bygren noted that fluctuations in the town's food supply may have health effects that spanning at least two generations. Grandfathers who lived their preteen years during times of plenty were more likely to have grandsons with diabetes—an ailment that doubled the grandson's risk of early death. Equally notable was that the effects were sex specific. A grandfather's access to a plentiful food supply affected the mortality rates of his grandsons only, not those of his granddaughters, and a paternal grandmother's experience of feast affected the mortality rates of her granddaughters, not her grandsons. This led Pembrey to suspect that genes on the sex-specific X and Y chromosomes were being affected by epigenetic signals. Further analysis supported his hunch and offered insight into the signaling process. It turned out that timing—the ages at which grandmothers and grandfathers experienced a food surplus—was critical to the intergenerational impact. The granddaughters most affected were those whose grandmothers experienced times of plenty while in utero or as infants, precisely the time when the grandmother's eggs were forming. The grandsons most affected were those whose grandfathers experienced plentitude during the so-called slow growth period, just before adolescence, which is a key stage for the development of sperm. The studies by Pembrey and other epigenetic researchers suggest that our diet, behavior, and environmental surroundings today could have a far greater impact than imagined on the health of our distant descendants.

Scientists are still coming to understand the many ways that epigenetic changes unfold at the biochemical level. One form of epigenetic change physically blocks access to genes by altering what is called the histone code. The DNA in every cell is tightly wound around proteins called histones, and it must be unwound to be transcribed. Alterations to this packaging cause certain genes to be more or less available to the cell's chemical machinery and so determine whether those genes are expressed or silenced. A second, well-understood form of epigenetic signaling, called DNA methylation, involves the addition of a methyl group to particular bases in the DNA sequence. This interferes with the chemical signals that would put the gene into action and thus effectively silences the gene. (Epigenetic changes involve gene expression.) Until recently, the pattern of an individual's epigenome was thought to be firmly established during early fetal development. Although it is still seen as the crucial period, scientists have lately discovered that the epigenome can change in response to the environment throughout an individual's lifetime. 'People used to think that once your epigenetic code was laid down in early development, that was for life,' says Moshe Szyf, a pharmacologist at McGill University in Montreal. 'But life is changing all the time and the epigenetic code that controls your DNA is turning out to be the mechanism through which we change along with it. Epigenetics tells us that little things in life can have an effect of great magnitude.'…. Through numerous studies, Szyf has found that common signaling pathways known to lead to cancerous tumors also activate the DNA-methylation machinery, knocking out one of the enzymes in the pathway that prevents the tumors from developing.

When genes that typically act to suppress tumors are methylated, the tumors metastasize. Likewise, when genes that typically promote tumor growth are de-methylated, those genes kick into action and cause tumors to grow. Szyf is far from alone in the field. Other researchers have identified dozens of genes, all related to the growth and spread of cancer, that become over- or under-methylated when the disease gets underway. The bacteria, *Heliobacter*, believed to be a cause of stomach cancer, have been shown to trigger potentially cancer-inducing epigenetic changes in gut cells. Abnormal methylation patterns have been found in many cancers of the colon, stomach, cervix, prostate, thyroid and breast. Szyf views the link between epigenetics and cancer with a hopeful eye. Unlike genetic mutations, epigenetic changes are potentially reversible. A mutated gene is unlikely to mutate back to normal; the only recourse is to kill or cut out all the cells carrying the defective code. But a gene with a defective methylation might well be encouraged to reestablish a healthy pattern and continue to function. Already one epigenetic drug, 5-azacytidine, has been approved by the Food and Drug Administration for use against myelodysplastic syndrome, also known as preleukemia, or smoldering leukemia. At least eight other epigenetic drugs are currently in different stages of development or human trials."

Waters further notes that: "Epigenetic researchers around the globe are rallying behind the idea of a human epigenome project, which would aim to map our entire epigenome. The Human Genome Project, which sequenced 3 billion pairs of nucleotide bases in human DVA, was a piece of cake in comparison: Epigenetic markers and patterns are different in every tissue type in the human body and also change over time. Jirtle says: 'A single individual doesn't have one epigenome, but a multitude of them.' One genome could be said to give rise to many epigenomes." Cellular differentiation processes in environmental compartments, crucial for embryonic development, rely almost entirely on epigenetic rather than genetic inheritance from one cell generation to the next.

Epigenome adaptation occurs as a biochemical process; it occurs at many levels within an individual organism. Environmental selection may also operate at many levels within an individual organism; however, organism survival occurs only at the level of the individual organism. Said otherwise, epigenome adaptation occurs as a biochemical process; however, phenotype (indirectly genotype) environmental selection for survival is at the organism level. The environment is deeply integrated into norm-of-reaction expressions of the phenotype, and into survival through adaptation.

DNA genes not only replicate themselves, they form the structural protein and functional protein regulation of life. A gene is a linear DNA segment of a chromosome that codes for a protein (polypeptide).

Previously noted, it may even code for more than 1 protein, or only a part of a large polypeptide. The genome is the entire component of DNA contained in an organism or a cell, which includes both the chromosomes within the nucleus and the DNA in the mitochondria. DNA genes code for proteins by being decoded in two steps. The first step, transcription, occurs near the DNA in the chromosome itself; it involves forming a DNA-similar RNA (single strand) intermediate (uracil replaces thymine). The second step, translation, occurs in a remote location; it involves the RNA intermediate in forming the protein. The gene (the DNA portion of the chromosome that makes a protein in the two steps) first acts as a template for the synthesis of RNA in transcription. The DNA double chain unzips a piece at a time for construction of the shorter RNA segments. Transcription of DNA (deoxyribonucleic acid) into a form of complementary (purine or pyrimidine) nucleotide bases of RNA (ribonucleic acid) is then followed by the second step in protein formation, RNA translation into a sequence of amino acids; the remote location in eukaryotes is outside the cell nucleus. Only a part of this DNA is transcribed to produce nuclear RNA (nRNA), and only a minor portion of the nuclear RNA survives RNA processing steps. In most mammalian cells, only 1% of the DNA sequence is copied into functional RNA. Organisms utilize DNA triplet code; three sequential nucleotide bases code for an amino acid. These triplets are called codons; each one codes for a specific amino acid in life's 20 amino acids used to construct proteins. The genetic code can be expressed as either DNA codons, or following transcription, complementary RNA codons. The genetic code consists of 64 triplets of nucleotides. Each of the 64 codons specifies one of 20 amino acids, or else serves as a punctuation mark signaling the end of a message. Having this many triplets produces some redundancy in the code, most of the amino acids being encoded by more than one codon; there is some flexibility as to which base occupies a particular position. This nucleotide exchange of equal value will produce the same amino acid and is neutral variation in the genome. Given 64 codons, punctuation marks, and 20 amino acids, there are 10 to the 83rd power possible genetic codes. The genetic code is almost universal. Not only are the same codons assigned to the same amino acids, they are also assigned to the same "start" and "stop" signals in the vast majority of genes in animals, plants, and microorganisms. Some exceptions have been found.

Transcription of DNA sequences into RNA is carried out by RNA polymerase enzymes. Transcription begins at special "promoter" locations (adjacent to a downstream exon) along the DNA chain. One of the most important stages in RNA processing is RNA splicing. In many genes, the DNA sequence coding for proteins, called "exons," may be interrupted by stretches of non-coding DNA, called "introns." Many organisms share the same introns and types of repeats. Previously discussed, the nuclear DNA includes all the exons and introns of the gene that is first transcribed into a complementary RNA copy of "nuclear RNA" or nRNA. In a second step, introns are removed from nRNA by the process of RNA splicing. The edited sequence is "messenger RNA," or mRNA. Exons coding for a particular protein are not necessarily arranged side-by-side on the same chromosome; they are constructed in the transcription process, assembled from numerous messenger RNA molecules. The mRNA leaves the nucleus and travels to the cytoplasm, where it encounters the ribosomes. The mRNA, which carries the gene's instructions, dictates the production of proteins by the ribosomes. The amino acids used to make the proteins are carried to the ribosome site by "transfer RNAs" or tRNA.

Closer examination of RNA splicing, where introns (the internal sequences interrupting protein-encoding parts on the long strands of precursor RNA) are processed (by the removal of these apparently functionless internal "spacer" sequences), reveals that this "functionless" spacer material does have some regulatory function. Introns can be found in viruses and eukaryotes (cells with a nucleus), but are rare in prokaryotes (no nucleus). RNA introns and single strands extruded from duplex DNA have the potential to form stem-loop structures. In general, nucleic acids can form intra-strand stem-loop structures if the complementary nucleotide bases are suitably located. A lollipop-shaped structure is formed when a single-stranded nucleic acid molecule loops back on itself to form a complementary double helix (stem), topped by a loop comprised of around half-a-dozen to a dozen purine or pyrimidine bases. For many taxa, if transcription is to the right, the top (mRNA synonymous) DNA strand has purine-rich loop potential; if transcription is to the left, the top (template) strand has pyrimidine-rich loop potential. Nucleotide base order serves either the encoding of a protein (e.g. structural, enzyme or regulatory factor) or further indirect control of a gene by recombination-moderated stem-loop structures. In **rapidly evolving genomes, nucleotide base-order dependent stem-loop potential is as important as other functions.** There appears to be circumstances under which nucleotide base order synergizes with, or antagonizes, nucleotide base composition in determining total stem-loop potential. There is room for conflict (nucleotide base order serves many local "strategies," whose demands may conflict) between the "desires" of a sequence to encode *both* a protein **and** stem-loop potential. The conflict would be particularly apparent in the case of genes under very strong positive phenotypic (Darwinian) selection, as in the case of genes affected by "arms races" with predators or prey (Forsdyke, D.R.:—*Reciprocal Relationship between Stem-loop Potential and Substitution Density in Retroviral Quasispecies under Positive Darwinian Selection.* Journal of Molecular Evolution 1995. Dec; 41(6):1022–37). The stem-loop potential of a DNA sequence is unlikely to be just a passive and indirect consequence of the action of various evolutionary pressures on DNA. There appear to be powerful genome-wide pressures which actively confer or inhibit the potential to form stem-loops. An environmentally-induced stress would increase the potential for stem-loop formation and adaptive change.

Other than being involved with epigenetic-type changes in gene expression at the level of the chromosome, stem-loop formation is a significant biochemical mechanism that indirectly influences the genetic expression to environment stress in several other ways. Also mentioned above, the **general function of stem-loops is moderating** inter- or intra-chromososomal **recombination** (shuffling already-existing allele variation). The former (inter-) varies chromosome combintions; the latter (intra-) can produce a chromosome change mutation (still reshuffles existing alleles). Said otherwise, stem-loop formation allows the potential for changes in existing allele combinations and for internal chromosomal rearrangement mutations. In addition, increased formation of stem-loops seems to decrease transcription, allowing gene transcription regulation (indirect gene inhibition of protein production). Stem-loops also seem to stabilize protein formation, by interacting with protein-folding potential. The indirect genetic significance of protein folding will be addressed shortly (recall protein folding implements phenotypic change). In the above cases, an extreme environmentally-induced stress would greatly increase the potential for stem-loop formation and adaptive change.

As in simple life forms (virus), stem-loops also form in eukaryotes, introns abound. Stem-loop formation in eukaryotes induces similar change by varying the genes expressed by the epigenome. Stem-loop formation moderates recombination frequency (mutation); changes nucleotide base order (mutation); changes rates of transcription; and modifies protein folding. All of this allows stem-loop formation significant potential for indirect random change of genetic expression in eukaryotes. **Stem-loops form as a response to environmental stress**; stem-loops shotgun severe stress solutions. The formation of stem-loops under stress may account for significant phenotypic adaptive change; that is, stem-loop formation would seem to be very likely to form phenotype changes outside the norm-of-reaction. Under environmentally-induced stress (simple life-virus or complex life), stem-loop formation is a genetically and environmentally-connected biochemical mechanism (via epigenetic changes in gene expression, mutation control, transcription control, and protein folding) for adaptation. The adaptation patterns of simple life require some upgrading in complex life. Everything seems to be upgraded in complex life; however, fundamental processes don't necessarily change that much. Whether these processes involve more common genome norm-of-reaction displays, or formation of new random expressions, some of which may become environmental adaptations, genetic expression is indirectly modified by the environment.

Nearly all phenotypic expression change is within the genotype's norm-of-reaction. In the Weak Regulatory Linkage chapter, Kirschner and Gerhart discuss an experiment by Jacques Monod growing the bacterium *E. coli* on different sugars. "The growth rate of *E. coli* on a mixture of two sugars was not the sum of the rates on either one alone. The bacterial culture would first grow for a while using one of the sugars, then pause, and then start growing again, using the other sugar.... To grow on a sugar, the bacterium had to have a particular enzyme to degrade it. Monod found that the bacterium did not have the enzyme initially. It first produced one kind of enzyme to metabolize one of the sugars, and then produced a different kind to metabolize the other...." Enzyme adaptation referred to the parsimonious (simple) bacterium adjusted its metabolism according to the food supply. The explanation for Monod's paradox of why the bacterium consumed the sugars sequentially rather than simultaneously was that one sugar preferentially stimulated the bacterium to produce an enzyme to degrade it, while also inhibiting the bacterium from producing the enzyme to metabolize the other sugar. Here was an exquisite form of physiological adaptation, the organism fastidiously changing the proteins it produced (part of the phenotype) in response to changing environmental conditions. (The environment influenced a core process phenotype change in metabolism—via gene regulation.) For further study, Jacques Monod chose to use the milk sugar lactose and the enzyme that metabolized it, lactase (now used for removing lactose from dairy products for lactose- intolerant people), also known as beta-galactoside. *E. coli*, in the gut of an infant, would encounter lactose after the infant nursed and would soon be ready to metabolize it at a furious rate. In the intervals between feeding, it would turn off the production of the enzyme and metabolize other sugars. Monod found that beta-galactoside was not stored in an inactive form, but each time was synthesized anew from available amino acids. Thus, enzyme adaptation was not a process of activation of an enzyme but a process of synthesis of an enzyme. In the end, the regulation was on the synthesis of the RNA transcript from the DNA. The enzyme itself was not adapting but the bacterium was, by producing more or less of the specific enzyme (gene regulation). The phenomenon was named enzyme induction.

Although this might seem like merely another example of physiological adaptation (like oxygen uptake responding to oxygen supply), it was special because it involved the synthesis of a protein that would ultimately be connected to the genes that encode that protein (i.e., gene regulation). At the time it was not known how an organism could selectively make one kind of protein of the many that were encoded in its DNA. It was possible that every protein had its induction mechanism, but it was much more likely that proteins were made continuously. Inducible proteins like beta-galactoside would be exceptional by being not synthesized at specific times. Monod theorized that the synthesis of each inducible protein would be inhibited by a specific inhibitory protein, the repressor, which the cell made continuously. Lactose, when present, would bind to the repressor and remove it from the DNA, in this way releasing the inhibition. In this view there were no real inducers per se, only repressors. Activation was then achieved by the antagonism of repression, or as Monod called it, depression. He was adamant that depression is the only possibility; however, his colleague Francis Jacob speculated correctly that true activators might exist that could bind to DNA. They would activate specific genes to produce the corresponding RNA, which would yield the protein. Biology's solution to the selective use of information was elegantly simple. The bacterial cell could adapt to changes in the environment by repressing and depressing genes in a simple stimulus-response circuit. (Gene regulation is perhaps the most powerful conserved core process, also responsible for much of the phenotypic variation on which selection acts. This example indirectly links the environment to a specific biochemical-genetic mechanism.) Jacob and Monod's experiments proved the model to satisfaction in bacteria. For example, they isolated cells defective in the gene encoding the repressor protein. These cells made beta-galactoside all the time, regardless of whether lactose was in the environment. This simple physiological circuit was proposed as a general solution for regulating the expression of different genes in different cells... they had discovered another core process, which would have unlimited applicability—the process of 'regulation of gene expression.'

Monod worried about one part of the model. Physiological responses are usually quantitative, showing smooth and continuous variation, not on-off extremes. This trait was true of lactose metabolism. Over a wide range, the more lactose in the medium, the more enzyme the bacterium made. What the two men had described was an on-off switch, a poor model for physiology. Jacob was able to resolve this theoretical difficulty: 'The insight came to me as I watched one of my sons playing with a small electric train. Although he did not have a rheostat, he could make the train travel at different but constant speeds just by turning the switch on and off more or less rapidly.' Physiological adaptation in bacterial enzyme synthesis was now solved in a wonderfully simple manner, a genetic switch, turning the gene on and off by binding a repressor to it and turning it on by removing the repressor with an inducer. The fraction of time the switch was on would determine the rate of synthesis. The details were soon filled in: the repressor protein of the beta-galactoside synthesis binds to a specific short sequence on the DNA, located next to the start of the beta-galactoside gene sequence, where it occludes the binding of machinery necessary to transcribe the RNA.... There are two (indirect) linkages in this system. First, the repressor binds to the DNA at the specific location and blocks RNA polymerase from synthesizing RNA on that gene only. Second, the lactose binds to the repressor and causes the repressor to change so that it no longer binds to the DNA. Though several molecules like lactose regulate their own metabolism, none binds to DNA directly; they all act indirectly through a repressor protein. In such an indirect system, the response can be easily modified and generalized. If the lactose repressor were to bind next to some other gene, that gene should also respond to lactose, and the response (RNA synthesis) offers many opportunities for changing regulation." "At the heart of enzyme induction in bacteria is the repressor, which in response to binding lactose somehow changes its shape, causing it to fall off the DNA. Here we have an example of signal transduction, where one kind of signal, the level of the metabolite, is transduced into another type of signal, the binding of a protein to DNA. This form of (indirect) linkage is widespread in all of biology. In understanding how the repressor binds to lactose, we are solving a much more generalized problem of how molecules talk to one another and transmit signals. Jacques Monod's first great insight, depression (repression), showed how the genome could respond intelligently to simple signals.

His second great insight, which he called allostery, explained how proteins do the heavy lifting of decision making. Allostery, from the Greek *allo* meaning other, and *stere* meaning solid, referred at first to the fact that proteins can have two kinds of interaction. One is the locus of the protein's function, and the other is the locus of regulation of that function. The protein has a functional part and a regulatory part. Although this seems unexceptional to us now, in 1965 it was a profound insight. It was profound not only because it contradicted the prevailing view of biochemists at the time, that each kind of enzyme had only a single kind of site for carrying out its chemical reaction, but also because it liberated proteins to engage in an unconstrained variety of regulatory interactions. Part of the profundity follows from the fact that at level of the atomic level of proteins, explanations of their behavior can no longer be vague and ad hoc. They must conform to the laws of chemistry and physics. Allostery was no hand-waving model, but a chemical model that showed how a molecular switch operates. (Monod published just after the first atomic level structures of proteins were completed.) In his model, permissiveness was explained at the molecular level. These insights prompted Monod to exclaim with characteristic bravado, 'I have discovered the second secret of life.'

Biochemists at the same time were working on another problem of regulation that involved not control of the synthesis of an enzyme, but instead the direct control of an enzyme's activity. Enzyme inhibitors were well understood, and several were well known drugs. For example, sulfa drugs directly inhibit an enzyme that makes a component of DNA, penicillin inhibits an enzyme that makes bacterial cell walls, and the AIDS drug AZT inhibits a viral enzyme used in replicating human immunodeficiency virus. All of these inhibitors act by impersonating the normal target of the enzyme, known as the substrate. The inhibitor occupies the site on the enzyme where chemical reactions occur and physically blocks the binding of the real substrate. Ever since Emil Fischer drew the analogy in 1894, enzymes and substrates have been compared to locks and keys that fit together. An inhibitor was a false key that fit well enough into the lock to keep other keys from entering, but not well enough to turn the tumblers and open the lock, that is, to undergo a chemical transformation. Thus, inhibitors were expected to bear many likenesses to the substrate. By 1960, paradoxically, several enzymes were already known to be inhibited by molecules that looked nothing like their substrates. Enzymes that stood at the beginning of a biosynthetic pathway were often inhibited by chemical entities produced at the end of the pathway, many steps removed. Hence, the whole process was called feedback inhibition. Feedback inhibition makes logical sense. If you owned a factory manufacturing automobiles and sales were so sluggish that finished automobiles piled up in the showrooms, you would cut back your purchase of raw materials such as steel, rubber and glass. It would make no sense merely to slow down the final steps of assembly, such as the paint job; while curtailing the production of finished autos, the process would still consume costly materials, energy and labor. (Control is needed close to DNA—the site of transcription.) Since the end product of a pathway did not generally resemble the substrate used by the enzyme of the first step, it could not impersonate the substrate. Regulatory control had to be (indirectly) exerted at an alternative or 'allosteric site.' Commenting on allostery, Francis Crick, codiscoverer of the structure of DNA, said in 1971, 'That meant that you could connect any metabolic circuit with any other metabolic circuit, you see, because there was no necessary relation between what was going on at the catalytic site and the control molecule that was coming in.' Separate sites were designed independently.

Monod and Jacob were aware that this model went beyond metabolic control and had significance for the evolution of circuits coordinating complex processes and hence for the evolution of complex organisms. Freeing the business side of the enzyme (the catalytic or primary binding site) from the regulatory side allows their independent evolution, without the constraint imposed on a single site to meet dual functions. The catalytic side is constrained by all of the specialized chemistry of catalysis. The regulatory side can be constructed to interact with almost anything that has regulatory relevance. Since much of evolution involves connecting conserved core processes in new ways, it is a distinct advantage to separate the functional part of a protein from its regulatory part, which can then evolve in an unconstrained manner. Today it is known that many proteins are modular, having separate functional and regulatory parts, and that through the regulatory part they communicate signals across very different pathways—from cell proliferation to protein synthesis, from metabolism to heart rate, from inflammation to cell death. The power of proteins to integrate new regulatory connections in a simple way has fueled much of their change during multicellular evolution. Proteins with different domains arise readily in evolution, a fact that creates a major deconstraint on the evolution of regulatory connections. Although the biological implications of separate domains on the same protein were profound, mechanistic insights awaited an understanding of how the allosteric site could actually control the catalytic site, despite separation from it. Evidence was building for the notion that protein molecules are not rigid and can have more than one folded conformation or shape. Monod's conceptual breakthrough on the mechanism of allostery was to argue that allosteric proteins have not one but two confirmations that differ in degree of activity of the active site. The protein was itself a molecular switch having active and inactive states with regard to enzymatic activity. He postulated that the protein could oscillate freely between the two confirmations. It was like a toy that could flip into a new state, where all aspects of the geometry were altered.... For allosteric enzymes, the two states differ not only in their catalytic activity but also in their ability to bind the regulator. If the inactive confirmation binds the regulator more tightly, the regulator is an allosteric inhibitor; binding it would hold the protein in its inherently inactive state. If the regulator binds more tightly to the active state of the enzyme, it is an allosteric activator; binding the regulator would hold the protein in the active state.

In the case of the repressor, which is an allosteric protein, the form of the protein that binds lactose tightly binds DNA weakly. And the form of the protein that binds DNA tightly binds lactose weakly. Therefore, when lactose is present, the protein is held in a state where it binds DNA poorly; the repressor stays off the DNA, and transcription begins.... Monod was particularly proud of his assertion that in a protein made of multiple subunits, each with its own enzymatic and regulatory site, all subunits pass concertly from one confirmation to the other. Such organization makes the response behavior of the protein all or none. In this way, the repressor's on-off behavior is converted into transcription on-off behavior. Monod asserted that everything in biology is either all or none, either on or off.

Intermediate levels of activity reflect a mixed population of molecules, some of which are entirely on and some of which are entirely off—just like the switch on the toy electric train. This simple idea has stood the test of time. As we move to the molecular level, the allosteric protein faithfully maintains the distinction between permissive (indirect) and instructive (direct) signals. In Monod's model, the inhibitor that binds to the regulatory allosteric site does not instruct the enzyme to change from an active to an inactive state; it merely binds preferentially to the preexisting inactive state, encouraging that state to persist and accumulate in that population. The inhibitor is really a selector of a preexisting response, not the creator of the response.... Much of allosteric enzyme regulation is essentially designed into the protein in advance of the regulator's arrival. It is not that permissive signaling signals systems are less complicated than instructive ones. It is that permissive signals are less complicated, thanks to the prepared complex responses of the receiver. Permissive signaling is (indirect) weak linkage, because the signal does not alter the actual process; it merely selects upon it (gatekeeper).

We are now better able to understand how conserved and constrained mechanisms facilitate variation around them. An allosteric protein is highly constrained. A typical protein has thousands of weak chemical interactions that collectively hold it in a single stable configuration. It took a great deal of metabolic energy to build the protein, from the synthesis of amino acids to the synthesis of RNA and its translation. An allosteric protein is poised on a knife-edge, with two stable configurations differing in activity and in how strongly a regulator bonds to an allosteric site. The protein shifts continually from one state to another. The allosteric protein is a design so constrained that it cannot endure mutational change without damage to this allosteric function. Yet this extensive internal constraint enables extensive deconstraint in the evolution of regulatory connections. Allostery makes a protein capable of (indirect) weak linkage. A regulatory signal does not have to generate the active or inactive state—those options are already built in. It simply selects one form or the other by binding more tightly to it. The evolution of such regulatory sites has few constraints, for the regulator has little to accomplish. It does not have to interact with the highly constrained and precise catalytic site. The regulatory site can be almost anywhere on the protein surface. We have delved so deeply into the workings of this form of physiology because two-state proteins are important well beyond metabolic control. The greatest novelty that has evolved in multicellular organisms is the passage of information, not the chemical rendering of chemical intermediates. Much of the transfer of information comes from switch-like molecules that can exist in two confirmations. For switch-like molecules, the core mechanism is allostery. Such molecules communicate much of the information for control of cell growth and cell differentiation…. Allostery promotes (indirect) weak linkage by separation regulation from function. Continual selection for the retention of (indirect) weak linkage facilitates the generation of phenotypic variation and deconstrains the selection for new functions and new regulatory connections."

In eukaryotes (complex life), Kirschner and Gerhart note that: "The great innovations of core processes were not magical moments of creation but periods of extensive modification of both protein structure and function…. Many new proteins of large size were produced as new combinations of small, functional proteins similar to those found in single-celled eukaryotes. Large combinations of pieces are novel to animals… this kind of protein evolution has been undeniably facilitated by the exon-intron structure of eukaryotic genes and by the capacity of eukaryotic genes to 'splice RNA transcripts.' (The splicing capacity functions continuously in the individual in the production of messenger RNA in the cells.) The new proteins participate in multicellular functions, such as adherence to the extracellular matrix, cell-to-cell communication, and tissue reorganization through formation of intercellular junctions.

The second kind of novelty (genetic change) arises from the duplication of old genes, followed by the divergence of protein coding sequences to give related but distinctive functions." (a color vision example is upcoming) "The genome of multicellular animals is set up in a way that facilitates the evolution of new genes, using a kind of weak (indirect) linkage along the protein-encoding domains in the genome. During Precambrian metazoan evolution, new genes were created by fusing together various pieces of other genes, especially new genes for components of signaling pathways and of the extracellular matrix…. Given the size of the genome and the large amount of noncoding DNA, one might imagine that the likelihood of making exact couplings of many pieces into one large gene, for one large protein, is small. However, we now know that most genes encode proteins in pieces. (Recall that) each short coding piece is an island surrounded by long stretches of noncoding DNA, called introns. During (eukaryote) transcription, the introns are spliced out precisely, yielding messenger RNAs coding for multidomain proteins. The machinery for splicing together the RNA is a complex, highly constrained, and extremely conserved aggregate of about two hundred proteins and RNAs. It recognizes general sequence features at the two boundaries of the intron flanked with coding sequences. At these boundaries, it perfectly cuts out the intron and splices together the ends of the remaining RNA. Still, this highly constrained and conserved splicing machine is tolerant of the length of intron sequence between two boundaries. If a new piece of coding sequence with its intron boundaries (and bits of introns) is placed within an existing intron elsewhere in the genome, it will be spliced properly and incorporated into the final messenger RNA in the correct frame. Hence its encoded protein domain will be incorporated into final protein. Since there are several mechanisms for moving blocks of DNA around the genome and for dropping those blocks into existing introns, the structure of the genome with its long intron sequences facilitates the formation of new multidomain protein structures. Thanks to the RNA splicing machinery, the sequences can be dropped in without precision and still incorporated precisely into new structures."

"Much of the 'junk DNA' that is no longer in use are remnants of old genes. When genes aren't in use, they are no longer kept mutation-free by (environmental) selection. Genes require environmental support or they will atrophy." (from Sean Carroll chapter on Making New from the Old- *The Making of the Fittest: DNA and the Ultimate Forensic Record of Evolution*). "(The) strong preservation of protein sequences at most (gene) sites, with the synonymous evolution of the corresponding DNA sequence, and diversity limited to a few sites in the protein is the predominant pattern of evolution in the DNA record. DNA sequences that encode the same protein sequence but that are substantially different (redundancy substitution of interchangeable base sequences) are unmistakable evidence of (environmental) selection, allowing mutations that do not change protein function (could also have base change in non-critical areas), while acting to eliminate mutations that would. (In theory, selection for a redundancy difference could happen if the physical presence of the each DNA could be considered as a phenotype available for environmental selection, e.g. methylation differences). The preservation of genes among different species over vast periods of time is thus definitive proof of (environmental) selection." Carroll documents numerous examples of genetic activity reinforcement from environmental stabilization throughout the text; this perspective is markedly different from the genotype-determined, orthogenic views of Kirschner and Gerhart. "Junk DNA" can be manufactured. Brosius, in the first essay of Vrba's *Macroevolution, Diversity, Disparity, Contigency* suggested retroposition (reverse transcribed RNA to DNA) as a mechanism to reintegrate retronuons into genomes (further discussed in epilogue). He attributes this to be the major source of junk DNA.

A single gene may cause varied expressions in different organisms. Kingsley, Schluter and their co-workers found changes in a single gene can produce major changes in the skeletal armor of fish living in the wild (*Science* March 25, 2005). This brings new data to the long- standing debate about how evolution occurs in natural habitats. "People have been interested in whether a few genes are involved, or whether changes in many different genes are required to produce major changes in wild populations. The answer is that evolution can occur quickly, with just a few genes changing slightly, allowing newcomers to adapt and populate new and different environments.... Sticklebacks are enormously varied... biologists have realized most populations are recent descendants of marine sticklebacks. Marine fish colonized new freshwater lakes and streams when the last ice age ended 10,000–15,000ya. Then they evolved along separate paths, each adapting to the unique environments created by large scale climate change. There are really dramatic morphological and physiological adaptations to the new environments. For example, sticklebacks vary in size and color, reproductive behavior, in skeletal morphology, in jaws and teeth, (and) in the ability to tolerate salt and different temperatures at different latitudes. Kingsley, Schluter and their co-workers picked one trait—the fish's armor plating—on which to focus intense research, using the armor as a marker to see how evolution occurred. Sticklebacks that still live in the oceans are virtually covered, from head to tail, with bony plates that offer protection. In contrast, some freshwater sticklebacks have evolved to have almost no body armor. It's rather like a military decision, either to be heavily armored and slow, or, to be lightly armored and fast. Now, in countless lakes and streams around the world, these low-armored types have evolved over and over again. It's one of the oldest and most characteristic differences between stickleback forms. It's a dramatic change: a row of 35 armor plates turning into a small handful of plates, or even no plates at all. Using genetic crosses between armored and unarmored fish from wild populations, the research team found that one gene is what makes the difference. Now, for the first time, they've been able to identify the actual gene that is controlling this trait, the armor-plating on the stickleback. The gene they identified is called *Eda*, originally named after a human genetic disorder associated with the ectodysplasin pathway, an important part of the embryonic development process. The human disorder, one of the earliest ones studied, is called ectodermal dysplasia. It's a famous old syndrome; Charles Darwin talked about it. Darwin talked about 'the toothless men of Sind,' a pedigree (in India) that was striking because many of the men were missing their hair, had very few teeth, and couldn't sweat in hot weather. It's a simple Mendelian trait that controls formation of hair, teeth and sweat glands. It's a very unusual constellation of symptoms, and is passed as a unit through families. Research had already shown that the *Eda* gene makes a protein, a signaling molecule called ectodermal dysplasin. This molecule is expressed in ectodermal tissue during development and instructs certain cells to form teeth, hair and sweat glands. It also seems to control the shape of bones in the forehead and nose. Now it turns out that armor plate patterns in the fish are controlled by the same gene that creates this clinical disease in humans. And this finding is related to the old argument whether Nature can use the same genes and create other traits in other animals."

Pleitropy is the case of a single gene exerting an effect on many aspects of an individual's phenotype. Kirschner and Gerhart note pleiotropy is subdued by compartmentalization (environmental epigenetic identity) in the map of the embryo; this would reduce any conflict and lessen the lethality of mutation. Sean Carroll, in *Regulating Evolution,* discusses an example of beneficial loss due to mutation of a regulatory switch affecting a compartmentalized body-building pleiotropic gene. "The evolutionary forerunners of the hind limbs of four-legged vertebrates are the pelvic fins of fish. Dramatic differences in pelvic fin anatomy have evolved in closely related fish populations. The three-spined stickleback fish occurs in two forms in North America—an open water form that has a full spiny pelvis, and a shallow water, bottom-dwelling form with a dramatically reduced pelvis and shrunken spines. In open water, the long spines help to protect the fish from being swallowed by larger predators. But on the lake bottom, those spines are a liability because dragonfly larvae that feed on the young fish can grasp them. The differences in pelvic morphology have evolved in the 10,000 years since the last ice age (period). Long-spined oceanic sticklebacks colonized many separate lakes and the reduced form evolved independently several times. Because the fish are so closely related and interbreed in the laboratory, geneticist can map the genes involved in the reduction of the stickleback pelvis. David Kingsley and others have shown that changes in the expression of a gene involved in the building of the pelvic skeleton are associated with the pelvic reduction. Like most other body-building genes, the Pitx1 gene has multiple jobs in the development of the fish. But its expression is selectively lost in the area of the fish that will give rise to the pelvic fin bud and spines. A mutation in a regional regulatory switch (enhancer switch) deactivated the pleiotropic regional body-building gene. There are no coding changes in the Pitx1 protein between different forms of the stickleback.... Mutations in regulatory sequences circumvent the pleiotropic effects of mutations in coding sequences and allow for the selective modification of individual body parts."

Dogs, a subspecies of wolf, have high variation within a single species. The differences between breeds of dogs can only be small changes in a single gene. How does a single gene cause multiple effects? How does ectodermal dysplasin vary the expression of other genes that make the armor of a stickleback, or hair, sweat glands and teeth in humans? Kirschner and Gerhart state: "Eukaryotic gene regulation is perhaps the most powerful conserved core process, responsible for much of the phenotypic variation on which (environmental) selection acts. In eukaryotes, various regulatory proteins (factors) bind to specific regions of DNA sequence near the gene to be controlled and directly activate transcription (1st step in formation of proteins) in response to extra-cellular signals…. The typical mammalian gene is regulated by dozens of such factors binding in the vicinity of the gene at particular DNA sequences they recognize. Each carries a message from some different signaling system, saving time, space, amount, cell type, and other information. These signals are not on-off switches but partial-on and partial-off switches. The multiple (internal environment) factors result in a certain level of RNA synthesis (1st step in formation of proteins) from a specific gene…. The (indirect) regulation of gene expression (through internal environmental factors produced by genes of a higher hierarchy) is one of the core processes most critical for generating phenotypic variation." In short, **developmental eukaryote genes** (selector genes) **indirectly control** (remotely regulate) **other genes in isolated environmental compartments through the production of signaling molecules** (internal environmental factors).

Kirschner and Gerhart, in their chapter on Facilitated Variation, note that: "The term *cis*-regulatory refers to the DNA sequences that are adjacent to a gene, through which the gene's transcription is controlled. According to this attractive view, the most important evolutionary change is that occurring in the regulatory regions of genes, by which random mutation creates or eliminates sites on the DNA for binding various transcription factors of the great variety available in the cell. Each (DNA regulatory) site is a few bases in length, perhaps six to nine, and not entirely unique in sequence since some positions carry alternate bases. When a (DNA regulatory) site changes and is newly bound by a factor, the expression of the (protein-encoding) gene changes its time, place, or amount, depending when and where the factor is present. The change of regulatory sequence is, of course, heritable. Thus, without too many special requirements, a new condition for expression can be added without losing the old. Many candidate factors are active in binding DNA only when the cell receives (internal or external environmental) signals or when the factors are carried forward in the cell lineage from earlier developmental stages. The newfound transcription of the gene therefore relates to some spatio-temporal aspect of development of the embryo. The changing of *cis*-regulatory DNA by mutation does not affect protein structure in most cases, because *cis*-regulatory sequences are not usually transcribed into proteins. Since changes of the *cis*-regulatory region do not cause detrimental amino acid changes, they are not eliminated by selection as a result of functional failures of proteins…. (But they are free to mutate, unless) improved binding sites would be preserved by positive (environmental) selection (environmental support). This model provides a direct and efficient means to express genes in new combinations and amounts, with little investment in mutational change—exactly what our facilitated variation theory seeks. Transcription with its means of regulation embodies these conserved features of ready modification and is one of the most important of the core processes that manifest weak linkage.... The model implies that the proteins make development happen, however that is done. It assumes that the linkage of gene expression in new combinations throughout transcriptional regulation will be effective in generating new kinds of development and hence new traits.... The cell has many other ways, other than transcriptional *cis*-regulation, to control the presence or absence of protein function and thereby to control a time, place, or amount of function."

An internal environmental factor molecule may be described as *cis*-acting when it affects other entities only if they are physically adjacent. If it affects a more remote location, it may be considered "the opposite" of *cis*-acting, a *"trans*-acting"molecule. It is common to describe transcription factors as either *cis*- or *trans*-acting. A *cis*-acting transcription promoter facilitates the transcription of adjacent polypeptide-encoding sequences whereas trans-acting promoters affect the transcription of regions of DNA *not* in close physical proximity (Wikipedia). Eukaryotes also have "enhancer" DNA sequences remotely located. An enhancer is a short region of DNA that can be bound with proteins (namely, the trans-acting factors, much like a set of transcription factors) to enhance transcription levels of genes (hence the name) in a gene-cluster. An enhancer does not need to bind close to the transcription initiation site to affect its transcription, as some have been found to bind several hundred thousand base pairs upstream or downstream of the start site. Enhancers can also be found within introns. An enhancer's orientation may even be reversed without affecting its function. Furthermore, an enhancer may be excised and inserted elsewhere in the chromosome, and still affect gene transcription. That is the reason that intron polymorphisms are checked though they are not transcribed and translated (Wikipedia). "Gene expression begins when transcription factor proteins attach to binding sites within the (DNA) enhancer sequence. The complex they form acts as an 'on' switch that triggers the enzyme polymerase to begin transcribing an RNA copy of the gene. The RNA transcript is then read by a ribosome, which translates its message into an amino acid chain that folds itself into a coded protein.... When multiple enhancers control the expression of a gene in different parts of the body, a change to one enhancer can alter the gene's activity in a specific place without affecting it elsewhere. A fruit fly gene called yellow, for example, produces black pigment in a fly's developing body and wings, but various species have evolved distinct pigmentation patterns through changes to their enhancer sequences.... A flip side of evolution, loss of features, is very common, though much less appreciated. (In *D. kikkawni*, the enhancer can no longer drive high levels of yellow expression in the rear of the abdomen because a few mutations have disrupted some of its transcription factor binding sites.) The loss of body characters perhaps best illustrates why the evolution of enhancers is the more likely path for the evolution of anatomy." (Sean Carroll-*Regulating Evolution.* Scientific American.)

As life gets bigger and more complex, more and more environmental interaction has been internalized into the genes themselves, and their interactions. Subtle alterations in the timing, location, and levels of protein synthesis have considerable consequences at both the molecular level and organism level. Protein production regulation may be influenced by, or affect, various levels of transcription regulation in complex animal life, and this action is thought to be achieved by the combined action of multiple transcription (internal environmental signal) factors, which bind to *cis*- or *trans*-regulatory DNA sequences. A major part of the gene regulatory apparatus seems to be organized in separate *cis*-regulatory modules. A given module defines specific aspects in the spatio-temporal pattern of a gene expression by the combined action of multiple transcription factors, which together define the rate of transcription and translation into protein. Previously noted, compartment body-building and selector/regulatory genes are influenced by mutational modification of these regional regulatory switches, producing numerous types of expressions in a phenotype that incorporate varying degrees of transcriptional control in the structural/functional gene expressions.

It is believed that neural crest cell embryonic induction in vertebrates allows even another level of indirect control in the genetic hierarchy of complex life development. "The development of the neural crest is a powerful combination of exploratory cell behavior and the compartment map. Explosive diversification can occur when there is a synergism between the independent and diversified compartments and a multiprogenitor population that migrates over them and explores their diversity (K&G)." Kirschner and Gerhart present numerous examples of how the properties of developmental systems may have actually contributed to the evolution of a major evolutionary novelty. One example of evidence for this comes from beak studies on Darwin's finches. "Embryos of the various species differ in beak development in a way correlated with the level in the beak of a certain growth factor protein, called Bmp4 (internal environmental factor—a bone modeling protein), which stimulates the deposition of bone (and probably beak materials). Neural crest cells produce Bmp4 in the beak region. In the large Galapagos ground finch, Bmp4 is produced earlier and at higher levels than in the pointed small-beak species. If this factor (internal environmental signal) is introduced into the beak neural crest cells of the chicken embryos, they develop broader, larger beaks than normal, similar to the beaks of ground finches. Other growth factors do not have this effect. Nonetheless, when the experimental beak changes size and shape, it is still integrated into the anatomy of the bird's head. It is not a monstrous aberration. The precise regulatory changes accompanying the changes of Bmp4 production in the finches must still be established (seems similar to bone formation in limbs). Significantly, what changed was the time and level of a signaling molecule, Bmp4. It is found in all metazoans, even jellyfish. Changing the level artificially in a different species, the chicken, had similar effects. What seems to happen is exactly what might have been predicted by facilitated phenotype variation. The changes are regulatory, affecting time, place, and amount of Bmp4. They perturb, in a quantitative way, via Bmp4, the adaptive cell behavior of the neural crest cells, themselves a conserved multipurpose adaptive agency of development. Crest cells do not produce an outright deformity, but in fact modify their development compatibly with the rest of the head." Kirschner and Gerhart further note that: "Transcription factor proteins (internal—like Bmp4 or ectodermal dysplasin) are themselves targets of change. Many contain repetitive runs of a few kinds of amino acids. These runs expand and contract at high frequencies, many thousands of times higher than other mutations, with the consequence of increasing or decreasing factor (regulatory) activity. Some of these changes correlate with skull shape in various breeds of dog; another change, entailing a loss of 17 repeated amino acids in a transcription factor protein present in limbs, correlates with the presence of a second dewclaw on the rear leg of the Great Pyrenees breed.... Furthermore, the addition of such a repeat sequence to one of the Hox (regulatory gene) proteins, as well as the loss of a sequence for receiving phosphate signals, correlates with a newfound suppression of appendages in the abdomen of insects, whereas this change did not occur in the many-legged crustacean-like ancestors. All of these are changes in coding sequence for new forms of functional regulation, not changes of *cis*-regulatory (or *trans*-regulatory) control." Said otherwise, in addition to *cis*-regulation and *trans*-regulation of compartmentalized body-building genes and regulatory genes; regulatory variations in their produced factors influence regional body-building genes to create novel expression. When mutation does occur, facilitated variation (sudden mutation supported by random exploratory cell behavior) and allosteric protein switches (newly-expressing an earlier mutation)... explain comprehensively how a minimal input of random mutation (change in a gene) can generate (new) phenotypic variation. Sean Carroll, in *Regulating Evolution*, states: "Mutations in regulatory sequences are not the exclusive mode of evolution—they are just the more likely path when a gene has multiple roles and only one of them is modified."

There is still a genetically-inherited process of transcriptional and post-translational gene moderation in eukaryotes to be discussed. Post-translational feedback affects rates of transcription. Epigenetic change does not involve DNA change and is environmentally-connected. Previously discussed, it may involve methylation of a DNA base (interferes with the chemical signals that activate the gene—effectively silences it). It may involve allosteric (switch) protien marker control of gene expression. The DNA in every cell is wound around histones (nuclear proteins) to form nucleosomes and the DNA must be unwound in order to be transcribed. These core nucleosides are involved with numerous post-translational feedback-type reactions (acetylation, methylation, phosphorylation and others). Cancer research has revealed some understanding in the epigenome's role. Alterations in DNA-histone arrangements involve exposure to environmental influences; i.e., nucleosome alteration enables epigenome modification. *Developmental Biology On-line* discusses an example of chromatin remodeling. Chromatin structure determines the state of activity of genes by gating (chemical gatekeeper example of environmental selection) the access of the transcription machinery to transcriptional regulatory regions. Chromatin structure also plays a role in other genomic activities, such as recombination and repair. Acetylation of the lysine residues at the N-terminus of histone proteins could remove positive charges, thereby reducing the affinity between histones and DNA. This would make it easier for RNA polymerase and transcription factors to access the promoter region, which would not otherwise be accessible. Histone acetylation could enhance transcription while histone deacetylation could repress transcription; a dynamic state of histone acetylation could be maintained by two enzyme families. DNA bound to acetylated histones is demethylated. Methylation on the DNA's cytosine base is believed to repress transcription. The human genome is globally hypomethylated; a certain few genes are regionally hypermethylated. Global hypomethylation is a hallmark of cancer. Proteins that inhibit histone acetylation inhibit demethylation. Chromatin modifying proteins cause regional hypermethylation, preventing access to demethylase. Increased demethylase is responsible for global hypomethylation and maintaining tumor invasion genes (hypomethylated and active). Inhibition of demethylase causes hypermethylation and silencing of tumor invasion promoting genes. Active demethylation is directed by chromatin structure (nucleosome protein markers). DNA methylation is a reversible reaction; the DNA methylation pattern is a balance of methylation and demethylation. The importance of environmental influence cannot be underestimated. **Inherited genes and environmental feedback are connected, indirectly, in the epigenome. The majority of gene control in** any organism occurs in the epigenome, the chromosome control center, natural selection genetic bias responds to indirect environmental feedback. Inherited genomes can change expression without suddenly mutating, or changing themselves (not by changing any gene itself, but changing which genes will be expressed), depending on the environmentally-variable selection of which genes are expressed by an environmentally-sensitive epigenome (natural selection genetic bias = inherited epigenome norm-of-reaction patterns of environmentally-sensitive expressions).

"Our (K&G) point is that the conserved core processes respond to genetic variation or environmental variation (an environmental variation acknowledgement?) by producing their special type of phenotypic change." Under environmental stress, the entire epigenome may even act as a single unit. The epigenome is the hierarchal level of interaction between genetics and environment; it is the plastic range of possibilities for the norm-of-reaction. It includes patterns of gene expression passing from one cell to its descendants (in gametes; i.e., sex cell formation), gene expression changes during the differentiation of one cell type into another (development), and environmental factors reversibly changing the way genes are expressed in adults, all occurring without a change in DNA sequence. The epigenome, and through it, even the genome, have an indirect heritable connection to environmental variation. Variation is not in DNA alone. Kirschner and Gerhart could argue that even epigenetic gene expression modification could be a form of facilitated variation (as with transcription factor run changes). However, it may be better to separate their sudden mutation-related facilitated variation discussion from epigenetic changes, which are the heritable changes in genome function that occur without a change in DNA sequence. Said otherwise, Kirschner and Gerhart's facilitated variation seems to be limited to the support of newly-formed gene mutations; it is mostly, but not exclusively, mutation-induced. Epigenetic expression of a mutation could also involve an expression of an older previously-silent mutation; however, it is mostly, but not exclusively, environment-induced. The result (a newly expressed mutation) is the same; however, both forms of new mutation expression are not a common occurrence, and are usually seen only under severe environmental stress.

Under normal circumstances; that is, almost always, epigenetic expression involves expressing the norm-of-reaction (without expressing any new mutations). Cellular differentiation processes in environmental compartments, crucial for embryonic development, rely almost entirely on epigenetic rather than genetic inheritance from one cell generation to the next. Kirschner and Gerhart's text provides a superb understanding of the inherited norm-of-reaction developing phenotype. It also provides a superb understanding of the evolving phenotype, which uses facilitated variation (sudden mutation supported by random exploratory cell behavior) or allosteric protein switches (newly-expressing an earlier mutation); epigenetic change utilizes feedback loops, involving internal and external environmental signals. Without this understanding, the environment's indirect influence would only remain as some vague idea. Nevertheless, the genes must be expressed for anything to work. **The epigenome controls gene expression response to nearly all environmental feedback. The epigenome is an environmental-genetic interface, and it is** (possible to short-circuit, but is) **a doubly-indirect link.** Conrad Waddington originally defined the term epigenetics as "the interactions of genes with their environment that bring the phenotype into being." **Present-day definition regards it as a plastic change in the phenotype without a change in the DNA.** The epigenetic modification of gene expression (in the norm-of-reaction) is sensitive to environmental feedback signals. Environmentally-related indirect feedback is first used to stabilize the ever-present random exploratory cellular behavior that forms the phenotype, changing the random exploratory cellular behavior into a physiologic response. The feedback mechanism used to stabilize the phenotype in a given environment likely involves protein allostery switch function. The stabilization of the phenotype then indirectly stabilizes the genetic mechanism of expression, at the epigenome interface. In all instances, both the genome and the environmental conditions-of-life influence the random exploratory cell behavior that forms the phenotype.

In severely stressful environments, any population confined to this area becomes extinct. Almost without exception, DEATH occurs due to lack of an adequate adaptation to manage the adverse environmental change. Sometimes, but only on rare occasion, the adverse environmental selection process does not cause extinction of the severely- stressed isolated population. In this rare case, SURVIVAL may involve one of three biological options: **the 1st OPTION is an increased rate of mutation and immediate expression of a new-mutation phenotype, which fortuitously has survival value for an organism in the stressful environment.** This is likely to be an exceptionally uncommon occurrence, even for the rare process of environmental adaptation. Most mutations are silent, or if immediately expressed, fatal. **Perhaps less uncommon, a 2nd OPTION, a pre-existing phenotype structure (or function) allows an organism to quickly adapt (improve interaction) to the hostile new environment.** In *The Making of the Fittest: DNA and the Ultimate forensic Record of Evolution,* Sean Carroll notes: "There is a general theme in evolution, that **one innovation creates opportunity to evolve additional innovations.**" A structure that functioned in one role may respond to environmental stress and be adapted to another role (exaptation/coption). Because there is little room in nature for neutral (phenotypic) variation in structure or function in life, this "pre-adaptive" approach would not be the most common route in the rare process of successful environmental adaptation. (Genetic support of a neutral phenotype should at least be linked to a structure or function that is supported by environmental selection.) **When the seldom-seen process of successful environmental adaptation does occur, it usually involves the 3rd OPTION, a life-saving environmentally-generated** and **environmentally-stabilized change in genetic expression, which makes a survival difference in some population members. In this last case, there is a genome-related response** (utilizing existing genotype variation, and possibly an expression of a previously unexpressed silent mutation in a regulatory area) **that results in new genetic expression** (new phenotype) **in response to an unusual environmental challenge. If this structure** (or function) **allows an organism to improve interaction with the hostile new environment,** (environmental and genetic) **stabilization of the new phenotype** (adaptation) **occurs.** An adaptation needs to be supported by positive environmental feedback, which first indirectly stabilizes random cellular exploratory functions and is in turn indirectly linked to genetic expression; the indirect modification of epigenome expression is likely a remote, doubly-indirect modification. (Layers of biochemical reactions operate variable switches to express previously-formed programs or pathways.) **This last case doesn't involve genome (DNA) change per se** (i.e., sudden mutation); **it is only the genome's new expression of a phenotype change** (modulation of gene expression or epigenome change). **The epigenome can change, according to an individual's environment, and this change is passed from generation to generation.** (Only adaptive phenotypes survive environments lethal to non-adaptive phenotypes.) **The generation of shared character states in survivors is a genetic-related response to environmental challenge (change).** Any of the above scenarios allow survival in the newly-created environment; this survival would have not been possible without a change in genetic expression that permitted environmental adaptation; timing is everything. The survivors have the resultant genetic bias of natural selection. Importantly, almost without exception, common environmental stress is already managed by expressing pre-existing adaptations, already preserved in the norm-of-reaction as natural selection genetic bias. **The expression of adaptations to common stress occurs during development, and along with adult acclimation, it is managed by epigenetic (plastic) change.**

Returning to protein (polypeptide) synthesis, following transcription, the translation into a sequence of amino acids occurs next. There are three types of RNA involved in translation's protein (polypeptide) synthesis. In addition to mRNA, and the previously-mentioned transfer RNA (tRNA) that carries the amino acids to the ribosome, there is ribosomal RNA (rRNA). Ribosomes are the sites where the cell assembles proteins according to genetic instructions; this is the site of translation. A bacterial cell may have a few thousand ribosomes; a human cell has a few million ribosomes. Cells that have high rates of protein synthesis have particularly great numbers of ribosomes. Cells active in protein synthesis also have prominent nucleoli, which make the ribosomes. Ribosomes are particles composed of about 60% rRNA and 40% protein (enzymes). Ribosomes are suspended in the cytosol and are also bound to the endoplasmic reticulum. These ribosomes are (respectively) free ribosomes and bound ribosomes. They translate the information encoded in messenger RNA (mRNA) into a polypeptide. Initiating factor proteins (internal environmental signals) enable the process; ribosomes cannot bind to mRNA by themselves. (An epigenetic response to initiating factor proteins will be addressed shortly.) The sequence of amino acids in a polypeptide is dictated by the codons in the messenger RNA (mRNA) molecules from which the polypeptide was translated. A number of ribosomes may be attached to the same messenger, each manufacturing its own chain of polypeptides. Transfer RNA carries the amino acids, as high-energy esters, to the ribosome. The genetic code is the same in all living organisms; it has been demonstrated that eukaryotic ribosomes are able to translate bacterial mRNAs correctly. In the ribosome, the tRNA base pairs with in a specific way with the mRNA (the template RNA). A particular sequence of nucleotides can specify a particular sequence of amino acids by means of transfer RNA (tRNA) molecules, each specific for one amino acid and for a particular triplet (codon) of nucleotides in mRNA. The family of tRNA molecules enables the codons in a mRNA molecule to be translated into the sequence of amino acids in the protein. At least one kind of tRNA is present for each of the 20 amino acids used in protein synthesis. Some amino acids use two or three different kinds of tRNA; most cells contain as many as 32 different kinds of tRNA. The amino acid is attached to the appropriate tRNA by an activating enzyme (1 of 20) specific for that amino acid as well as for the tRNA assigned to it. Each kind of tRNA has a sequence of 3 unpaired nucleotides, the anticodon, which can bind, following the rules of base pairing to the complementary triplet of nucleotides (codon) in a messenger RNA (mRNA) molecule. As DNA replication and transcription involve base pairing of nucleotides running in opposite direction, the reading of codons in mRNA also requires that the anticodons bind in the opposite direction (Montgomery—*Biochemistry*).

Proteins and polypeptides are formed by peptide linkages (bonds linking the amine end of 1 amino acid and a carboxylic acid end of another) between the amino acids. There may be from 50 to 1000 amino acids in a protein (polypeptide). Because of electrical charge (different ions on the large polypeptide result in a large molecule with areas of different electric charge), proteins fold into precise structures. A change in one amino acid (mutation) can change the local electric charge and can change the shape of the protein or polypeptide. The folding produces secondary structures (alpha helix and beta sheets), that may involve nearly half of the total protein mass. Proteins (polypeptides) will stick together in a specific way to form a composite structure that functions as a single entity. Kirschner and Gerhart stated that altered protein states (allostery) play a major role in (indirect) weak phenotype linkage flexibility, which is a crucial part of the facilitated phenotype variation process. Proteins are the basic components of protoplasm in the cells of animals and plants; they are the "building blocks" of the structural components in the body. Collagen is very important in animals, making-up to 25% of their structural proteins (Wikipedia). Because proteins can change shape in response to the environment, cells (like fibrocytes) can change shape in response to the environment. Proteins are involved in every cellular process. Many hormones and nearly all enzymes are proteins. Enzymes enhance chemical reactions at life-friendly temperatures; enzymes (formed from DNA coding sequences) make things happen within a cell's stored chemistry. Proteins are involved in transport (hemoglobin), storage (casein), contraction (actin and myosin), protection (antibodies) and defensive toxins. Protein molecules contain nitrogen (+ some sulfur). Nitrogen-fixing bacteria (e.g. on bean/legume roots) and blue-green algae convert inert atmospheric nitrogen and bind it with hydrogen, creating ammonia (NH_3). NH_3 is a more biochemically-available form of nitrogen. But NH_3 can be toxic to complex plant life; NH_3 is oxidized to a friendly form by nitrite (NO_2^-) forming and nitrate (NO_3^-) forming bacteria. Green plants absorb nitrates and reduce them to manageable amounts of ammonium ions (NH_4^+). Ammonium ions are used to make amino acids in the chemistry of respiration.

Not all cellular DNA is in the nucleus; some is in the mitochondria. Mitochondria also contain RNA as well as enzymes for protein synthesis. Mitochondrial DNA and RNA are inherited maternally and bear a closer resemblance to bacterial nucleic acid than to animal nucleic acid. Mitochondria conduct cellular respiration. The major function of mitochondria is to convert food energy to the chemical energy of the cell, adenosine triphosphate (ATP). Human mitochondrial DNA has information contained in approximately 16,500 nucleotides (paired); these code for 2 ribosomal and 22 transfer RNAs. These in turn code for the synthesis of 13 proteins, all components of the oxidative phosphorylation system. Mitochondrial DNA does not contain information for all mitochondrial proteins; most are coded by nuclear genes. Most mitochondrial proteins are synthesized in the cytosol, from nuclear-derived mRNAs. They are then transported to the mitochondria, where they contribute structural and functional elements to the organelle (Montgomery—*Biochemistry*).

Mitochondrial DNA is often used in molecular DNA analysis; it is abundant (human mtDNA is present at 100–10,000 separate copies per cell) and lends well to extraction techniques. Much of the molecular lineage of life is traced through mitochondrial DNA; it is a valuable tool for evolutionary research. Richard Dawkins, in his book on *The Ancestor's Tale* discusses the molecular clock in the chapters on the Lungfish's Tale and Velvet Worm's Tale. "As long as the mutation rate at a neutral (no change in phenotype) genetic locus remains constant over time, the fixation rate (gene is permanent feature of the population) will also be constant.... Count the number of letters by which the starfish gene differs from the pangolin gene. Assume that half the differences accumulated in the line leading from ancestor to starfish and the other half in the line leading from ancestor to pangolin. That gives you the number of ticks in the (molecular) clock.... In a Geiger counter, the timing of the next tick is unpredictable.... But...the average interval over a large number of ticks is highly predictable. The hope is that the molecular clock is predictable in the same way as a Geiger counter, and in general, this is true. The tick rate varies from gene to gene in the genome. This was noticed early, when geneticists could look only at the protein products of DNA, not DNA itself. Cytochrome-c evolves at its own characteristic rate, which is faster than histones but slower than globulins, which in turn are slower than fibrinopeptides. In the same way, when a Geiger counter is exposed to a very slightly radioactive source such as a lump of granite, versus a highly radioactive source, such as radium, the timing of the next tick is always unpredictable but the average rate of ticking is predictably and dramatically different as you move from granite to radium. Histones are like granite, ticking at a very slow rate; fibrionpeptides are like radium, buzzing like a dementedly randomized bee. Other proteins, such as cytochrome-c (or rather the genes that make them), are intermediate. There is a spectrum of gene clocks, each running at its own speed, and useful for dating purposes, and for cross-checking with each other. Why do different genes run at different speeds? What distinguishes 'granite' genes from 'radium' genes? Remember that neutral (expression) doesn't mean useless, it means equally good. Granite genes and radium genes are both useful. It is just that radium genes can change at many places along their length and still be useful. Because of the way a gene works, portions of its length can change with impunity without affecting its functioning. Other portions of the same gene are highly sensitive to mutation, and its function is devastated if these portions are hit by a mutation.... Maybe the cytochrome-c gene has a mixture of granite bits and radium bits; fibrinopeptide genes have a higher proportion of radium bits, while histone genes have a higher proportion of granite bits…. Tick rates really do vary between genes, while the rate for any given gene is pretty constant even in widely separated species…. Truly neutral mutations, as in junk DNA or in 'synonymous substitutions' seem to tick in generation time as opposed to real time; creatures with short generation times show accelerated DNA evolution if you measure it in real time. Conversely, mutations that actually change something, and therefore fall foul of (environmental) selection, tick away more or less constantly in real time."

In *The Making of the Fittest: DNA and the Ultimate forensic Record of Evolution* Sean Carroll notes (in The Immortal Genes chapter): "In the DNA record, there is more information than just the history of a particular gene—there is information about the species that carries it, and about all the preceding species that also carried it, right back through eons of life's history. **Because of the power of** (environmental) **selection to preserve information** (natural selection genetic bias) **that would otherwise be erased in time, genomes contain a record of the history of life.**" Molecular clock research provides strong evidence that DNA change (mutation) is not the driving force of evolution. The above Dawkins source reveals that the overall rate of genetic change is independent of the rate of morphological evolution. Environmental adaptations (phenotypes displayed by genotype) may be stable for long periods of time, even though there may be significant changes in the DNA. Lungfish and Coelacanths underwent rapid DNA changes 400mya, but have not changed morphology for the last 200my. Yet, their DNA change is greater than ray-finned fish; ray-finned fish have changed morphology greatly during the same recent 200my time period. And, neither do organisms that show the least morphological change have the least amount of DNA change. This refutes genetically-determined evolution as the source of variation. That is, this essentially eliminates the gene itself or mutation itself as the driving force of evolution; neither the gene nor the gene mutation include the all-important environmental components in the phenotype. The driving force of evolution is adaptation to an isolated, changing environment; i.e., environmental stress may result in environmental adaptation, which incorporates an environmentally-integrated change in phenotype. Constant environments involve little change in phenotypes (norm-of-reaction change).

While mutation may not be the driving force of evolution, it is the biologic foundation of genetic variance; that is, mutation is the driving force of genetic variation (in genotype) within an individual and its offspring. And while mutation has already been discussed in detail, Sean Carroll's chapter on Making New from the Old in *The Making of the Fittest: DNA and the Ultimate forensic Record of Evolution* discusses interesting changes (as promised- a detailed gene duplication example) that exemplifies most (non-regulatory) gene mutation modifications over time. "All the Old World (African and Asian) apes monkeys have trichromatic color vision (see full visible light spectrum) and three cone opsin (color-specific protein) genes, while the New World monkeys, as well as rodents and other mammals have dichromatic vision (see only yellow and blue colors) and two opsin genes.... We can deduce that full color vision arose in an ancestor of the old world primates, after a separation of Old World and New World lineages. Furthermore, because Old World primates possess a third cone pigment, this opsin gene must have originated after this split as well. This tells us that human color vision dates from a deep, Old World ancestor and was not invented independently during the recent course of hominid evolution. The existence of two cone opsins in other mammals (squirrels, cats, dog, etc.) suggests that the presence of two color opsins and dichromatic vision was a condition of a common ancestor of mammals.... We have to consider the vision status of other vertebrates besides mammals. Birds have fabulous color vision; so do reptiles and many fish, such as the goldfish. Members of these groups have at least four opsin genes... when considered in the context of the whole evolutionary tree of vertebrates, the non-primate mammals are impoverished with respect to color vision and opsin genes throughout the vertebrates, we can deduce that the pattern of opsin genes in our history was one of initial abundance, then a loss in the ancestors of mammals, and then expansion again in an ancestor of Old World primates.... The most likely explanation has to do with the evolution of nocturnality in mammals. Early mammals were small and lived a cryptic, nocturnal lifestyle in ecosystems dominated by bigger animals, such as the dinosaurs. The evolution of nocturnality shifted the dependence of color vision in bright light to vision in dim light and darkness, and full color vision was lost (color gene atrophy from mutation—because environmental selection did not support full color vision gene function; night vision was supported by environmental selection).

We can say for certain when, relative to primate and mammal evolution, our third opsin (protein) gene evolved.... The 'tuning' of opsins (proteins) in the adaptation to specific environments is a general phenomenon in color vision.... Rats, mice, squirrels, rabbits goats and other mammals have one MWS/LWS opsin (protein) whose maximal absorbance is at wavelengths from about 510–550 nm. This opsin is encoded by a single gene. In contrast, humans have two opsins (1 for MWS, 1 for LWS), encoded by two genes on our X-chromosome that lie together as a head-to-tail tandem pair. These opsins are very similar (98%) to each other at the level of their DNA code. Their position as next-door neighbors in our DNA and their great similarity are telltale signs that they arose by the duplication of a single MWS/LWS gene in a primate ancestor. Gene duplications are a common form of change in DNA—many of our genes are members of multicopy families that have expanded in the course of evolution. The expansion of gene number increases the information that (environmental) selection can act upon, and a common pattern of duplicated genes is for their functions to become different. This is exactly the case with our two X chromosomes. Our pair of opsins and those of other trichromatic primates are most stimulated by light with wavelengths of about 530 (green) and 560 (red) nm (these points of greatest stimulation are referred to as absorption maxima). Advances in understanding the functional properties of opsins have revealed that it is very easy to shift the adsorption spectrum of individual opsins by changing particular amino acids. The maintenance of the 530 and 560 nm absorption maxima throughout the trichromatic primates suggests that there is (environmental) selection pressure to maintain this precise separation. There are just 15 amino acid differences between the green and red pigments. Biologists have been able to pinpoint which of these differences are responsible for the different functions of each pigment by making precise replacement of one amino acid with another and measuring the effects of these replacements on the spectral properties of each opsin. Three sites, at amino acid positions 180, 277, and 285, appear to account for most of the 30nm difference in the absorption peaks of our green and red pigments." (Unsurprisingly, improvement in color vision incurred the cost of night vision degradation and olfactory function degradation.)

Kirschner and Gerhart, in the chapter on The Sources of Variation, just do not seem to acknowledge any environmental influence on gene expression when they state: "No mechanism is known to direct a specific environmental stress toward the alteration of a specific gene or set of genes, as a way to ameliorate the stress.... Genetic variation and selection are completely uncoupled (not directly connected)." And while it is true that the environment does not directly instruct a gene on how to vary its expression, the above discussed indirect influence of the environment should be acknowledged. Kirschner and Gerhart discuss differential modifications of genetic code expression in single organisms and internal environmental factors (signals) in great detail; however, they see these functions as expressions of the organism itself, rather than a response to an increasingly internalized environment. Kirschner and Gerhart also seem to return to the confusing idea that novel phenotypic change is an indication of a sudden mutation occurrence. They discuss an experiment that demonstrated that starvation of a population increased the mutation rates of all genes (not just the required gene). They point out that mutation rates, but not the mutations themselves, are affected by the environment; a stressful environment increases mutation rates. Mutations can be caused by both chemical and physical agents, although the action of physical agents (e.g. ionizing radiation—alpha, beta, gamma and non-ionizing UV) can usually be explained by a chemical mechanism. Examples of chemical mutagens (mutation agents) include acridine orange, bromine (and some of its compounds), ethyl methanesulfonate, diethylsulfate, methylmethane-sulfonate, nitrogen mustard, nitrous acid, nitrosoguanidine, peroxides, proflavin and sodium azide. Generally speaking, mutations are a bad thing for an organism.

In the case of the mutation-induced immediate adaptation, recall that that all genes are composed of four different types of purine or pyrimidine nucleotide bases; any agent that specifically reacts with only one of the four bases could potentially cause mutations in every gene. Mutations may be produced in many ways. Nucleotide bases may be deleted or new ones may be inserted. More frequently, existing bases are chemically modified so that on replication, improper base pairing will cause a different base to appear at the modified position. Sometimes an entire region of DNA (thousands or millions of nucleotides) is accidentally deleted or duplicated. If it happens as a single event, this is counted as a single mutation, just as is the former example. Mutations are random events. During evolution, environmental selection eliminated large numbers of harmful mutations (and sometimes ones that were not harmful). Most of the time, the insertion of the wrong amino acid would cause an inability in the function of a protein. If the amino acid replacement occurs at a less important position, activity may not be affected at all (neutral genotype variation). On occasion, there may even be a beneficial improvement in activity; a small number of helpful mutations changed the phenotype and gave life survival value. The time of a mutation occurrence and the time of the expression of the mutation usually are not the same. Mutation affects the genotype, and may or may not affect the present phenotype. The majority of the mutations are not expressed in the phenotype. When they are expressed, environmental selection will interact with the mutation-induced phenotypes. The majority of expressed mutations are eliminated by environmental selection. Said otherwise, most immediately-expressed mutations are damaging; it would not be very likely for a mutation and an immediate adaptation to a stressful environment to occur.

An occurrence far more common than mutation is seen in uncommon environments. It is an increase in the frequency for epigenetic generation of norm-of-reaction phenotypic expressions, sometimes even incorporating random combinations from the genome that are outside the norm-of-reaction, making something new. And this includes expression of a previously neutral mutation that can be supported by random cellular exploratory function and environmental feedback. An increase in the rate of epigenetic expressions, forming an environmentally-integrated phenotype seems to be (relatively) a more common solution to a stressful environment problem. To be stable, the expressed phenotype needs interactive environmental support. The phenotype interacts with the internal environment. The phenotype interacts with the external environment. **The epigenome is a hierarchy of protein switches that control phenotype expression** (norm-of-reaction and/or expressing new or previous mutations); **it is the interface between the genome and the environment.** Nothing happens unless the epigenome allows expression of the genes for an observable change to occur. **A gene or gene mutation must be expressed by the epigenome. Only then can transcriptional control be modified, producing numerous expressions in phenotypes that incorporate varying degrees of gene expressions.**

Unlike mutation agents that alter the molecular structure of DNA, it is important to remember that internal and external environmental agents can promote or inhibit the epigenetic expression of genetic information coded in the DNA. Epigenetic promoting agents (not to be confused with the DNA transcription promoter locations previously mentioned, but the factors or environmental signals that Kirschner and Gerhart discussed) include a variety of substances, such as hormones, protein growth factors, and plant products. These substances influence genetic expression by binding to epigenetic receptors. The actions of epigenetic promoting agents (say an internal environmental signal) are mostly reversible in the adult. In addition to epigenetic promoting agents, there are epigenetic initiating agents (usually harmful external environmental agents, not to be confused with the initiating factors previously mentioned, which are internal environmental signals). Harmful external environmental agents can cause more damaging epigenetic changes, and may even alter the DNA structure and cause somatic mutations over the lifetime of an individual. Irreversible changes similar to a mutation can occur, but phenotypic change is not necessarily obligatory. Harmful external environmental agents (like smoke) can program the cells so that exposure to a promoting agent causes a response, as in the formation of cancerous cells (Montgomery— *Biochemistry*). A variety of (epigenetic initiating agent) compounds are considered as carcinogens; they result in an increased incidence of tumors, but they do not necessarily show <u>mutagen</u> activity. Examples include arsenite, benzene, carbon tetrachloride, chloroform, cyanides, dichloromethane, diethylstilbesterol, hexachlorobenzene, polychlorinated biphenyls, and compounds of cadmium, lead, mercury and nickel.

Epigenetics invalidates the genetically-determined view of evolution. Environmentally AND genetically determined evolution is essential to the WHEN and WHERE of environmental biology. WHY and HOW the environment impacts the life contained within it has been a topic of considerable discussion in this text. With norm-of-reaction development, adult acclimation, and adaptation, epigenetics addresses relevant environmental impacts upon the epigenome. Signals from the internal and external environment generate the plastic range of responses from the natural selection-biased norm-of-reaction. Recall the indirect environmental stabilization of random exploratory cellular behavior; stabilizing the random exploratory cellular behavior produces physiologic patterns that form the phenotype. The genetic control of phenotype formation is indirect; this indirect control is coordinated by feedback signals from the forming phenotype. The epigenome is the genome's interface between genes and the cell physiology that forms the phenotype. Cellular differentiation processes in environmental compartments, crucial for embryonic development, rely almost entirely on epigenetic rather than genetic inheritance from one cell generation to the next. The indirect environmentally-connected cell changes that occur with the map of the embryo formation are heritable cellular epigenetic changes. These epigenetic changes endow cells with an address and the means to explore. The changes that occur with the outcome of the exploration are also indirect environmentally-connected heritable cellular epigenetic changes. Anything that appears to not follow Mendelian principles (environmentally-related inheritance) would likely be an epigenetic change in gene expression. **The epigenome is indirectly affected by the environment and the epigenome controls genome expression.**

The following epigenetic discussion from Wikipedia (online encyclopedia) is a very good, but still somewhat genetically-determined, summary of the topic. It links (online) references for further reading. Epigenetics is a term in biology used to refer to features such as chromatin and DNA modifications that are stable over rounds of cell division but do not involve changes in the underlying DNA sequence of the organism. These epigenetic changes play a role in the process of cellular differentiation, allowing cells to stably maintain different characteristics despite containing the same genomic material. Epigenetic features are inherited when cells divide, despite a lack of change in the DNA sequence itself and, although most of these features are considered dynamic over the course of development in multicellular organisms, some epigenetic features show trans-generational inheritance and are inherited from one generation to the next. Specific epigenetic processes include paramutation, bookmarking, imprinting, gene silencing, X chromosome inactivation, position effect, reprogramming, transvection, maternal effects, the progress of carcinogens, many effects of teratogens, regulation of histone modifications and heterochromatin, and technical limitations affecting parthenogenesis and cloning.

In epigenetics, paramutation is an interaction between two alleles of a single locus, resulting in a heritable change of one allele that is induced by the other allele. Paramutation violates Mendel's 1st law, which states that the process of the formation of the gametes (egg or sperm), the allelic pairs separate, one going to each gamete, and each gene remains completely uninfluenced by the other. In paramutation, an allele in one generation heritably affects the other allele in future generations, even if the allele causing the change is not transmitted. RNA is a molecule of inheritance, just like DNA; it can be packaged in egg or sperm and cause paramutation in the next generation. The RNAs transmitted in such a case are RNA's such as piRNAs (RNA-induced silencing complex), siRNAs (small interfering RNA- interferes with expression of a specific gene), miRNAs (micro-RNA), or other regulatory RNAs. Micro-RNA's are shorter versions of stem-loop structures, transcription products that are partially complementary to messenger RNA. The binding of miRNA triggers the degredation of the mRNA transcript through a process similar to RNA interference, though in other cases it is believed that the miRNA complex blocks the protein translation machinery, or otherwise prevents protein translation without causing the RNA to be degraded. Also targeted is methylation of the genomic site corresponding to a particular mRNA. The miRNA's function in association with a complement of proteins collectively termed the miRNP. The molecular basis of paramutation is being unraveled. Paramutation may share common mechanisms to other epigenetic phenomena, such as gene silencing, genomic imprinting, and transvection (genetics). Paramutation is RNA-directed; stability of the chromatin states associated with paramutation and transposon silencing requires the mop1 gene, which encodes an RNA-dependent RNA polymerase. This polymerase is required to maintain a threshold level of the repeat RNA, which causes the paramutation. The mechanism is yet to be fully understood, but like any other epigenetic change, it involves a covalent modification of the DNA an/or the DNA-bound histone without changing the sequence of the DNA itself.

In genetics and epigenetic bookmarking there is a biological phenomenon believed to function as an epigenetic mechanism for transmitting cellular memory of the pattern of gene expression in a cell, throughout mitosis, to its daughter cells. This is vital for maintaining the phenotype in a lineage of cells so that, for example, liver cells divide into liver cells and not some other cell type. It is characterized by non-compaction of some gene promoters during mitosis. In terms of mechanism, it is believed that at some point, prior to the onset of mitosis, the promoters of genes that exist in a transcription-competent state become "marked" in some way. This mark persists both during and after mitosis, and the marking transmits gene expression memory by preventing the mitotic compaction of DNA at this locus, or by facilitating reassembly of transcription complexes on the promoter, or both. In some cases, bookmarking is mediated by binding of specific factors to the promoter prior to the onset of mitosis, but in other cases could be mediated by patterns of histone modification or presence of histone variants that are characteristics of active genes, and which are believed to persist throughout mitosis. In the case of specific genes, for example, the stress-inducible hsp70 gene, bookmarking may also function as a mechanism for ensuring that the gene can be transcribed early in G1 phase if a stress were to occur at that time. If this gene promoter were compacted it would take time to decompact it in G1, during which time the cell would be unable to transcribe this cytoprotective gene, leaving it vulnerable to stress-induced cell death. In this case, bookmarking appears to be important for cell survival.

Genomic imprinting is a genetic phenomenon by which certain genes are expressed in a parent-of-orgin specific manner. Imprinted genes are either expressed only from the allele inherited from the mother, or in other instances, from the allele inherited from the father. Forms of genomic imprinting have been demonstrated in insects, mammals and flowering plants. The phrase "imprinting" was first used to describe events in the insect *Pseudococcus nipae*. In Pseudococcids (mealybugs), both the male and female develop from a fertilized egg. In females, all chromosomes remain euchromatic and functional. In embryos destined to become males, one haploid set of chromosomes becomes heterochromatinised after the 6th cleavage division and remains so in most tissues; males are thus functionally haploid. In insects, imprinting describes the silencing of the paternal genome in males, and thus is involved in sex determination. In mammals, genomic imprinting describes the process involved in introducing functional inequality between two parental alleles of a gene. Experimental manipulation of mouse embryos in the early 1980s showed that normal development requires the contribution of both the maternal and paternal genomes. Gynogenetic embryos (containing 2 female genomes) show relatively normal embryonic development, but poor placental development. In contrast, androgenetic embryos (containing 2 male genomes) show very poor embryogenic development but normal placental development. Further investigation identified that these phenotypes were the result of unbalanced imprinted gene expression. NOEY2 is a maternally-expressed located on chromosome 1 in humans. Loss of NOEY2 expression is linked to an increased risk of ovarian and breast cancers; in 41% of breast and ovarian cancers the protein transcribed by NOEY2 is not expressed, suggesting that it functions as a tumor suppressor. Therefore, if a person inherits both chromosomes from the mother, the gene will not be expressed and the individual is put at risk for breast and ovarian cancer. Imprinting is a dynamic process. It must be possible to erase and re-establish the imprint through each generation. The nature of the imprint must therefore be epigenetic (modifications to the structure of DNA but not the sequence). In germline cells, the imprint is erased, and then re-established according to the sex of the individual; i.e. in the developing sperm, a paternal imprint is established, whereas in the developing oocytes, a maternal imprint is established. This process of erasure and reprogramming is necessary such that the current imprinting status is relevant to the sex of the individual. In both plants and mammals there are two major mechanisms that are involved in establishing the imprint; these are DNA methylation and histone modifications.

Gene silencing is a general term describing epigenetic processes of gene regulation. The term gene silencing is generally used to describe the "switching off" of a gene by a mechanism other than genetic modification. That is, a gene which would be expressed (turned on) under normal circumstances is switched-off by machinery in the cell. Genes are regulated at either the transcriptional or post-transcriptional level. Transcriptional gene silencing is the result of histone modifications, creating an environment of heterochromatin around a gene that makes it inaccessible to transcriptional machinery (RNA polymerase, transcription factors, etc.). Post-transcriptional gene silencing is the result of mRNA of a particular gene being destroyed. The destruction of the mRNA prevents translation to form an active gene product (in most cases, a protein). A common mechanism of post-transcriptional gene silencing is RNAi. Both transcriptional and post-transcriptional gene silencing are used to regulate endogenous genes. Mechanisms of gene silencing also protect the organism's genome from transposons and viruses. Gene silencing may thus be part of an ancient immune system protecting against such infectious DNA elements. X-inactivation is a process by which one to the two copies of the X chromosome present in female mammals is inactivated. The inactive X chromosome is silenced by packaging in repressive heterochromatin. X-inactivation also occurs so that the female, with two X chromosomes, does not have twice as many gene products as the male, which only possesses a single copy of the X chromosome. The choice of which X chromosome is random in placental mammals (such as mice and humans), but once an X chromosome is inactivated, it will remain inactivated throughout the lifetime of the cell. Unlike the random X-inactivation in placental mammals, inactivation in marsupials applies exclusively to the paternally-derived chromosome.

The position effect is the effect on the expression of a gene when its location in a chromosome is changed, often by translocation. This has been well described in *Drosophilia* with respect to eye color and is known as position effect variegation. The phenotype is well-characterized by unstable expression of a gene that results in the red eye coloration. In the mutant flies the eyes typically have a mottled appearance of white and red sectors. These phenotypes are often due to a chromosomal translocation such that the color gene is now close to a region of heterochromatin. The heterochromatin can spread stochastically (in jumps) and switch off the color gene resulting in the white eye sectors. Position effect is also used to describe the variation of expression exhibited by identical transgenes that insert into different regions of a genome. In this case, the difference in expression is often due to enhancers that regulate neighboring genes. These local enhancers can also affect the expression pattern of the transgene. Since each transgenic organism has the potential for a unique expression pattern.

Reprogramming refers to erasure and remodeling of epigenetic marks, such as DNA methylation, during mammalian development. After fertilization, some cells of the newly-formed embryo migrate to the germinal ridge and will eventually become germ cells (sperm and oocyte) Due to the phenomenon of genomic imprinting, maternal and paternal genomes are differentially marked and must be properly reprogrammed every time they pass through the germline. Therefore, during the process of gametogenesis, the primordial germ cells must have their original biparental DNA methylation patterns erased and re-established based on the sex of the transmitting parent. After fertilization the paternal and maternal genomes are once again demethylated and remethylated (except for differentially-methylated regions associated with imprinted genes). This reprogramming is likely required for totipotency of the newly-formed embryo and erasure of acquired epigenetic changes. In-vitro manipulation of pre-implantation embryos has been shown to disrupt methylation patterns at imprinted loci and plays a crucial role in cloned animals.

Transvection is an epigenetic phenomenon that results from an interaction between an allele on one chromosome and the corresponding allele on the homologous chromosome. Transvection can lead to either gene activation or repression. It can also occur between non-allelic regions of the genome as well as regions of the genome that are not transcribed. Pairing-related phenomena have been observed in *Drosophilia*, other insects, nematodes, mice, humans, fungi and other plants. In light of these findings, transvection may represent a widespread form of gene regulation. Transvection is believed to occur through a variety of mechanisms. In one mechanism, the enhancers of one allele activate the promoter of a paired second allele. Other mechanisms include pairing-sensitive silencing and enhancer bypass of a chromatin insulator through pairing-mediated changes in gene structure.

A maternal effect, in genetics, is the phenomenon of where the genotype of the mother is expressed in the phenotype of its offspring, unaltered by paternal genetic influence. The phenotype of the individual reflects the genotype of its mother, rather than the genotype of the individual. This maternal effect is usually attributed to maternally-produced molecules, such as mRNAs, that are deposited in the egg cell. Maternal genes often affect early developmental processes. An example in *Drisophila melanogaster* morphogenesis is axis formation, in which mRNA such as *Bicod* and *nanos* is of maternal orgin and loaded into the egg prior to fertilization. "Maternal effect" should not be confused with maternal inheritance, in which some aspect of an offspring is inherited solely from the mother. This is often attributed to the inheritance of mitochondria or plastids, each of which contains its own genome. Maternal inheritance is distinct from maternal effect because in maternal inheritance, the individual's phenotype reflects its own genotype, rather than the genotype of a parent. In contrast, paternal effect is when a phenotype of an individual results from the genotype of the father; the genes responsible for these effects are components of sperm that are involved in fertilization and early development.

Carcinogenesis (meaning literally, the creation of cancer) is the process by which normal cells are transformed into cancer cells. Cancer cells are caused by a series of mutations. Each behavior alters the behavior of the cell somewhat. Cell division is a physiological process that occurs in almost all tissues and under many circumstances. Normally, the balance between proliferation and programmed cell death (usually apotosis) is maintained by tightly regulating both processes to ensure the integrity of organs and tissues. Mutations in DNA that lead to cancer disrupt these orderly processes by disrupting the programming regulating the processes. A new way of looking at carcinogenesis comes from integrating the ideas of developmental biology into oncology. The cancer stem-cell paradigm proposed that some or all cancers arise from transformation of adult stem cells. These cells persist as a subcomponent of the tumor and retain stem cell properties. Furthermore, the relapse of cancer and the emergence of metastasis are also attributed to these cells. The cancer stem-cell hypothesis does not contradict earlier concepts of carcinogenesis. It simply points to adult stem-cells as the site where the process begins.

Carcinogenesis is caused by this mutation of the genetic material of normal cells, which upsets the normal balance between proliferation and cell death. This results in uncontrolled cell division and tumor formation. The uncontrolled and often rapid proliferation of cells can lead to benign tumors; some types of these may turn into malignant tumors (cancer). Benign tumors do not spread to other parts of the body or invade other tissues, and they are rarely a threat to life unless they compress vital structures or are physiologically active, for instance as in producing a hormone. Malignant tumors can invade other organs, spread to distant locations (metastasis), and become life threatening. More than one mutation is necessary for carcinogenesis. In fact, a series of several mutations to certain classes of genes is usually required before a normal cell will transform into a cancer cell. Only mutations in those certain types of genes which play vital roles in cell division, apoptosis (cell death), and DNA repair will cause a cell to lose control of its cell proliferation. Cancer is ultimately, a disease of genes. In order for cells to start dividing uncontrolaby, genes which regulate cell growth must be damaged. Proto-oncogenes are genes which promote cell growth or mitosis (a process of cell division), and tumor suppressor genes discourage cell growth, or temporarily halt cell division from occurring in order to carry out DNA repair. Typically, a series of mutations to these genes are required before a normal cell transforms into a cancer cell.

Proto-oncogenes promote cell growth in a variety of ways. Many can produce hormones, a "chemical messenger" between cells which encourage mitosis, the effect of which depends upon signal transduction of the receiving tissue or cells. Some are responsible for the signal transduction system and signal receptors in cells and tissues themselves, thus controlling sensitivity to such hormones. They often produce mitogens, or are involved in transcription of DNA in protein synthesis, which creates the proteins (enzymes) responsible for producing the products (biochemical) cells use for interaction. Mutations in proto-oncogenes can modify their expression and function, increasing the amount or activity of the product protein. When this happens, they become oncogenes, and thus have a higher chance to divide excessively and uncontrollably. The chance of cancer cannot be reduced by removing proto-oncogenes from the genome, as they are critical for growth, repair, and homeostasis of the body. It is only when they become mutated that the signals for growth become excessive.

Tumor suppressor genes code for anti-proliferation signals and proteins that suppress mitosis and cell growth. Generally, tumor suppressors are transcription factors that are activated by cellular stress or DNA damage. Often DNA damage will cause the presence of free-floating genetic material as well as other signs, and will trigger enzymes and pathways that lead to the activation of tumor suppression genes. The function of such genes is to arrest the progression of cell cycles in order to carry out DNA repair, preventing mutations from passing on to daughter cells. Canonical tumor suppressors include the p53 gene, which is a transcription factor activated by cellular stress, such as hypoxia or ultraviolet radiation damage. However, a mutation can damage the tumor suppressor gene itself, or the signal pathway that activates it—"switching it off." The invariable consequence of this is that DNA repair is hindered (inhibited). DNA damage accumulates without repair, inevitably leading to cancer.

In general, mutations in both types of genes are required for cancer to occur. For example, a mutation limited to one oncogene would be suppressed by normal mitosis control and tumor suppressor genes. A mutation of only one tumor suppressor gene would not cause cancer either, due to the presence of many "backup" genes that duplicate its functions. It is only when enough proto-oncogenes have mutated into oncogenes, and enough tumor suppressor genes are deactivated or damaged, that the signals for cell growth overwhelm the signals to regulate it, and cell growth quickly spirals out of control. Often, because these genes regulate the processes that prevent most damage to genes themselves, the rate of mutations increase as one gets older (DNA damage forms a feedback loop). Usually oncogenes are dominant alleles, as they contain gain-of-function mutations, while tumor suppressors are recessive alleles, as they contain loss-of-function alleles. Each cell has two copies of the same gene (1 from each parent). Under most cases, gain of function mutation in one copy of a particular onco-gene is enough to make that gene a true oncogene; loss-of-function mutation needs to happen in both copies of a tumor suppressor gene to render that gene completely non-functional. Mutation of tumor suppressor genes that are passed-on to the next generation can cause an increased likelihood for cancers to be inherited. Members within these families have increased incidence and decreased latency of multiple tumors. The mode of inheritance of mutant tumor suppressors is that the affected member inherits a defective copy from one parent and a normal copy from the other. Because inheritance of tumor suppressors acts in a recessive manner, the loss of the normal copy creates the cancer-prone phenotype.

Many cancers originate from viral infection; this is especially true in animals such as birds, but less so in humans. Viruses are responsible for 15% of human cancers. Viruses that are known to cause cancer, such as HPV causing cervical cancer, Hepatitis B causing liver cancer, and EVB causing a type of lymphoma, are all DNA viruses. It is thought that when a virus infects a cell, it inserts part of its own DNA near the cell growth genes causing cell division. The group of changed cells that are formed from the first cell dividing all have the same viral DNA near the cell growth genes. The group of changed cells is now special because one of the normal controls on growth has been lost.

Many mutagens are carcinogens, but some carcinogens are not mutagens. Examples of carcinogens that are not mutagens include alcohol and estrogen. These are thought to promote cancer through their stimulating effect on the rate of cell mitosis. Faster rates of mitosis increasingly leave less opportunity for repair enzymes to repair damaged DNA during DNA cell replication, increasing the likelihood of a genetic mistake. A mistake made during mitosis can lead to daughter cells receiving the wrong number of chromosomes, which leads to aneuploidy and may lead to cancer. Cells, depending on their location, can be damaged through radiation (e.g. sunshine), chemicals (e.g. cigarette smoke), inflammation (e.g. bacterial infection or other viruses). Each cell has a chance of damage, a step on the path to cancer. Cells often die if they are damaged, through failure of a vital process or the immune system; however, sometimes damage will knock out a single cancer gene. In an old person, there are thousands, tens of thousands, or hundreds of thousands of knocked-out cells. The chance that any one of them would form a cancer is very low. When the damage occurs in any area of changed cells, something different occurs. Each of the cells has the potential for growth. The changed cells will divide quicker when the area is damaged by physical, chemical, or viral agents. A vicious cycle has been set up. Further damaging the area will cause the changed cells to divide, and then it is more likely they will suffer cancer gene knock-outs. This model of carcinogenesis is popular because it explains why cancers grow. It would be expected for cells that are damaged by radiation would die, or at least be worse-off because they have fewer genes working; viruses increase the number of genes working.

In the 19th Century, teratology related to botanical biological deformities. Currently, it's most instrumental meaning is that of the medical study of teratogenesis, congenital malformations, or grossly deformed individuals. With greater understanding of the origins of birth defects, the field of teratology now overlaps with other fields of basic science, including developmental biology, embryology and genetics. It was previously believed that the mammalian embryo developed in the impervious uterus of the mother, protected from all extrinsic factors. However, after the thalidomide disaster of the 1960's, it became apparent and more accepted that the developing embryo could be highly vulnerable to certain environmental agents that have negligible or non-toxic effects on adult individuals. Along with this new awareness of the in-utero vulnerability of the developing mammalian embryo, came the development and refinement of the *Six Principles of Teratology*, which are still applied today. It is these principles that guide the study and understanding of teratogenic agents and their effects upon the developing organisms. Susceptibility to teratogenesis depends upon the genotype of the conceptus and the manner in which this interacts with adverse environmental factors. Susceptibility to teratogenesis varies with the developmental stage at the time of exposure to an adverse influence. There are critical periods of susceptibility to agents and organs systems affected by these agents. Teratogenic agents act in specific ways on developing cells and tissues to initiate sequences of abnormal developmental events. The access of adverse influences to developing tissues depends upon the nature of the influence. Several factors affect the ability of a teratogen to contact a developing conceptus, such as the nature of the agent itself, route and degree of maternal exposure, rate of placental transfer and systemic absorption, and composition of the maternal and embryonic/fetal genotypes.

There are four manifestations of deviant development; they are Death, Malformation, Growth Retardation, and Functional Deficit. Manifestations of deviant development increase in frequency and degree as dosage increases from the No Observable Adverse Effect Level (NOAEL) to a dosage producing 100% Lethality (LD100). A wide range of different chemicals and environmental factors are suspected or are known to be teratogenic in humans and in animals. A selected few include: Ionizing radiation: atomic weapons, radioiodine, radiation therapy. Infections: cytomegalovirus, herpes virus, parovirus B-19, rubella virus (German measles), syphilis, toxoplasmosis, Venezuelan equine encephalitis virus. Metabolic imbalance: alcoholism, endemic cretinism, diabetes, folic acid deficiency, hyperthermia, phenylketonuria, rheumatic disease and congenital heart block, virilizing tumors. Drugs and environmental chemicals: 13-cis-retinolic acid, isoretinoin (Accutane), temazepam (Restoril, Noemission), nitrazepam (Mogadon), nimetazepam (Ermin), PCB's, Dioxin, coumarin, cyclophosamide, diethylsibesterol, diphenylhydantoin (Phenyltoin, Dilantin, Epanutin), ethanol, ethidium bromide, ethretinate, lithium, methimazole, organis mercury, penicillamine, tertacyclines, thalidomide, trimethadione, uranium, methoxyethyl ethers, and valporic acid. Under suspicion: Agent Orange, nicotine, aspirin, NSAIDs, cyclopamine.

In biology, histones are the chief protein component of chromatin. Histones act as spools around which DNA winds; they play a role in gene regulation. Without histones, the unwound DNA in chromosomes would be very long. For example, each human cell has about 1.8 meters of DNA, but wound on the histones, it has about 90 millimeters of chromatin, which, when duplicated and condensed during mitosis, result in about 120 micrometers of chromosomes. This enables the compaction necessary to fit the large genomes of eukaryotes inside cell nuclei; the compacted module is 30,000 times shorter than an unpacked molecule. There are classes of histones; H1 (sometimes called the linker histone, also related to histone H5), H_2A, H_2B, H_3, H_4, and Archaeal histones. Two each of the class H_2A, H_2B, H_3, and H_4, so-called *core histones*, assemble to form one octameric nucleosome core particle by wrapping 146 base pairs of DNA around the protein spool in 1.65 left-handed super-helical turn. The linker histone H1 binds the nucleosome and the entry and exit sites of the DNA, thus locking the DNA into place and allowing the formation of higher order structure. The most basic such foundation is the 10 nm fiber or beads on a string confirmation. This involves the wrapping of DNA around nucleosomes with approximately 50 base pairs of DNA spaced between each nucleosome (also referred to as linker DNA). The assembled histones and DNA is called chromatin. Higher order structures include the 30 nm fiber, these being the structures found in normal cells. During mitosis and meiosis, the condensed chromosomes are assembled through interactions between nucleosomes and other regulatory proteins.

In general, genes that are active have less bound histone, while inactive genes are highly associated with histones during interphase. It also appears that the structure histones have been conserved through evolution, as any deleterious mutations would have been severely maladaptive. Histones are found in the nuclei of eukaryotic cells and in certain Archea, namely Euryarchea, but not in bacteria. Archael histones may well resemble the evolutionary precursors to eurykarotic histones. Histone proteins are among the most highly conserved proteins in eurykarotes, emphasizing their important role in the biology of the nucleus. There are very few differences among the amino acid sequences of the histone proteins of different species. Linker histone usually has more than one form within a species and is also less conserved than the core histones.

The nucleosome core is formed of two H_2A-H_2B dimers and a H_3-H_4 tetramer, forming two nearly symmetrical halves by tertiary structure (C2 symmetry ; one macromolecule is the mirror image of the other). The H_2A-H_2B dimmers and H_3-H-4 tetramers also show pseudodyad symmetry. The 4 "core" histones (H_2A, H_2B, H_3 and H_4) are relatively similar in structure and are highly conserved through evolution, all featuring a "helix turn helix turn helix" motif (which allows the easy dimerisation). They also share the feature of long "tails" on one end of the amino acid structure—this being the location of post-transcriptional modification. Histones undergo posttranslational modifications, which alter their interaction with DNA and nucleoproteins. The H_3 and H_4 histones, with their long tails protruding from the nucleosome, can be covalently modified at several places. Modifications of the tail include methylation, acetylation, phosphorylation, ubiqutination, sumoylation, citrullination, and ADP-ribosylation. The core histones (H_2A and H_3) can also be modified. Combinations of modifications are thought to constitute a code. Histone modifications act in diverse biological processes, such as gene regulation, DNA repair, and chromosome condensation (mitosis). In all, histones make 5 types of interactions with DNA. Helix-dipoles from alpha-helices in H_2B, H_3 and H_4 cause a net positive charge to accumulate at the point of interaction with negatively charged phosphate groups on DNA. Hydrogen bonds between the DNA backbone and the amine group on the main chain of histone proteins. Nonpolar interactions between the histone and deoxyribose sugars on DNA. Salt links and hydrogen bonds between side chains of basic amino acids (especially lysine and arginine) and phosphate oxygens on DNA. Non-specific minor groove insertions of the H_3 and H_2B N-terminal tails into two minor grooves on the DNA molecule. The highly basic nature of histones, aside from facilitating DNA-histone interactions, contributes to the water solubility of histones.

Heterochromatin is a tightly-packed form of DNA. Its major characteristic is that transcription is limited. As such, it is a means to control gene expression, through regulation of the transcription initiation. Chromatin is found in two varieties: euchromatin and heterochromatin. Originally, the two forms were distinguished cytologically by how darkly they stained—the former is lighter, while the latter stains darkly, indicating tighter packing. Heterochromatin is usually located in the periphery of the nucleus. Heterochromatin mainly consists of genetically-inactive satellite sequences, and many genes are represented to various extents, although some cannot be expressed euchromatin at all. Heterochromatin also replicates later than euchromatin in S-phase of the cell cycle, and is only found in eukaryotes. Both centromeres and telomeres are heterochromatic, as is the Barr body of the second inactivated X-chromosome in a female. Heterochromatin is believed to serve several functions, from gene regulation to the protection of the integrity of chromosomes; all of these roles can be attributed to the dense packing of DNA, which makes it less accessible to protein factors that bind DNA or its associated factors. For example, double-stranded DNA ends would usually be interrupted by the cell as damaged DNA, triggering cell cycle arrest and DNA repair. Heterochromatin is generally clonally-inherited; when a cell divides, the two daughter cells will typically contain heterochromatin within the same regions of DNA, resulting in epigenetic inheritance. Variations cause heterochromatin to encroach on adjacent genes or to recede from genes at the extremes of domains. Transcribable material may be repressed by being positioned (*in cis*) at these boundary domains. This gives rise to different levels of expression from cell to cell, which may be demonstrated by position effect variegation. Insulator sequences may act as a barrier in rare cases where constitutive heterochromatin and highly active genes are juxtaposed.

All cells of a given species will package the same regions of DNA in constitutive heterochromatin, and thus in all cells any genes contained within the constitutive heterochromatin will be poorly expressed. For example, all human chromosomes 1, 9, 16, and the y-chromosome contain large regions of constitutive heterochromatin. In most organisms, constitutive heterochromatin occurs around the chromosome centromere and near telomeres. The regions of DNA packaged in facultative heterochromatin will not be consistent within the cell types of species, and thus a sequence in one cell that is packaged in facultative heterochromatin (and the genes within poorly expressed) may be packaged in euchromatin in another cell (and the genes within no longer silenced).

However, the formation of facultative heterochromatin is regulated, and is often associated with morphogenesis or differentiation. As an example of facultative heterochromatin, is X-chromosome inactivation in female mammals: one X-chromosome is packaged in facultative heterochromatin and silenced, while the other X-chromosome is packaged in euchromatin and expressed. *Saccharomyces cerviseae,* or budding yeast, is a model eukaryote and its heterochromatin has been defined thoroughly. Although most of its genome can be characterized as euchromatin, *S. cerviseae* has regions of DNA that are transcribed very poorly. These loci are the so-called silent mating type loci (HML and HMR), the rDNA (encoding ribosomal DNA), and the sub-telomeric regions. Fission yeast (*Schizosaccharomyces pombe*) uses another mechanism for heterochromatin formation at its centromeres. Gene silencing at this location depends upon components of RNAi pathway. Double-stranded RNA is believed to result in silencing of the region through a series of steps.

Parthenogenesis is a form of asexual reproduction in which females produce eggs that develop without fertilization. Parthenogenesis is an asexual form of reproduction found in females where growth and development or a seed occurs without fertilization by males. The offspring produced by parthenogenesis are always female in species where the XY chromosome system determines gender. As with all types of asexual reproduction, there are both costs (low genetic diversity and succeptability to adverse mutation) and benefits (no need for a male in reproduction) that are associated with parthenogenesis. Asexual reproduction existed alone for many epochs from the beginning of life on earth. When sexual reproduction arose (behavior adaptation—presumably through mutation), it introduced a means to expand genetic diversity through the partial contribution from the male, providing more options for the survival of the species in which it began (environmental adaptation). Many species followed this reproductive path successfully, some to the exclusion of the asexual pattern from which it arose, some enabling both, and some retaining the capacity to revert to asexual reproduction, if necessary, and yet others abandoned sexual reproduction and reverted to the asexual. (This behavior adaptation must have occurred early in eukaryote history; the description is that of a classic adaptation.)

Parthenogenesis is distinct from artificial animal cloning, a process where the new organism is identical to the cell donor. Parthenogenesis is truly a reproductive process, which creates a new individual or individuals from the naturally-varied genetic material contained in the eggs of the mother. A litter of animals resulting from parthenogenesis may contain all genetically unique siblings without any twins or multiple numbers from the same genetic material. Parthenogenesis occurs naturally in some species, including lower plants, invertebrates (water fleas, aphids, some bees, some scorpion species, and parasitic wasps), and vertebrates (some reptiles, fish, and very rarely—birds + sharks). This type of reproduction has been artificially introduced in other species.

Cloning is the process of creating an identical copy of something. In biology, it collectively refers to procedures used to create copies of DNA fragments (molecular cloning) or organisms. Molecular cloning refers to the procedure of isolating a defined DNA sequence and obtaining multiple copies *in vivo*. Cloning is frequently employed to amplify fragments containing genes, but it can be used to amplify any DNA sequence, such as the promoters, non-coding sequences and randomly-fragmented DNA. It is utilized in a wide array biological experiments and practical applications, such as large-scale protein production. Occasionally, the term cloning is misleadingly used to refer to the identification of the chromosomal location of a gene associated with a particular phenotype of interest, such as in positional cloning. In practice, localization of a gene to a chromosome or genomic region does not necessarily enable one to isolate or amplify the relevant genomic sequence. In essence, in order to amplify the sequence in a living organism, that sequence must be linked to an origin of replication, a sequence element capable of directing the propagation of itself and any linked sequence. In practice, however, a number of other features are desired and a variety of specialized cloning vectors exist that allow protein expression, tagging, single-stranded RNA and DNA production and a host of other manipulations.

Cloning of any DNA fragment essentially involves four steps: fragmentation, ligation, transfection, and screening/selection. Although these steps are invariable among cloning procedures, a number of alternative routes can be selected; these are summarized as a "cloning stragedy." Initally, the DNA of interest needs to be isolated to provide a relevant segment of suitable size. Subsequently, a ligation procedure is employed, whereby the amplified fragment is inserted into a vector. The vector (which is frequently circular) is linearized by means of restriction enzymes, and incubated with the fragments of interest under appropriate conditions with an enzyme called DNA ligase. Following ligation, the vector with the insert of interest is transfected into cells. A number of alternative techniques are available, such as chemical sensitivation of cells, electroporation and biolistics. Finally, the transferred cells are cultured. As the aforementioned procedures are of particularly low efficiency, there is a need to identify the cells that have been successfully transfected with the vector construct containing the desired insertion sequence in the required orientation. Modern cloning vectors include selectable antibiotic resistance markers, which allow only cells in which the vector has been transfected to grow. Additionally, the cloning vectors may contain color selection markers that provide blue/white screening (alpha-factor complementation) on X-gal medium. Nevertheless, these selection steps do not absolutely guarantee that the DNA insert is present in the cells obtained. Further investigation of the resulting colonies is required to confirm that the cloning was successful. This may be accomplished by means of PCR restriction fragment analysis and/or DNA sequencing.

Cloning also encompasses situations whereby organisms reproduce asexually. Cloning a cell means to derive a population of cells from a single cell (vegetative reproduction). In the case of unicellular organisms such as bacteria and yeast (simple life), this process is remarkably simple and essentially only requires the inoculation of the appropriate medium. However, in the case of cell cultures from multi-cellular organisms (complex life), cell cloning is an arduous task, as these cells will not readily grow in standard media. A useful tissue culture technique used to clone distinct lineages of cell lines involves the use of cloning rings (cylinders). According to this technique, a single-cell suspension of cells which have been exposed to a mutagenic agent or drug used to drive selection is plated at high dilution to create isolated colonies; each arising from a single and potentially clonally distinct cell. At an early growth stage when colonies consist of only a few cells, sterile polystyrene rings (cloning rings), which have been dipped in grease are placed over an individual colony and a small amount of trypsin is added. Cloned cells are collected from inside the ring and transferred to a new vessel for further growth.

In summary, epigenetics plays a prominent role in development and adaptation. Recall that bone development is the starting point for limb development. Internal gradients in the developing organism play the definitive role in bone formation, and bone formation plays the definitive role in the remaining development of the limb. Young and Badyev, published an online article *Evolution of ontogeny: linking epigenetic remodeling and adaptation in skeletal structures*, in the Society for Integrative and Comparative Biology online, May 22, 2007 (Oxford Journal). "Earlier or increased expression of Ihh, BMP-2, or BMP-4 (experimentally-induced by environment change) can result in premature ossification, thereby inhibiting developmental response to environmental variation. Alternatively, delayed ossification may reflect upregulation of FGF-2 (BMP antagonist), prolonging exposure to epigenetic signals." They "showed that variation in timing of ossification can result in similar phenotypic patterns through epigenetically-induced changes in gene expression, and proposed, relative to exposure in unpredictable environments, that both genetic accommodation of environmentally-induced developmental pathways and flexibility in development across environments evolve through heterochronic (developmental timing) shifts in bone maturation." They suggested that "such heterochronic shifts can not only buffer development under fluctuating environments (while maintaining epigenetic sensitivity crucial for normal skeletal formation), but also enable epigenetically-induced gene expression to generate specialized morphological adaptations." Unlike pre-Darwinian Lamarackism, epigenetics emphasizes the heritable transfer of traits; epigenetics integrates well with related natural selection genetic bias and of the alteration of the DNA genome by random mutation (Waters). Random epigenetic expressions, caused by organism isolation in a stressful environment, can be inherited from one generation of organisms to the next, and they arise at the time of formation from regulatory mechanisms, encoded by the entire genome (natural selection genetic bias). They will likely be transient and eventually reversible, unless they are environmentally-adaptive changes that are supported through time by either negative or positive environmental selection. Some environmentally-related genetic accommodation may follow during this time. So, in the big picture, the phenotype is initiated by both heritable genetic transfer of DNA coding and/or epigenetic switch patterns from the norm-of-reaction, and it is mostly formed by simple physiologic processes that utilize environmental feedback. And in a stressful environment, either plastic (epigenetic) norm-of-reaction genetic variation, or mutation, or new epigenetic expression of an old mutation, could enable instant formation of a random phenotype that may or may not be supported by selection. If it is supported by environmental selection, it becomes an adaptation to the environmental stress. Only adaptive phenotypes survive environments fatal to non-adaptive phenotypes. Gene frequency is changed by selection. The biochemistry of genetic mechanisms is a part of the biochemistry of metabolism. Metabolism, the total sum of all chemical processes in living organisms, includes not only the biochemistry of genetic mechanisms, but it also includes other areas of protoplasm production (non-genetic cellular dynamics), maintenance (vital function), detoxification (rendering waste products harmless), and respiration (energy production). Metabolism requires a power source. The metabolism of the more familiar surface life is dependent on the biochemistry of photosynthesis.

33

Primary Production

The primary productivity of an ecological <u>community</u> is the amount of biomass produced through photosynthesis by plants (per unit area and time), which are the primary producers. Primary productivity is usually expressed in units of energy (e.g., joules/sq meter/day) or in units of dry organic matter (e.g., kg/sq m/year). Green plants contain layers of the pigments chlorophyll a and b, and carotene in cellular structures called chloroplasts. In addition, photosynthetic bacteria and algae can also have phycobilins in smaller structures (chromatophores). This pigment diversity permits light absorption at varied wavelengths, increasing light-trapping efficiency. In the photosynthesis reaction, chlorophyll acts as an enzyme to transfer the light energy of the photon into chemical energy (sugar) and O_2 (using H_2O and CO_2 as raw materials). Interestingly, the reaction in photosynthetic bacteria is enhanced by the presence of iron (electron acceptor).

Photosynthetic primary production is mostly at sea. Photosynthetic *Prochlorococcus marinus* (a cyanobacterium or blue-green algae) is the single largest producer of organic matter (and oxygen) on earth. *Prochlorococcus* cells dominate the tropical and temperate oceans (40 to 40 latitudes); they are the most abundant organism in the oceans. This organism accounts for up to 80% of oceanic primary production. Its tiny size (0.005 mm) optimizes uptake of nutrients such as nitrogen or phosphorus, even when concentrations are below the detection limit. *Prochlorococcus marinus* is the smallest organism or genome known to sustain life by photosynthesis. *Prochlorococcus marinus* is the primary control of global warming (CO_2 reduction), even though the water's surface delays the process of ultimate limestone lock-up in ocean sediments. *Prochlorococcus marinus* has a single chromosome, 1884 protein-making genes and 40 transfer RNA genes. There are two ecotypes, physiologically and genetically adapted to grow under different light intensities. Typical concentrations are 100,000–300,000 cells per ml over the first upper 100m (shallow ecotype). *Prochlorococcus* cells can be found as deep as 150–200m at depths reached by only 0.1% of the irradiances in surface (deep ecotype). Even in the clearest water, this is as dark as the darkest night on land. *Prochlorococcus marinus* lacks phycobilisomes that are characteristic of other cyanobacteria, and contains chlorophyll b as its major accessory pigment. The specific pigment complement includes unique divinyl derivatives of chlorophyll *a* and *b* (chlorophyll *a2* and *b2)* as well as a-carotene and zeaxanthin. *Prochlorococcus* is the only photosynthetic organism known to contain this particular combination of pigments. This enables it to absorb blue light efficiently at the low-light intensities and blue wavelengths characteristic of the deep euphotic zone. These low-light adapted strains also must cope with low oxygen conditions as well. Though they derive their energy from photosynthesis, they also use dissolved organic carbon as a source of reduced carbon and energy (Kenyon College Website—*Prochlorococcus marinus*).

Aquatic algae are the #2 primary producers (mostly marine—primarily brown, but some red, green, and other blue-greens). Algae strongly support oxygen (O_2) production and carbon dioxide (CO_2) reduction. What does this reveal about the ocean's importance? The forest is the #3 primary producer; the rain forest (also important to regional biodiversity) plays a prominent role in primary production, oxygen (O_2) production and carbon dioxide (CO_2) reduction. Forests presently cover 10% of the earth's surface, or 30% of the land area. During the growing season of spring and summer, conifers of the sub polar region will match photosynthesis production for the rain forest's entire year (rain forest is only 3% of land surface; taiga and boreal forest is 17% of land surface); deciduous temperate forests (area is midway in-between preceding forests) and grasslands (grassland ecosystems cover 25% of earth's land area) are also highly significant sources of primary production. Understanding primary production is the essential to understanding the ecosystem; it is essential to understanding the conditions of life.

34

Respiration

Understanding respiration is also essential to understanding the ecosystem; it is essential to understanding the conditions of life. For complex surface life, respiration is the opposite of photosynthesis; sugar is oxidized to carbon dioxide (CO_2) and water (H_2O). The machinery of respiration is located on the inner membranes of the mitochondria, or in the case of more primitive life forms, in cell membranes. The purpose of respiration is to break down outside energy sources to form adenosine-tri-phosphate (ATP) energy packets and to dispose of the electrons produced by the breakdown. In aerobic respiration, the terminal electron acceptor is oxygen (O_2). Oxygen supercharges (aerobic) respiration. The terminal electron waste disposal product is H_2O. The food waste product depleted of electrons is CO_2. Respiration is generally the same, but not exactly the same for all plants, animals and bacteria. In high light intensities and temperatures, photorespiration in plants causes inefficiency in photosynthesis, and it can also be a source of CH_4 and NH_3. Plants and animals commonly use the Krebs Citric Acid Cycle for aerobic metabolism of sugar, producing CO_2 and H_2O. This cycle also provides intermediates for other pathways to make nucleic acids, fatty acids and amino acids (and proteins). Animals can use the Embden-Meyerhoff Hexose-Monophosphate Shunt for limited non-aerobic respiration, producing lactic acid; plants may produce alcohol (fermentation) as an intermediate prior to forming an organic (e.g., lactic, acetic) acid. Bacteria often use all of the above, plus may have additional anaerobic pathways for respiration, notably producing anaerobic by-products like ammonia (NH_3), hydrogen sulfide (H_2S) and methane (CH_4), the terminal electron waste disposal products. The terminal electron acceptors respectively include nitrate (NO_3), sulfate (SO_4) and carbon dioxide (CO_2). Despite Michael Behe's claim that the origination of a pathway to AMP is not possible under Darwinian evolution, many others don't think it is a coincidence that the main energy source is ATP with AMP being the purine base adenine, and the 2nd main energy source is GTP, with GMP being the other purine base guanine. Chemosynthetic primitive life, with the ability to form carbohydrate chain structure, already had the ability to form the ribose-5-phosphate chain foundation of AMP and GMP during the varied conditions of life experienced from around 4 billion years ago up to the present time. It is almost certain that looking for original pathways in highly-evolved aerobic complex life today would encounter visualization barriers. Examination restricted to computer-controlled autos of today would make one wonder how there could have been a workable horseless carriage just over a century ago.

Decomposers break down dead plants, animals and detritus (small bits of nonliving organic material). The bacteria generally break down animal flesh and the fungi generally break down plants. Different bacteria and fungi work together or alternate in the breakdown processes; team effort is often necessary. Decomposers, as a group, work under a variety of conditions (still respiration). Bacteria will grow in the presence or absence of oxygen. Anaerobic respiration is incomplete, so not all carbon goes to CO_2 (like sooty flame) or bacterial protoplasm. Bacteria often create carbon-rich anaerobic Terry Hilleman- 386 sediments and soil faster than oxygen can diffuse into medium. A micro-underground world exists below soils where enough water is present. Bacteria, fungi, algae, protozoa, nematodes and mites comprise a major biomass community of the micro-decomposers. Biomass in this area has been estimated to equal that of 12 horses per acre. The biomass of bacteria and nematodes is staggering. There can be millions of these microorganisms in a few cubic centimeters.

Most bacteria are placed into one of three groups based on their response to gaseous oxygen. Bacteria generally function primarily in oxidation or in reduction. Aerobic bacteria thrive in the presence of oxygen and require it for their continued growth and existence. Other bacteria are anaerobic, and cannot tolerate gaseous oxygen, such as those bacteria which commonly live in soil and in underwater sediments (e.g. *Clostridia*, which can also cause tetanus, or highly toxic bacterial food poisoning). The third group comprises the facultative anaerobes, which prefer growing in the presence of oxygen, but can continue to grow without it. Bacterial heterotrophs (living off of dead organic matter) are important in the processes of biodegradation and decomposition under both aerobic and anaerobic conditions. There are dual-role bacteria, intermediates between producers and decomposers . They play a role in overall production; they also function under conditions unfavorable for most green plants. When the photosynthetic or chemosynthetic functions are in use, these forms of energy production are autotrophic (self-feeding). However, some of the photosynthetic bacteria can also grow anaerobically, functioning as heterotrophs in the dark. Some of the chemosynthetic bacteria can also use sunlight for energy or can function in the dark as heterotrophs. Heterotrophy, usually by some means of limited non-aerobic respiration (same as animals) or fermentation (same as yeast or lactic acid bacteria), is organic-dependendent. Recall that deepwater *Prochlorococcus marinus* gets some of its energy from photosynthesis, but also uses dissolved CO_2 as an energy source.

Careful examination at ground level often reveals a green bacterial layer at (or just below) the moist surface; photosynthesis occurs here. Below the green layer is a red layer of sulfate-fixing bacteria; this is an area of sulfate (and nitrate) formation via oxidation. Further below the red layer is a black layer of sulfide bacteria; this is an area of sulfate reduction. Hydrogen sulfide (H_2S) production occurs here. A similar community exists in ponds, lakes and oceans. The near-surface green layer is in the water column. Sometimes the bottom's surface sulfate-fixing cloud of bacteria can be seen floating in oxygenated water just above the bottom floor; it may even be found near the thermocline just above black anaerobic deep water. In the more common bottom black sulfide (sub-bottom) layer, smelly, toxic, hydrogen sulfide (H_2S) is produced. This black layer can also contain inorganic carbon, carbon dioxide (CO_2) and anaerobically locked-up carbon as methane (CH_4).

Sulphur compounds, mainly hydrogen sulphide (H_2S), can serve as sources of electrons for bacterial chemosynthesis. As in photosynthesis, inorganic carbon (CO_2) is reduced to organic carbon (CH_4), while the oxidation of sulphur serves as the source of energy instead of light. This oxidation could go all the way to sulphate ($SO_4=$) by the acid-producing (sulphur) bacteria but it often stops at an intermediate oxidation state. Said otherwise, chemobacteria can create carbohydrates capturing close-by carbon in a chemosynthetic cousin of the carbon cycle. Any sulfate produced could be reduced once again by sulfate-reducing bacteria, providing additional energy.

Chemosynthetic bacteria obtain energy by chemical oxidation or reduction of simple organic compounds. Examples include N_2 or (NO_2-) or (NO_3-), or to (NH_4+); NH_3 to N_2; H_2S to S, or ($SO_3=$), or ($SO_4=$); or, vice versa; and FeS_2 to $Fe(OH)_3$, ($SO_4=$) and H+, as well as Fe_2O_3 and H_2S to FeS_2. Ammonia (NH_3 and NH_4OH) is toxic to most plant life, so nitrogen for protein production must first be converted to nitrate (NO_3)- form (by nitrogen-fixing or cyanobacteria). Oxygen must be present in the environment for (O_2) oxidation to occur. There are exceptions. Black sulfide layer activity has been recently discussed (chemobacteria can create carbohydrates capturing close-by carbon in a chemosynthetic cousin of the carbon cycle). Anaerobic photosynthetic sulphide oxidation occurs primarily in estuaries. Jannasch, H.W., in *Interactions between the Carbon and Sulphur Cycles in the Marine Environment*, has an interesting perspective.

"Since hydrogen suphide is a product of sulphate reduction and that uses photosynthetically-produced organic matter as reductant, chemosynthesis by sulphur-oxidizing bacteria could be considered in the flow of energy as a form of secondary production. Microbial sulphur oxidation appears twice. It appears first as an aerobic and chemosynthetic process. It appears second as an anaerobic photosynthetic process. In estuaries, bacterial anaerobic reduction of CO_2 requires light as a source of energy and uses hydrogen sulphide (H_2S) as a source of electrons. In a way, if the above terminology is used, this bacterial photosynthesis represents (as does the green plant photosynthesis) a form of primary production. Green plant (O_2- producing) photosynthesis uses H_2O rather than H_2S for the electron source. The distinction between primary and secondary production of organic carbon is important if the interactions between the carbon and sulphur cycle are linked to the flow of energy, be it light or chemical energy." It could be that this anaerobic photosynthetic pathway laid the groundwork for oxygen-generating photosynthesis, over 2bya. Understandably, the organism capable of this breakthrough molecular process gained a natural selection advantage in its environment. One type of salt-loving extremophile (a halophyle) uses sunlight to make energy, but not the way plants do. This extremophile has light-harvesting pigment (bacteriorhodopsin) in its cell membrane; this pigment reacts with light and enables the cell to make ATP for respiration.

Mid-ocean rift, thermal sea-vent chemobacteria extremophiles mimic activities of their photosynthetic cousins from above, making carbohydrates from hydrogen, oxygen and carbon; they can also create carbohydrates capturing close-by carbon in a chemosynthetic cousin of the carbon cycle. Recall that reduced sulphur compounds, mainly hydrogen sulphide (H_2S), can serve as sources of electrons for bacterial chemosynthesis. As in photosynthesis, the presence of inorganic carbon (CO_2) permits reduction to organic carbon, while the oxidation of sulphur serves as the source of energy instead of light. This oxidation may go all the way to sulphate by the acid-producing (sulphur) bacteria, or may stop at an intermediate oxidation state (often preferred). Other types of sea vent extremophiles simply use sulfate reduction (to sulfides) for respiration; both parallel the activity seen in other green-red-black layers. Chemobacteria extremophiles in thermal sea vents and non-marine hot springs (both at very high temperature) can also use any locally-abundant sulfur (S), rather than oxygen, as an electron acceptor in respiration. In this case, the reduction of sulfur itself generates hydrogen sulfide (H_2S) rather than water, as a by-product of respiration. This H_2S can be used by other extremophiles (recall thermal vents in the mid-ocean rift, in the area of the hot under-water water cycle, can contain certain chemobacteria capable of creating carbohydrates capturing close-by carbon in a chemosynthetic cousin of the carbon cycle). If the available sulfur is organically bound, the sea vent/hot spring extremophiles may produce hydrogen gas, which is also of use to other extremophiles (deep-rock extremophile discussion follows). Sea-vent and hot-spring extremophiles are often anaerobic, but can tolerate oxygen to some degree. Their ability to generate hydrogen from organic waste products, while tolerating oxygen allows great flexibility. Hot thermal vents on the seafloor may then support complex ecosystem communities of white crabs, prawns, barnacles, starfish and fish. Tubeworms (e.g. *Riftia*), form the base of the animal community; their hemoglobin-red plumes, rimmed with blood vessels, extend into hot vent waters. O_2, CO_2 and H_2S bind to carrier molecules in the blood. These three compounds are circulated to specialized tissue, containing densely packed symbiotic bacteria, similar to the free living forms found in the surrounding water. The chemosynthetic (sulfur) bacteria exchange ATP and all the organic compounds *Riftia* needs for growth for a constant source of O_2, C, S, energy, and a protected environment. A similar symbiotic system is duplicated by surrounding mussels.

Cold seeps are areas of the ocean bottom (sometimes associated with brine pools) that release H_2, H_2S, or CH_4 gas (bacteria oxidize gas as it exits cracks), or oil. "Bushes" of tubeworms may grow on methane-ice deposits. Hemoglobin-enabled tubeworms (e.g. *Riftia* and others), and various clams (*Calyptogenea* and *Vesicomyids*) harbor sulfide-fixing bacteria in their mantle tissues. Respiration rates are lower than rates found in hot thermal vents, but similarly complex seafloor communities are present in the cold seep area (e.g. numerous crabs, snails, and ice-methane-worm *Hesiocaeca methanicola*).

Chemobacteria extremophiles can have many different pathways for their very slow respiration. Chemobacteria extremophiles that live deep in the earth only need water to live in and heat of the earth (energy) to reduce (or oxidize) rocks and minerals. H_2 gas is released as water seeps through rock (produced from iron oxide reacting with water); molecular hydrogen can be reacted with carbon, oxygen or sulfur to sustain deep rock extremophile respiration. In caves, extremophile (sulfur) bacteria derive their energy from inorganic hydrogen sulfide; previously discussed, H_2S is carbohydrate-convertible when small amounts of carbon are present (chemobacteria can create carbohydrates capturing close-by carbon in a chemosynthetic cousin of the carbon cycle). Limestone extremophiles can reduce the rock and release CO_2; the resultant carbonic acid dissolves limestone, forming caves.

35

Ecosystem Energy Flow

Understanding the energy flow of primary production to the other organisms in the ecological community is essential to understanding the conditions of life. Both deep and surface food chains require energy to power the biochemistry of life. Primary production (photosynthesis) captures solar energy in the form of chemical bonds (sugar). In respiration, living things use enzymes (proteins) to form and break these bonds. Both plants and animals can modify the sugar to form new compounds with potential energy (always lose large amounts of energy, usually 90%, at each tropic level— primarily as detritus). If a grazer consumes a green plant, most of the green plant's potential energy will be lost as grazer waste. The rest will be degraded as heat of metabolism (mostly respiration) or incorporated into grazer protoplasm. Technically, the first grazers only appeared after grasses showed up about 35mya; before that, browsers on land ate leaves from plants. The production of living material per unit area (or volume) per unit time by herbivores is known as secondary production.

When a predator eats a grazer, most of the grazer's potential energy is lost as waste, the rest goes to predator metabolism (mostly respiration) and predator protoplasm. When a big predator eats a small predator, most of the small predator energy is lost as waste; some small predator energy goes to big predator metabolism (mostly respiration and lost as heat) and some small predator energy goes to big predator protoplasm. When a big predator dies and falls to ground, decomposers attack big predator flesh (bacteria and fungi). Most of the big predator protoplasm energy is lost to decomposition, as the big predator protoplasm is degraded (decomposer metabolism, mostly respiration- and lost as heat), and some of the big predator protoplasm energy is incorporated into decomposer protoplasm (bacteria and fungi). If a worm eats the dead big predator, along with the bacteria and fungi, the energy of all three goes to worm metabolism/respiration (heat), protoplasm, and waste. In the photosynthetic-productive ecosystem, the bulk of potential energy is lost as waste and ultimately utilized by detritus decomposers, or lost to soil, freshwater or marine deposits.

The oceans comprise 95% of the living area on the planet (excluding the deep rock extremophile bacteria). In the deep blue sea, the grazers and predators are really on the move. The Deep Scattering Layer (DSL) migration of grazers and predators can rise from the mid-ocean depths (from nearly a kilometer down—¼th of average ocean depth) to the surface at speeds of up to 7 meters per second. It's the largest animal migration on the planet and it happens twice a day (dark and dawn). It may also approach near-shore areas. Zooplankton, e.g. pelagic Foraminiferans, like the protozoan *Globigerina*, eat cyanobacteria and algae; micro-zooplankton like these are the dominant consumers of phytoplankton in both the open ocean and coastal waters. The speediest members of the DSL migration are the Copepods. Most Copepods are grazers; their numbers increase with primary production. They are the most numerous "bugs" in the sea; they make up 70% of all macro-zooplankton. On average, one Copepod inhabits about every single liter of sea water. Predators, like juvenile fish, Cnidarians (jellyfish), Ctenophores (comb-jellies) and Tunicates (salps), eat the grazers. Copepod and salp waste can represent a significant source of carbon in bottom sediments. Different sizes of other fishes and cephalopods also join in the food chain/web feeding frenzy before all retreat to the safety of the water column middle layers to avoid still larger daytime predators and/or stay cool. The average DSL depth decreases gradually, e.g. from 800m to 300m, with increasing latitude, correlating with light intensity from solar irradiation. At higher latitudes, shrimp-like krill (#2 ocean "bugs") become more abundant. Krill eat phytoplankton, but copepods are also listed as a favorite on the krill menu.

36

Common Ground

The tree of life is composed of three great branches. Extremophiles (e.g. sea-vent/hot spring sulfur forms, methanogens and halophiles) are the most primitive (-like); they are the Archaea. Archaea are the most abundant life on earth (rock extremophiles). Many use hydrogen (H_2), hydrogen sulfide (H_2S) and inorganic carbon (CO_2) as a source of energy. As noted previously, these most ancient forms of life may have their own carbohydrate metabolism. Some of the Archea, like the "primitive" *Prochlorococcus marinus* blue-green algae (bacteria), are photosynthetic. Many marine cyanobacteria can be found in limestone (calcium carbonate) or lime-rich substrates, such as coral algae and the shells of mollusks. Many Archaea have tough outer cell walls, these walls contain different kinds of amino acids and sugars than those found in true bacteria. One type of acidophyle can tolerate pH levels near zero; there are alkaphiles that tolerate an environmental pH near thirteen. Archaea cell membranes are chemically distinct from membranes of true bacteria, with differing lipid structures and chemical links. True bacteria comprise the 2nd branch; they have evolved advanced forms of DNA and/or RNA; a small percentage of the bacteria are the pathogens that challenge our immune systems. Some bacteria are aerobic; some are not. In the absence of O_2 many soil bacteria use anaerobic respiration for their somewhat-efficient metabolism. Ecologically, the most important soil bacteria are the nitrifying bacteria, which together convert NH_3 to NO_2, and NO_2 to NO_3, and the colorless sulfur bacteria, which oxidize H_2S to S and S to SO_4. Bacteria are known as simple life, but they aren't that simple. The 3rd branch, Eucarya (complex life), include the more familiar life forms, which have a nucleus (plants, animals, and single cell life forms). The first eukaryotic cell was the greatest leap in the technology of life since life's first beginning (in a hot ocean thermal vent?). The eukaryotic cell has many functions that may be random; however, random functions are environment-sensitive. Environmental feedback stabilizes a function and the genetic pathway to this function.

The function is genetically-heritable but environmentally-integrated; this is the basis of environmentally AND genetically determined adaptation.

Kirschner and Gerhart, in the chapter on Conserved Cells, Divergent Organisms, state that: "The controlled fluid environment inside the multicellular epithelial organism was a novelty that promoted communication between animal cells via secreted and received signals. Communication, of course, also occurs in single-celled eukaryotes, and prokaryotes (no nucleus), but not to the same extent as in metazoans (animals). Other eukaryotes such as multicellular algae, slime molds, and mushrooms, lacking intercellular junctions, nonetheless achieve significant intercellular signaling for elaborate multicellular functions, such as the fruiting bodies of mushrooms. Plants, for example, have sizable channels that allow large signaling molecules, even RNA molecules, to pass between cells. The controlled internal milieu (environment) of animals, though, must have provided the context for the elaboration of a greatly expanded set of signals and receptors, and indeed animals have evolved many kinds of cell-cell signaling." The internal areas of a single cell is an environment, as is the compartment of a developing metazoan, and the external (surrounding) conditions of all organisms. Single-cell eukaryotes first formed in a created environment of changing conditions. Their origi can be traced to a time following the first snowball earth (2.3bya), near the time (2.2bya) when oxygen had suddenly appeared in the atmosphere and the ocean. When oxygen first arrived, it was an anaerobic world; at this time, only small traces of oxygen were present. The compartmentalized cell design of the eukaryotes is well-suited to aerobic metabolism; nevertheless, many eukaryotic forms retain an essentially anaerobic metabolism. The diversification of eukaryotes seems to be associated with the rise in atmospheric oxygen during the Proterozoic era (2,300mya–544mya); it was during this time that many eukaryotes acquired mitochondria and chloroplasts. Complex, many-celled animal life (metazoan) only arrived after the time of the last snowball earth event (580mya), as atmospheric oxygen levels began to approach the atmospheric oxygen levels of today. Just as the blood of modern animals reflects the (saline) environment of ancient oceans (when multicellular life began to create its own environment), the internal environment of the eukaryotic cell reflects the conditions of life at the rise of the eurykarotic cell.

The genetic code is the same in all living organisms; the genetic code guides the way organisms form and it controls their daily function. It has been demonstrated that eukaryotic ribosomes are able to translate bacterial mRNAs correctly. Eukaryotic ribosomes are much larger than prokaryotic (no nucleus) ones and most of their proteins are different. There is no obvious functional reason why mechanisms and components need to be so similar.

These similarities can be explained by descent from a common ancestor. Life's vast diversity has come about by only a slight difference in the same genes. Researchers at the human genome project have yet to find a gene structure that is unique only to humans. Every one of the human genes came from another species at another time. Humans share 50% of their genes with the common housefly. A lot of the human protein sequences are identical (100%) to those of all mammals. Human beings have far fewer genes than originally thought, about 30,000, perhaps just a few hundred more than a mouse. Humans share 98.5% of their genes with chimpanzees. If the genotype of so many different life forms is so similar, what explains the huge differences in phenotypes? **Epigenetic expression of the genes creates the phenotype.** Phenotypes are heritable genotype expressions with integrated environmental components; environmental selection interacts with these components, not with the genotype. Environmental selection causes biased survival of certain phenotypes, indirectly genotypes; this biased survival is the bias of natural selection. Natural selection is the (genotype) preservation of successful phenotype adaptations to previously-experienced environments. Adaptations are new environmental expressions (phenotypes), integrated into the inheritable genome. Signals from the external and internal environment may initiate flexible, but regulated, modifications in epigenome expression. It is naïve to think that because an adaptation is inheritable, that genetics alone creates the adaptation.

In Futuyma's Evolutionary biology text, regarding major points about the origin of mammals, he states: "The origin of mammals, via mammal-like reptiles, is doubtless one of the most fully, and beautifully documented example of evolution of a major taxon (group). In no other case are there as many steps preserved in the fossil record... each of many mammalian characteristics (e.g. posture, limb flexibility, agility, tooth differentiation, skull changes associated with jaw musculature, secondary palate, reduction of the elements that became middle ear bones) evolved gradually, so that in retrospect, a trend can be traced back to synapsids (mammal-like reptiles—250+mya) or beyond. As in every other origin of a major taxon (group), evolution has been mosaic, with different characters 'advancing' at different rates. At no point in this history has a new bone come on the scene; all the skeletal elements are modified from those of amphibians and even crossopterygians (coelacanth—lobefin fish), by fusions or changes in size and shape."

It was different species of plants that were the first described and placed into the taxonomic system; classification was based on differences between sexual components. Botany is fundamentally about sex. Flowering plants have co-evolved with insect pollinators and have thrived because they are better at sexual reproduction than spore producers. In *Seeds*, Wolfgang Stuppy notes that flowering plants gained a competitive evolutionary advantage over the spore-bearing competition 140mya; today, the latter is reduced to only 3% of all land plants. Seed-formation in flowering land plants has a significant common role in this great success.

"Nothing in biology makes sense, except in the light of evolution"— Theodosius Dobzhansky, 1973.

37

Biodiversity

The process of evolution requires the sort of environmental unrest that makes change both necessary and advantageous. Evolution needs something to work with, some purchase to give it leverage on life; it looks for the bits that stand out from the rest, rejoicing in diversity and limitless variation. Anything that subtracts from that variety is bad for business. Every loss of diversity represents a loss of organic vigor and a corresponding reduction in the possibilities of interconnection and cooperation; the loss is progressive. As diversity fades, so do the chances for change, and eventually, the system breaks down altogether. Diversity is the key to success. (Watson—*Dark Nature*) Biodiversity is "the variety of life, in all its forms, levels and combinations." It is genetic variation at all levels. This would include genetic variation within one individual, or between different individuals, species, genera, families and still higher taxonomic levels; that is, this would include the entire community of genetically variable populations within an ecosystem. Biodiversity enhances stability within the individual, species, genera, family and higher taxa; it enhances stability in the ecological community. To understand biodiversity is to understand the biologic genetic variation at different levels within the community of the ecosystem. Environmental biology is therefore strongly connected to biodiversity. Heterozygosis of paired alleles provides biodiversity within the organism. Genetic diversity within a species encompasses the differences in DNA composition between individual members. All species consist of a population of individuals that vary in genetic and epigenetic degree from one to another. Whether the diversity is minor, or reflect greater differences (often found with the isolation-based formation of a subspecies), there are some considerations. As a species undergoes environmental selection in the local environment, genetic "adaptation" of the phenotype will be directed towards conditions of life in that particular local environment (assuming isolation). In this "adaptation," directional selection sorts (stabilizing selection for pre-existing adaptive phenotypes from the norm-of-reaction) for phenotypes (developing organisms) with the best match to the conditions of life. Organism function in that environment will usually improve; the genetic diversity of that population of phenotypes may decrease. As long as conditions of the environment do not change, the displayed phenotype (inheritable adaptation) selected for that particular environment will work very well. "Organisms seal their own fate through continually increasing commitment to a given lifestyle." (Levinton—chapter on Patterns of Morphologic Change in Fossil Lineages). No environment is permanent; change is inevitable. In the long term, species with more genetic diversity (genotype + phenotype) have better potential for survival. This is expected from species that are dispersed over large landscapes of variable habitats. Within this large area of changing landscapes, environmental selection (as a gatekeeper) will have indirectly directed genetic variation towards diversity, i.e., towards "adaptation" (expression of pre-existing adaptations from natural selection genetic bias) to the many different habitats that are present. Increased genetic variation enables organisms to better adjust to changing environmental selection pressures. This diversification advantage is related to Darwin's observation in *Origin of Species* when he said: "isolation of large populations provided the greatest (genetic AND environmental) variation for the creation of new species."

Accurate quantitative measurement of a region's biodiversity isn't easy. It is often expressed as species diversity, and in particular, species richness. Although this is far easier than trying to measure than the genetic diversity in an individual organism's DNA, it is still a challenge. There is still profound ignorance of the number of species that exist worldwide. As of 2002, 1.75m (million) species of plants, animals and microorganisms had been identified; estimates for the number of species in existence range from 3m to 100m. In recorded number of species, insects account for more than half of all species in the world; about 40% of these are beetles. Beetle species outnumber all non-insect animal and plant species. Vertebrates make up less than 4% of the animal species. Further bias comes from unexplored areas, such as ocean depths, tropical tree canopies and soils of remote areas; here, examination and classification has only just begun. Many species, such as archaea, bacteria, fungi, protists, nematodes, and various arthropods are tiny and easily overlooked. Biomass in the soil can be equal to 12 horses per acre. Until recently, the molecular methods to identify bacterial diversity did not even exist. Noel Fierer, a microbial ecologist from the University of Colorado examined 98 soil samples from Arctic tundra to Amazon rain forest; the results were surprising. The Amazon rain forest is a bacterial desert compared to the dry grasslands of the Great basin in Nevada, which is lush with microbes. Rain forest soil is acidic; grassland soil is near neutral. The acidity in the environment may explain the difference in diversity of soil bacteria between these two environments. Woods Hole marine biologists have found more than 20,000 species of bacteria in a liter of seawater. Craig Venter, the first to sequence the human genome, has discovered tremendous biodiversity in the oceans, involving discovery of millions of new genes, thousands of new protein families and the characterization of thousands of new protein kinases from ocean microbes, using whole environment shotgun sequencing and new computational tools. The J. Craig Venter Institute (JCVI) launched the Sorcerer II Global Ocean Sampling (GOS) Expedition in 2004. Researchers believe these data will lead to better understanding of key biological processes which could eventually offer new ideas for alternative energy production, and could offer solutions to deal with climate change along with other environmental issues. The GOS data-set is 90× larger than other marine genomic datasets, thus making it the largest ever released in the public domain. The GOS analysis also nearly doubles the number of previously known proteins. This enormous amount of data allowed the researchers to better understand the genomic structure and evolution of microorganisms, as well as the function of important protein families such as protein kinases, which are key regulators of cellular function in all organisms. Although invisible to the naked eye, ocean microbes (like *Prochlorococcus marinus*) make up the vast majority of life in the hydrosphere; they are primarily responsible for creation and maintenance of Earth's atmosphere. It is important to understand the role and function of these organisms to ensure the survival of the planet and all aerobic life upon it. The data not only provide an unprecedented level of new genes and protein family discoveries, but are also pivotal in providing a compelling analysis of evolution, including the function of these genes and proteins within the larger context of organisms interacting with their environment. Venter stated: "Given the findings, it's clear that we've only begun to scratch the surface of understanding the microbial world around us." Upon closer examination, using DNA analysis as a guide, many small organisms originally identified as a species may turn out to be more than one species. Subspecies, formed by the same (isolating) processes that give rise to speciation, represent cases in which divergence hasn't proceeded far enough to create a separate species; this can also create additional confusion in assessing species biodiversity.

Species richness is considered as the standard for overall biodiversity when different areas are being compared. Could this also introduce a bias? Ninety percent of all species are tropical; this is the greatest area of biodiversity loss and extinction at the species level today. Notable past extinctions have been caused by climate change, chilling tropical environments and killing large numbers of cold-intolerant populations. What happens to biodiversity when there is a major environmental limiting factor occurring at higher latitudes, where there are fewer species, but greater numbers of individuals of each species? Higher latitude populations are more cold-tolerant and will be less affected by a cold environment. However, a cold climate is not alone as an agent of environmental extinction; past mass extinctions have been caused by other climate change. Global warming increases temperatures far more at Polar Regions than in the tropics; this will also eliminate habitats. What happens to biodiversity when the loss of habitat and individuals is at higher latitudes, with fewer species, but greater numbers of population individuals? How does the number of individuals affect population and community biodiversity? What happens to biodiversity if individuals are unaffected generalists or sensitive specialists? Is there any hope of avoiding a looming mass extinction caused by humanity? Is there a concern for wildlife and biodiversity before species extinction is threatened? What makes a species special? If beetle species outnumber all non-insect animal and plant species, do beetles get equal rights of endangered species protection? Should all wildlife, or only selective forms of wildlife, have rights? In the upcoming text, there will be some discussion of these topics related to tallgrass-prairie habitat loss. Biomass production is mainly dependent upon primary production and food web dynamics. Areas high in production and high in limiting factors (such as waters "turning over" near Arctic and Antarctic ice), display large numbers of individuals with limited species biodiversity. Deep sea basins generally have higher species diversity than more heavily-populated waters above, but energy input for this area is lower, so numbers of individuals are necessarily lower. Deep sea basins tend to be more stable environments than those found in the waters above. Previously noted in the discussion of extinction, **biodiversity is directly proportional to environmental stability; i.e., species richness is directly proportional to habitat stability over time.** Habitat continuity over time stabilizes and enhances biodiversity. Recall that the primary cause of extinction is destruction of habitat. The primary cause of biodiversity loss is destruction of habitat. **Biodiversity is inversely proportional to a location's environmental extremes and directly proportional to habitat diversity; i.e., species richness is directly proportional to favorable conditions of life and habitat diversity.**

The conditions of life give life meaning; the dynamic interaction of heredity and the environment produce the biodiversity in life. Biodiversity represents biological genetic (genotype and phenotype) variation in the individual and in the ecosystem community; it is the variety of life, in all its forms, levels and combinations. Biodiversity may also include the entire ecosystem, which even includes diversity of the physical environment, such as habitat (landscape) diversity; however, this is not really biological genetic (genotype or phenotype) diversity. It is ecosystem habitat diversity; and even though the conditions of life give life meaning, I do have a problem including habitat diversity (landscape diversity) within biodiversity. Habitat diversity is an environmental description; it is diversity (variation) in environmental selection. Biological (genetic) variation AND the environmental variation indirectly form the phenotype, but the physical environment is not part of the biologic ecosystem; it is not alive. Living organisms can be a physical aspect of the habitat, but the conditions of life are the environment of life, not life itself. This is just one more example of how biological phenomena continually get confused by inappropriately interchanging wording with the conditions of life. It is helpful to keep biodiversity and habitat diversity, as well as their connection to one another in clear focus. One thing for sure, protecting against the extinction of an endangered species will require the protection of its habitat, and everything in it. Plants and lowly invertebrates are essential to food webs; they will also need protection in order to protect any "special" species.

38

Habitat Diversity

Habitat diversity encompasses all conditions of life for the ecosystems within the area of interest. It is the physical diversity of the ecosystem; **the conditions of life give life meaning.** Habitat diversity is an environmental description; it is diversity (variation) in environmental selection. Habitat diversity is environment variation (a range of variability in all conditions of life, or variation in environmental selection at any one time). Physical habitat diversity must consider landscape variety; this requires large areas of physical habitat with both new and old growth. (Living organisms can be a physical aspect of the habitat, but the conditions of life are the environment of life, not life itself.) An area with a monotonous landscape (e.g. a flat sand desert) has far less habitat diversity than an area with innumerable, different, varied landscapes (e.g. tropical mountain rain forest). A habitat must be of adequate size. Habitat thresholds of adequate area (size) must be maintained; falling below habitat thresholds will not allow populations to be maintained. Habitat integrity must also be maintained; breaking up habitat integrity into small pieces changes the structure of the ecosystem, destabilizing populations. A 90% loss of habitat area could result in greater than a 90% loss of individuals within the resident biological community. A 90% destruction of the habitat could result in greater than a 50% biodiversity or community species loss, depending upon population distribution. Habitat diversity is a primary consideration for both biogeography and environmental biology. So, habitat diversity is very concerned with location (considers environmental conditions within it); polar habitats have more discriminating environmental selective factors than tropical habitats. Temperate climate habitats have degrees of environmental selective factors generally in-between these two extremes. Time changes environment variation; habitat diversity is constantly in a state of flux. Recall that within a large area of changing landscapes, both environmental variation and genetic variation direct life towards adaptation to the many different, changing habitats that are present.

The Biosphere, our planet (earth) can be divided into mega-habitats: the lithosphere (earth's crust—includes soil + bedrock; and upper mantle-plastic molten rock lying underneath); the hydrosphere (surface liquid water covering >70% of the earth's crust); and the atmosphere (10km layer of gasses above lithosphere and hydrosphere). Life often prefers to live on boundaries of contrasting locations (edge effect); for example, we live in the atmosphere at the lithosphere-atmosphere boundary. Most textbooks on ecology, biogeography or environmental science further group the above, according to more or less similar habitats. Lithosphere areas with similar habitats are systematically classified as macro-habitats (biomes) that are based on similar climax communities, which are in turn, based on the climate and the substrate. Examples include: the desert ecosystem (1/3rd of the earth's land area— < 10" or 25cm rain/yr); grassland ecosystem (¼th of earth's land area—prairies or steppes); savannah ecosystem (grassland with some trees); Mediterranean scrub ecosystem (California chaparral); temperate deciduous forest ecosystem; tropical seasonal forest ecosystem; tropical rainforest; northern conifer forest ecosystem (17% of land surface is boreal forest—which includes both taiga or black spruce/ sphagnum moss bog and coastal rainforest); and, tundra ecosystem (permafrost). The rain forest only comprises 3% of the land, but is the habitat of most of the land species. (All forests presently cover 10% of the earth's surface, or 30% of the land area.) Some call the tundra a grassland, even though forbs may be more prevalent. Some call Antarctica a cold desert, yet it is covered with nearly all of the earth's fresh water (ice and snow). The real hydrosphere is systematically grouped into areas with similar communities of macro-habitats (biomes). Examples include the: pelagic marine ecosystem (ocean surface and mid-layers); benthic marine ecosystem (mostly abyssal plain- or marine snow surface; and shelf/near-shore ecosystems—mud, sand, rock +/or coral reef); shoreline ecosystem (mangrove swamp, grass wetland, sandy beach, muddy shore or rocky shore); estuary (brackish water) ecosystem; and freshwater ecosystems (inland swamps, lakes, ponds, grass wetland, bogs, fens, rivers and streams). Recall that between 1913 and 1924, Richard Hesse wrote *Tiergeographie auf oekologischer Grundlage*, which was translated into English by Hesse, Allee and Schmidt as *Ecological Animal Geography* in 1937. This is the classic text on world-wide environmental considerations for animal life; this is the classic text on <u>environmental selection (factors that condition animal existence being favorable in varying degrees at different places on the earth's surface.)</u> Other books can give very good insight into the environmental considerations of a freshwater ecosystem; for example, Ruttner's *Fundamentals of Limnology* gives a comprehensive environmental insight into the conditions of life for a lake, providing a wonderful understanding of an isolated, but diversified habitat.

The systematic grouping of similar communities of macro-habitats (biomes) is useful for discussing general patterns and processes. In broad terms, the general structure of the ecosystems and the kinds of habitats present are similar. But not all freshwater lakes are exactly alike; even lakes of the same surface size, such as deep alpine and nearby shallower lakes in broad erodable valleys (created by damming feeder streams) can have great differences in habitat diversity. And not all deserts are exactly alike; there are extensive areas of different landscapes within the desert biomes of the freezing Gobi and the hot Sahara. Within these varied landscapes are many habitats; some of the habitats are unique. Habitats gain far greater significance as the limiting or positively supportive factors within them play an increasing role in environmental selection. For the environmental biologist, the next step is to understand why and how the conditions of life impact the life itself. This involves the identification of these environmental factors and their subsequent impact upon and within the life contained by the particular habitat. Recall that environmental creation includes and follows environmental extinction events, and that environmental creation can also occur in a more steady-state environment; it may be subtle. Environmental creation does not necessarily involve major environmental change. It only requires isolation and changing conditions of life. And this could even occur in a micro-habitat.

No one talks much about Iowa; no other state of comparable size has lost more habitats or landscape areas of habitats. No other state of comparable size has lost more wildlife (number or biodiversity). Could the same claims even apply in a comparison to the larger states? Between 1830 and 1900 (one lifetime) the tallgrass prairie was converted to cropland. It was one of the most astonishing alterations of nature in human history. Iowa was 85% tallgrass prairie; only 1/1000 remains (0.1%). Many species were lost as habitat thresholds fell. The tallgrass prairie isn't even in one piece; remnants of old cemeteries, school yards, and isolated reserves are all that remain. This represents a severe loss of habitat integrity. While only 20% of the vertebrate species are gone, insects formed the animal base of the prairie food web. Perhaps no one noticed their disappearance; they were only insects (mostly beetles). Prairie vertebrates that depended on the insects for nourishment no doubt noticed (so did flowering plants that depended upon specific insects for pollination). The amphipods that made up much of the base of the prairie wetland food chain are also gone, poisoned by agricultural chemicals. The waterfowl have noticed their disappearance; there is little for them to eat so they cannot stay in the amphipod-poor prairie wetlands. Everything is interconnected. These most important remaining wildlife habitat areas of the tallgrass prairie, these shallow wetlands, are over 99%

gone from Iowa. Iowa wetlands continue to decrease yearly; they continue to decrease yearly in all other states as well. Wetlands in the U.S. are currently being destroyed at a rate of 35 acres per hour. Since the first European settlers colonized North America, well over 50% of U.S. wetlands have been lost. Wetland habitats are home to 43% of the federally-listed endangered and threatened species. Iowa losses extend beyond the disappearance of wetlands in the tallgrass prairie. On the drier prairie, ecological specialists like the prairie chicken depended on fire. When the fires died out, so did the prairie chicken. They did not really go extinct (yet). They are only gone (extirpated) from Iowa. Local extirpations reduce biodiversity. The passenger pigeon is extinct. While some believe that they were hunted to extinction, hunting and (vermin-image) eradication treatment by farmers did not cause as much problem as the real cause, habitat destruction (Dinsmore, Fremling). The passenger pigeon could not tolerate habitat fragmentation; farmers and woodcutters (cutting for lumber and steam-powered fuel) destroyed the expansive hardwood rookeries that were indispensable for accommodating millions of nesters at one time. Generalists like the raccoon live on, devalued by humanity (fur coat protests) and reduced to vermin status (and eradication treatment) from their overpopulation (as a result of high adaptability). Overpopulation is the opposite extreme of extinction; it can also be a major problem in the ecosystem.

39

Overpopulation

Environmental overpopulation occurs when the carrying capacity of the environment is surpassed; a population crash is likely to occur. Before the population crash occurs, widespread environmental damage is prevalent. A historical experiment in overpopulation environmental damage occurred near Hawaii a century ago. Laysan is a small island that is part of the Northwest Hawaiian Island chain. It was first described by early voyagers as a green wonderland, with a coastal sandalwood forest, native palms, several rare species of land birds, and a remarkable assemblage of sea birds. University of Iowa museum director and professor of zoology, Charles Nutting, first visited Laysan Island in 1902 as a scientific advisor on a U.S. Government expedition to explore the Pacific waters around Hawaii. Nutting decided to recreate this incredible paradise in an ecological exhibit at the museum. It took 9 years to arrange a return to the island; museum taxidermist Homer Dill led the return expedition in 1911. Instead of finding a paradise, Dill found the island swarming with rabbits and the entire island's wildlife facing ecological devastation. Dill recreated what he saw, a time capsule of a collapsing environment. A university student returned to the island in 1912 and shot over 5000 rabbits, hardly making a dent in the rabbit population before running out of ammunition. The Laysan Island exhibit opened in 1914; it was the first attempt to recreate an entire ecosystem. It is the only exhibit of its kind still in existence. A University of Hawaii expedition arrived on Laysan in 1923; by then, it was only a desert wasteland, and it was too late to save much of anything. The rabbit population had mostly starved themselves to death and crashed; very few remained alive. Three species of birds were extinct and two more bird species were very close to extinction. Today a vast sandy flat on the north end of the island still has sand castings of sandalwood roots that were once part of a sizable coastal forest. Most native plant species are still absent. Without a full complement of plants, the island is likely drier than it was, meaning the freshwater seeps produce less water for a shorter time. What were the circumstances that could have caused this environmental collapse?

"Laysan is noted for its bird life; early expeditions to the island in the 19th Century estimated the avian population in the seabird colonies at several hundred thousand or even several million. However, the island's ecosystem was all but destroyed by human influence around the turn of the 20th Century. A period of guano harvesting lasting about a dozen years was followed by 'King of Laysan' Max Schlemmer's introduction of rabbits. After he brought the rabbits to the island shortly after 1900 (to provide a source of food and the potential of industry—a rabbit canning business), the rabbits began to eat the island bare of greenery, munching seedlings even as they sprouted. Complaints about this developing problem and about Japanese bird poachers led President Theodore Roosevelt to declare the Northwestern Hawaiian chain a bird sanctuary in 1909. Schlemmer allowed the Japanese to continue exporting illegal wings, so he was removed from the island. The rabbits continued to flourish and multiplied. They devastated the native vegetation; without plants to hold the earth together, much of the soil and sand became loose, and blew about in horrific dust storms. When the (University of Hawaii) Expedition arrived for a month in 1923, only four species of plants remained (of the two dozen plant species originally known to have been present). One of the goals of this expedition was to eradicate any remaining rabbits. Killing the rabbits was easy and quickly accomplished (at) this time. When the last of the starving rabbits were finally exterminated, the bird population had been reduced to about a tenth of its former size; three endemic taxa had become extinct…. Only after nearly every rabbit starved to death and the few remaining ones were eradicated by the biologists in 1923, did the ducks began to recover. The Laysan Duck and the Laysan Finch survive to this day, but are still each an endangered species. The impact of a few decades of exploitation of Laysan for its natural resources is still felt today." (Wikipedia) What can be learned from this?

Perhaps if Max Schlemmer had been allowed to stay, he could have relocated any surplus rabbits to a neighboring Hawaiian Island. This would have bought a little more time and postponed the environmental collapse on Laysan Island. To Max's advantage, it would have even doubled his canning business resource. And as both of these island populations doubled, Max could have relocated excessive numbers of rabbits to a third Hawaiian Island; well, actually, he would need to utilize a fourth Hawaiian island (in addition to the third island). This would have tripled, no, actually quadrupled, his canning business resource potential. Max could have been a very busy, but also a very rich man, as long as he could continue to relocate the rabbits to a neighboring Hawaiian Island. And as surplus rabbits were relocated to the fifth, sixth, seventh, and eighth Hawaiian Islands, Max might have begun to think about finding a bigger island, like Australia; now, there's an idea! Australia was huge (really a continent), big enough to handle any excess rabbit immigration; there was hardly anything else there! And the only down side of this rabbit relocation was a little loss of tourism to Hawaii (as all islands would be now full of rabbits), but that was in the future, and the future can take care of itself. At least this way, the rabbits wouldn't starve to death! There would be no population crash, as long as any rabbits in any overpopulation were relocated. Although there is no record of social problems in the dying rabbit population, there had to have been severe rabbit social unrest as the majority of the population died in the agony of starvation. Without question, after the rabbits overpopulated all of the Hawaiian Islands, the cost and effort of relocating the starving rabbits to Australia would have been far better for the rabbits than letting them experience social breakdown and starve to death. There would have been simply no other option! It is truly a tragedy that no one helped Max help the rabbits; ignoring their suffering was cruel and inhumane! (I must apologize for this tongue-in-cheek scenario, but when I think of those helpless hungry bunnies....) Overpopulation is a terrible problem for any rabbit population.

There is evidence of population-pressure problems seen in the behavior of some overpopulated individuals. Unlike the rabbits that will eat themselves into starvation and total collapse, there may be an evolutionary benefit in some anti-social behaviors, as it will reduce the numbers of overpopulation in a confined environment. For example, Scientific American had an article by John Calhoun on *Population Density and Social Pathology* in 1962. As rat population densities (crowding) increased within pens that provided unlimited food and water, aggression, cannibalism, infanticide, infant neglect and homosexuality became rampant among the overpopulated individuals. This was the first publication that linked overcrowding (overpopulation of a limited environmental space) and social pathology. In another example, there is another small rodent known as a lemming; lemmings are usually found in or near arctic tundra biomes. Lemmings are herbivorous, like rabbits, and like rabbits, they populate rapidly, dispersing in different directions for food or shelter. Rather than having their numbers following a linear growth curve to a carrying capacity or following Gaussian curves of regular predator-prey oscillations (discussed later), lemmings are one of the few vertebrates that reproduce so rapidly that their population fluctuations are chaotic. Said otherwise, they reproduce even faster than rabbits. When overpopulation of the lemming *Lemmus lemmus*, leads to a scarcity of food and overcrowding of habitat, many thousands of the animals will migrate in search of food. Lemmings don't hibernate and so must migrate even during harsh northern winters. The migrations cross mountains, swim across rivers and lakes, and if it is present, eat all the vegetation in their path. Eventually, some reach the sea; perhaps attempting to swim across it, they are drowned. While people may have seen a Disney film showing lemmings committing mass suicide by jumping off a cliff into the sea; in the case of overpopulation, lemmings may be more likely to kill each other than themselves. And while the survival strategies are not the same in the first population-pressure example (confined environment) as in the latter population-pressure example (unconfined environment), this self-destructive migration behavior also has a population-stabilizing end result. Overpopulation causes terrible problems for any population.

As resources dwindle in an environment, fewer and fewer individuals can be supported by that environment. Because the environmental carrying capacity is diminished by the loss of resources, overpopulation of an area can even occur without an increase in population number. Even the insects bug-out when the environment does not provide adequate resources to sustain their growing population. The longest insect migration ever recorded was performed by the desert locust; in 1950 individual swarms were tracked over 2 months, from the Arabian Peninsula over 5000 km to the west coast of Africa. There is also evidence of a swarm of locusts that flew at least 4500 km, from Africa to the east coast of South America and many Caribbean islands in 1988. While any creature may undergo random exploratory behavior while looking for food, with increasing global warming, locusts will increasingly swarm and migrate extensively to new areas. Desert locust hatch coincides with the arrival of the rainy season. Coinciding with the warmest year in recorded history, 2004 saw both decreased rainfall and the worst locust swarms in West Africa in 15 years. Most of the harvest was lost, resulting in the need for extensive humanitarian aid to hold off starvation throughout the entire region. Research has proven that the swarming is caused by (pheromone release- initiated by) overpopulation, and the need for more food or water. With swarming, the locust overpopulation devastates both the area it leaves as well as any area it newly occupies. Overpopulation causes terrible problems for any population, and it causes terrible problems for other populations.

"The migratory locust is the most widespread locust species. It occurs throughout Africa, Asia, and Australia. It used to be common in Europe but has now become rare there. Because of the vast geographic area it occupies, which comprises many ecological zones, numerous species have been described. However, not all experts agree on the validity of some of these subspecies. The main ones are: *Locusta m. migratoria* (West and Central Asia + eastern Europe), *Locusta m. migratorides* (mainland Africa and Atlantic islands), *Locusta m. capito* (Madasgar), and *Locusta m. manilensis* (eastern Asia). They transform behaviorally and physically under the effect of overpopulation (stress) and thus are called polymorph insects. There are two main phases, the solitary phase and the gregarious phase. As the density of the population increases, the locust transforms progressively from the solitary phase to the gregarious phase, in intermediate phases: Solitare = solitary phase to transiens congregans (intermediate form), to gregarious phase, to transiens dissocians (intermediate form), to solitare = solitary phase." (Wikipedia)

Darwin's fatal competition for limited resources mostly occurs between members of the same population; it primarily increases fitness within that population, unless essential resources are depleted from the environment. As individuals of an environmentally well-adapted population increase in number, competition for resources increasingly becomes more of a limiting environmental factor. Any environmental advantage for an individual places other members of the population, without the environmental advantage, at a life-threatening disadvantage, jeopardizing their survival as numbers continually increase and resources continually decrease.

In the wilderness, the problem of animal overpopulation is solved by predators. Predators tend to look for signs of weakness in their prey, and therefore usually tend to first eat the old or sick animals. This controls overpopulation and ensures a strong stock among the survivors. Predators balance the ecosystem. In the absence of predators, animal species are bound by the resources they can find in the environment, but this does not necessarily control overpopulation. In fact, an abundant supply of resources can produce a population boom that ends up with more individuals than the environment can support. In this case, starvation, thirst, and sometimes violent competition for scarce resources may effect a sharp reduction in population during a very short lapse (a population crash).... Rodents are known to have such cycles of rapid population growth and population decrease. Some animal species seem to have a measure of self control, by which individuals refrain from mating when they find themselves in a crowded environment. This voluntary abstinence may be influenced by stress or pheromones. In an ideal setting, when animal populations grow, so do the number of predators that feed on that particular animal. Animals that have birth defects or weak genes (such as runt of the litter) also die off, unable to compete with stronger, healthier animals. On occasion, an animal that is not native to an environment may have advantages over native ones, being unsuitable for the local predators. If left uncontrolled, such an animal can quickly overpopulate and ultimately destroy its environment. (Wikipedia) **Without some form of population check or means of dispersal to unpopulated areas** (to relieve a confined environment's population pressure), **accommodating numbers beyond the carrying capacity of the habitat will lead to environmental destruction and an inevitable population crash.**

40

Humanity

When Thomas Malthus wrote *An Essay on the Principle of Population* in 1798, noting that human beings reproduce at a rate far outstripping food supply, and their population grows until resources are exhausted, is this really true? Do the views of Malthus have any relevance to today's population growth? Could it be that the logical consequence of overpopulation and irresponsibility in environmental resource consumption is a decrease in the quality of life? Or will technology continue to work miracles? For humanity, the environmental limitation of increasing population may be a valid concern. Geometric human population growth has impacted the human and the non-human environment. Man's presence excessively disturbs the environment; increasingly overwhelming populations increasingly overwhelm the environment. Failure to perceive overpopulation as a problem for humanity does not resolve a potential threat; it only worsens the problem. There is current ecological concern for our small, fragile, beautiful, and rare planet, our life support. Minnesota geologist Roger Hooke estimates that current human earth-moving activities now rival natural geological processes. Can we really manipulate nature enough to change fundamental ecological laws? This is unlikely; we have been very good at buying time. Limiting factors may be temporarily out-maneuvered, but they never go away. **The conditions of life will control the destiny of humanity.**

There is global good news for mankind's conditions of life in recent time. Public health and medicine have increased mankind's survival today to levels that were unimaginable in the time of Malthus. Over the past 50 years, science/technology breakthroughs increased the world gross national product (GNP) tenfold (10×). Since World War II, the average real income in developing countries has doubled (because of outsourcing). Poverty rates have decreased more in the past 50 years than in the previous 500 years. Malnutrition declined by 33%. Food production has experienced the green revolution and gene revolution; there is enough food (today). Mankind is coming to realize that food security has more to do with culture, government (and corruption), equitable distribution, and economy, as with food availability. Access to safe drinking water went from less than 10% to 75%. Child death rates halved. Life expectancy increased 30%. (The HDI is used by the U.N. to rank member states according to life expectancy, education, and standard of living.) In 1960, ¾ of the world population lived below a Human Development Index (HDI) of 0.5. In 2000, less than 1/3 were below HDI 0.5 (half of the poor ¾ developed).

Although there is enough food production today, reserves are becoming conspicuously low. Australia, a major producer (#2 worldwide) of the world's wheat supply, is experiencing the worst drought in 100 years. Prices are on the increase. From March 2007–March 2008, wheat has risen in price 130%. In the same time period, soybean products have increased in price 87%, rice has increased in price 74% and corn has increased in price 31% (BBC news). The U.S. is using corn to produce ethanol to reduce dependence on foreign oil. In the same time period, oil went from $57 a barrel to $117 a barrel; this is a main reason for the food price increase. Petroleum is the energy source of agricultural mass production and distribution; it is the raw material for agricultural chemicals. Looming shortages of petroleum and food, combined with an ever-increasing population and ever-increasing standard of living, is creating a supply and demand problem. BBC news said there was only a 40 day food reserve in early 2008. They also estimated a 50% increase in food demand by 2020. While the Green Revolution and Gene Revolution may have provided for the current tenfold population increase, there is no assurance that food production can meet the increasing demand. Cropland and fisheries are on the decline. Food production may have already hit the wall.

The Millennium Assessment of 2005 was compiled by 1,300 researchers and 95 nations over 4 years. It reports that humans have changed most ecosystems beyond recognition in a dramatically short space of time. The way society has sourced its food, fresh water, timber, fiber and fuel over the past 50 years has seriously degraded the environment. Of 24 evaluated ecosystems, 15 are being damaged; said otherwise, over 60% of indexed world ecosystem services have been degraded. Species extinction is now 100 to 1,000 times above the normal background rate. The considerable gains in economies and food production have continued to grow; but, successes achieved have placed global prosperity in the future at risk. The improved economy and food production resulted in a population explosion in the underdeveloped countries. This population explosion has been responsible for extensive degradation of the environment and essential environmental resources. This is an unstable situation. Progress achieved in addressing goals of poverty and hunger eradication, improved health, and environmental protection is unlikely to be sustained if most of the ecosystem "services" on which humanity relies continue to be degraded. In other words, the short-term gains already achieved could become a great long-term loss for humanity, because of overpopulation and environmental destruction. In a finite environment, improvement in the conditions of life cannot be followed by increased population without negating the improvement in the conditions of life. The Human Development Index is now sharply declining in areas with irresponsible population increase. Sub-Saharan Africa has the highest birth rate, the highest rate of population increase and the lowest use of contraceptives of any major region in the world. Very recently, 12 of the 18 countries that have suffered negative Human Development Index reversals between 1990 and 2003 were in sub-Saharan Africa, with Southern Africa "hit hardest." South Africa has plunged by 35 places to 120 on the global Human Development Index (HDI), Zimbabwe fell 23 places and Botswana dropped 21 places. Reversals were also noted for Lesotho, Swaziland and Zambia. Irresponsible overpopulation decreases the quality of life. The Millennium Report of 2005 is essentially an audit of nature's economy; the audit shows that most of the accounts have been driven into the red. If resource economy runs into the red, ultimately there are significant consequences for mankind's capacity to achieve dreams of poverty reduction and prosperity. Two services, fisheries and fresh water, are now well beyond levels that can sustain current, much less future, demands. Predation (harvesting) by human hunters and fishermen can only hit wild populations so hard, without losing those populations (causing extinction or near-extinction). When overpopulation's demands exceed sustainable harvest levels, continued harvesting depletes the resource. This loss of a valuable resource has happened many times before in the past. Long lines, gill nets, bottom and/or water-column trawlers, fish traps, and sport fishing have combined effects and have depleted, or almost depleted, over 75% of global fish stocks; this is reminiscent of market hunting a century ago. According to BBC news, if current trends continue, scientific studies indicate that there will be virtually nothing left to fish from the seas by the middle of this century. Most damaging of all is the loss of diversity resulting from fishing-related habitat destruction. Farm-raised fish represent a hope for future food from the sea; however, aquaculture waste is a major source of nutrient pollution. In addition to over-fishing methods damaging the ecosystem, nutrient pollution and eroded topsoil suspended in stream water adversely change both the freshwater and marine ecosystems, further endangering the life within.

The nutrient pollution of today is mostly man-made. More than half of all the synthetic nitrogen fertilizers (first made in 1913) ever used on the planet were deployed after 1985 (damaging water supplies and world ecosystems). Nutrients, like nitrates (NO_3-) and phosphates ($PO_4\underline{=}$), occur naturally in water, soil and air. They are used in fertilizer to aid the growth of agricultural crops; excessive applications to agricultural land often amounts to double the appropriate amount that can be utilized by the crops. This is wasteful and expensive. Worse yet, the result is that the excess fertilizer runs into streams, rivers, lakes, bays and the oceans. Although nutrients are vital to the growth of plants within the streams, rivers, lakes, bays and oceans, excess fertilizer in the water causes serious excess nutrient problems. Excess amounts of phosphates and nitrates cause rapid growth of algae (phytoplankton), creating high/dense populations, or algae blooms. The turbidity produced from the bloom greatly decreases photosynthesis in the water below. These blooms become so dense as to block sunlight very near the surface (think of the darkness of a forest floor); this reduces the amount of sunlight available to any submerged aquatic vegetation. The increased bloom metabolism decreases the dissolved oxygen. Dead cells and wastes throughout the water column undergo aerobic decomposition, this leads to further oxygen loss throughout the water column. The loss of sunlight can even kill the grasses and other submerged vegetation on a lake or shallow coastal bottom; algae may also grow directly on the surface of underwater vegetation. Then the dead submerged vegetation will decompose; the high rate of kill leads to additional anoxia of the deeper layers (anaerobic decomposition). Excess numbers of surface layer algae that are not consumed in the food chain/web will ultimately sink and also be (anaerobically) decomposed by bacteria, still further depleting bottom waters of oxygen. There is a problem for aerobic aquatic life survival when oxygen decrease of the upper layers is not offset by increased photosynthesis. As nutrient levels increase, aerobic life dies in increasing numbers (anaerobes survive). Most fish populations are highly sensitive to depressed levels of oxygen.

Nutrient pollution has led to anoxic, excess eutrophication of inland waters and coastal dead zones. There is a permanent year-round dead zone at the mouth of the Mississippi River, in the Gulf of Mexico; it's a 7000 square mile area of low-oxygen, caused by fertilizer nitrate overuse in the Mississippi/Missouri/Ohio River drainage. Lake Erie, the shallowest of the Great Lakes, also has a similar fertilizer toxicity problem, with a "dead zone" 100 miles long each summer. Elsewhere, the largest coral reef (largest living organism) on the planet is dying. Agricultural sediments and nutrient pollution are combining with thermal pollution to kill the Great Barrier Reef in Australia. In addition, a starfish called the crown of thorns, in overwhelming numbers, is physically eating the coral base of this reef. There is evidence that the increasing starfish numbers are enabled by regional over-fishing and nutrient pollution from regional agriculture. Nearly half of all tropical coral reefs are dead or dying from thermal pollution, nutrient pollution, or inappropriate contact. Coral reefs and mangroves are the tropical fish nurseries; both are key habitats, essential for regional fish population viability. If these key habitats disappear, the fish will also disappear. Twenty-five years ago the Black Sea ecosystem production (commercial fishing) was 25 times that of the Mediterranean. The top 500' (133m) of brackish water above the thermocline was rich in the right amounts of nutrients carried by the rivers into the drainage basin. The Danube River now carries chrome, copper, mercury, lead, zinc, and oil to the Black Sea at 20× the levels that the Rhine carries the same pollutants into the North Sea. Since 1970, the Danube nitrate (NO_3-) and phosphate ($PO_4\equiv$) levels have respectively increased 6× and 4×. The Dniester River (from Ukraine) has increased nitrate (NO_3^-) and phosphate ($PO_4\equiv$) by 700× during the same period. Eutropification and toxification have caused Black sea fisheries to fail abruptly (1986–1992 < 10% catch) in the new dead zone above the thermocline.

Fresh water concerns relate to both quality and quantity. Ocean water cannot be used to drink or grow crops. BBC news notes that it takes 1–2 liters of water to grow a gram of wheat. Contaminated water cannot be used to drink or grow crops. The fresh water available for use by humanity is surface water and underground water. Serious problems occur when these supplies become unavailable and/or contaminated. Lack of fresh water can be life-changing. The source of surface water and ground water is precipitation; many large populations are dependent upon glacier and/or snow melt from mountainous areas as their primary source of water during the summer. As the glaciers disappear with global warming, this crucial environmental resource will literally dry-up. The Himalayan glaciers alone are the main source of water for 40% of the world's population. The Himalayan glaciers form the Indus, Ganges, Brahmaputra, Salween, Mekong, Yangtze, and Yellow Rivers. Summer snow melt water shortages already impact populations beyond Asia. Snow melt that was a reliable summer-long source is now melting earlier each year in the spring and disappearing earlier each summer. This is already a serious problem in South America and the drought-impacted Southwest United states. Southern California's population is presently dependent on snow-melt for 75% of their water; there has been very little snow in California lately. For some reason, there may even be a developing drought in the Southeastern United States; these southern droughts are beginning to impact the area's population.

The edges of the equatorial rain belt are seasonal, with summer rains (e.g. Tropic of Cancer). The North Atlantic is usually a summer high pressure area with calm seas. In an extremely cold summer, this high pressure area drifts further south. The southern movement of high pressure also moves the band of summer rains further south. Many areas that normally get summer rains do not get rain when the above circumstances occur. The deprived areas experience drought, crop failure and famine. Northern Central America, in the 9th century AD, had the worst drought in 7000 years. It was the result of the southern shift of summer rains, which was caused by the southern movement of high pressure in the North Atlantic. This southern shift of summer rain produced a major climatic impact on civilization; the Mayan culture, America's most advanced to date, starved and collapsed. By today's standards, the human population density at this time was very low. What are the logical consequences of drought (and the inevitable crop failure that follows) in an area of overpopulation? The environmental effect of drought has caused many civilizations to collapse when food production fails. Starving people eat seed grain and slaughter breeding stock (violent anarchy and/or war is likely). Drought caused the fall of the Egyptian old kingdom (2200 BC); the same drought brought catastrophe to West Asia and Europe. A huge Egyptian lake a in a bowl-shaped valley called the Faiyum Depression just dried-up; the Nile ceased to flow. Dead Sea studies show that around 4,000 years ago, its level dropped 100m in one century. The 2200 BC climatic change (drought) was likely due to severe cold spells more characteristic of an ice age than of the warmer conditions that prevailed before the event; in other words, this pattern parallels the southern shift of summer rains previously discussed. This spell of severe cold caused a shift to a drier climate in southeastern Europe. The first permanent settlement in the United States was impacted by a far lesser drought; but this was enough to be a major cause of the Jamestown starvation/ Algonquin conflict (1604–1624 AD). The Mississippi valley has been unusually wet for last 700 years; was the Little Ice age (1200AD–1900AD) responsible for this? Is there a major drought ahead for this area? Previous droughts were more frequent and could last for centuries (Fremling— *Immortal River*). The only droughts anyone in this area can relate to are insignificant; they involve the little drought cycles of the 1870's, 1900's, 30's, 50's, and 70's. The drought periods of the 80's were even less severe, but did contribute to the farm crisis of the 1980's. Population levels today are far higher than in the past (10× historical average of the last 2 millennia); once again, what are the logical consequences of the next significant drought? Climate changes resulting in drought are inevitable. Droughts have occurred many times in the historical past; they will certainly reappear in the future.

Ninety percent of all precipitation occurs at sea. On land, plants form the major contribution to the local water cycle. Transpiration cycles 75% of the annual rain on land in the tropics. In temperate climates, 97–99% of plant uptake is lost as transpiration (e.g. smoky mountains). Cutting climax community vegetation breaks down (short-circuits) the local water cycle; that is, cutting native vegetation eliminates transpiration, creating bare soil (only captures 20% of rainwater for groundwater compared to 100% vegetation capture). The loss of forested areas caused a permanent regional climatic change in Lebanon (desertification). The Epic of Gilgamesh is the earliest recorded story of desertification caused by the extensive destruction of forestlands. Lebanon went from more than 90% forest (the famous Cedars of Lebanon) to less than 7% over a 1,500-year period. Trees and their roots are such an important part of the water cycle that rainfall downwind of the deforested areas decreased by 80%. Over time, millions of acres of land in the (once) Fertile Crescent area turned to desert or scrubland, and remain relatively barren to this day. A similar process of desertification happened in the Southwestern United States for the (13th century) Anasazzi. The American Southwest is in a drought today; it is nearly as dry as it was 700 years ago. If the present forest destruction continues worldwide, desertification could even occur in the Amazon rain forest. In other words, deforestation of a native forest causes a regional drought—no matter where it occurs.

Inappropriate irrigation depletes underground aquifers of waters at a rate that cannot be sustained, and it also dries-up stream flows. This eventually creates salt deposition in the soil, due to salts dissolved in the underground water and flowing surface water. In addition, many soils have salt just below the surface, which is a residue from evaporation of ancient seas and lakes. In some locations, like Australia, just removing present native vegetation will cause the salt to appear in the surface soil (Diamond—*Collapse*). Saline soils are unsuitable for food production. Australia is experiencing the worst drought in 100 years; it is caused by a combination of removing native vegetation, overgrazing, inappropriate agriculture and over-irrigation. The Murray-Darling River Basin is the main irrigation area of Australia, and covers an area the size of France and Spain. The prolonged drought has reduced many of the rivers to a trickle, crippling Australia's farming sector and forcing many cities and towns to enact drastic water restrictions as reservoirs dry up. If it doesn't rain in sufficient volume soon (unlikely), there will be no water allocations for irrigation purposes in the basin until May 2008. Although there will be drinking water for city populations, Australia's agricultural foundation and economy are seriously threatened. Global warming will only worsen the Australian drought crisis.

Cutting native vegetation causes loss of topsoil (organic-topsoil holds water best). The dust bowl (1930's) was a combination of drought plus inappropriate agriculture. The loss of natural vegetation and topsoil led to the dust bowl problem; however, it did not occur in areas with adequate topsoil (e.g. Iowa). More land was converted to agriculture since 1945 than in the 18th and 19th Centuries combined (24% of all land is now under intensive agriculture; another 16% is used for grazing). Agricultural production is at a historical maximum, but there is little room for any increased food production as formerly productive cropland begins to decline. Agriculture becomes too intensive as populations swell in number. Areas unsuitable for crop production are planted in preference to experiencing a food shortage. As marginal areas are planted, topsoil is lost at an alarming rate. If sustainable farming techniques are not practiced, topsoil can be lost at an alarming rate even in non-marginal lands. When the topsoil goes away, the ability to grow crops will also go away. Silt (topsoil) is the major polluter of streams worldwide. In streams and adjacent offshore areas, massive quantities of waterborne silt (topsoil) can shut down photosynthesis within a foot (30cm) of the surface and greatly worsen other water pollution problems. Agriculture, the basis of civilization, is totally dependent upon topsoil. Topsoil is necessary to support viable crops with needed organics, mineral nutrients, retain moisture and maintain cohesiveness. Once gone, it cannot be regenerated quickly enough to make a difference. Once food production fails, society fails. It happened in the Fertile Crescent and it is happening in Haiti today; it is happening worldwide. It appears as if past lessons learned, like the topsoil breakdown from overgrazing marginal land, were really never learned, or were ignored, or were forgotten.

The logical consequences of irresponsibility in environmental resource consumption are directly related to an experience called the "tragedy of the commons" (Garrett Hardin, Science 162:1243, 1968). The tragedy of the commons also exemplifies the logical consequences of irresponsibility in overpopulation. Fourteenth-century Britain was organized as a loosely aligned collection of villages, each with a common pasture for villagers to graze horses, cattle and sheep. There are no problems as long as the number of animals is small in relation to the size of the pasture; i.e. the carrying capacity of the pasture is not exceeded. Each household attempted to gain wealth by putting as many animals on the commons as it could afford. As the village grew in size, more and more animals were placed on the commons; overgrazing ruined the pasture. No stock could be supported on the commons thereafter. As a consequence of population growth, greed, and the logic of the commons, village after village collapsed. Even if greed were not an issue, continued village population growth would eventually have had the same result. Unfortunately, greed is always an issue. Greed, environmental resources, and overpopulation require better management.

41

Do the Math

The tragedy of the commons strongly suggests that it is not possible for an area with an out-of-control overpopulation to achieve sustainability; it graphically illustrates the consequences of irresponsibility in environmental resource consumption. Agriculture (24% of all land is now utilized for maximum-intensive agriculture, and is rapidly losing essential topsoil in nearly every location), pastoral lands (16% of all land is used for grazing), fisheries, fresh water and energy resources are now pressing or passing beyond realistic sustainability limits. The over-utilization of these important resources is already degrading the environment. Sustainability is a worthwhile goal, but sustainability cannot happen with an overpopulation that surpasses sustainable limits (even with reduced consumption). Just over a century ago, world population was less than one billion people; this seems to have been the world's maximum carrying capacity throughout the Little Ice Age (1200–1900AD) and basically at all time previous to that time. A half-century ago world population was 3 billion people. It would have been 9 billion today if the 1st and 2nd world had not begun to control overpopulation in the past half-century. Nevertheless, today's population is over 6 billion people. Seventy percent of people living today will still be here in 2050; so will be their many, many offspring. Irresponsible human overpopulation is no longer sustainable; the numbers just do not add up. Irresponsible human overpopulation can only lead to an unstable situation, a world of steadily increasing suffering and death. It is not at all surprising that the areas with the greatest growth in population density tend to be the areas with the greatest problems. Recall that **as individuals of an environmentally well-adapted population increase in number, competition for resources increasingly becomes more of a limiting environmental factor.** Overpopulation overtaxes resources. **All biologic and physical changes modify the degree of an individual's environmental compatibility, which continually varies the outcome of environmental selection.**

Quantitatively, environmental selection at a given time varies directly with the sum (+/or intensity) of environmental limiting factors; the population number (of individuals) varies inversely with the sum (+/or intensity) of environmental limiting factors.

The human environmental impact on an area varies directly with the total human population of the area; it is an additive effect. The human environmental impact on an area varies directly with the average resource consumption and waste production of the human population members; it is also an additive effect. Overpopulation and resource availability are inversely related. Overpopulation and the standard of living are inversely related. Overpopulation occurs in the poorest areas with the fewest resources. The logical consequence of increased overpopulation in these overpopulated areas is increased environmental damage. The logical consequence of increased resource consumption (whether it is in overpopulated areas or less-populated areas) is increased environmental damage. The logical consequence of increased environmental damage is increased population damage. Areas with an out-of-control overpopulation can never support an increased standard of living (Diamond—*Collapse*). Talk of sustainability in an out-of-control overpopulation is an impossible dream (Diamond—*Collapse*). It does appear as if developing overpopulated countries are primarily concerned with their economic growth; there appears to be little evidence of concern for the environment in the most populated areas. It will do little good for developed nations (with small declining populations) to increase environmental responsibility, while far greater numbers in overpopulated developing countries are disproportionately increasing their populations and increasingly compromising the environment.

It is understandable that every couple wants a family; the problem is with sustainability of large families in areas with existing overpopulation. India may overtake China as the world's most populous nation, surpassing the 2 billion mark in 2025. With less than half that number (< 1b today), the continuing depletion of aquifers, where water tables are falling annually by 1–3 meters over much of the country, illustrates the impossibility of sustainability if the population doubles in number. Today there is also a shrinking cropland area per person problem in India; more than half of the children in India are already malnourished and underweight. At the time of this writing, the population rate of increase in India is more than three times that of China. Most of the population growth is in the (poor) rural areas; some population control is beginning to occur in urban areas, and urban area economies are greatly improved. No crystal ball is needed for a view of the future in India if the population continues to increase at the current rate. Dwindling cropland also threatens food security in Pakistan, Nigeria, Zimbabwe, Somalia, Ethiopia and the Sudan.

Exceeding the land's carrying capacity will not allow the resident population to produce enough food to support their population. There is a progressively worsening, unprecedented, extreme environmental crisis, due to increasing irresponsible overpopulation, developing in a region that includes both Nigeria and Ethiopia. Spanning the entire continent, an enormous swath of equatorial Africa called the Sahel (portion of sub-Saharan Africa stretching from coast to coast) is undergoing very rapid desertification. The irresponsible overpopulation of this area is contributing to this desertification, due to the clearing of forest for agriculture as well as for firewood. In 1900, 40% of Ethiopia was covered by forest; now only 4% is forested. Today 1/3rd of the earth's land mass is desert; the deserts are rapidly increasing in size. Desertification can happen nearly anywhere, even in the rain forests of Africa and South America.

Humankind is presently degrading their environment, both at their expense and at the expense of other living creatures. The just-mentioned drought in Africa is causing harsh times for both humans and wildlife. The present conditions of life are now strongly influenced by human overpopulation. The conditions of all life could even take a far more significant turn for the worse, if the members of society in the overpopulated areas do not respond responsibly. Society in the overpopulated areas is at an increasing risk of disastrous collapse; for example, in past overpopulated societies, collapse was predictable and inevitable. **The environment ultimately determines the carrying capacity, the standard at which human overpopulation occurs**. Diamond discusses this in his book on *Collapse*; present Viking populations in Greenland were unsustainable when the climate changed and it became colder. At that time, the Vikings were the toughest guys on the planet; nothing could stand against them, except a little cold weather (Little Ice Age—1200–1900AD). Climate dominates the environment; for mankind, the climate is now about as good as it gets. But global warming is expected to rise another 3 degrees Celsius (5°F) by the end of this century.

The worst example of an unsustainable population growth tragedy is yet to come; it will certainly come and it will happen at the location of the highest population growth on the planet. No matter what is done, the numbers lead to only one outcome, that of looming, predictable, environmental disaster, directly followed by the most severe human disaster in recorded history. Something bad will happen to people that have already suffered too much. This amount of suffering will get worse, much worse. At this point, only the degree of death and suffering can be managed. The only possible avenue of damage control is population control, today. Present population control patterns appear to make this only hope unlikely; however, it is the most realistic possibility of offsetting the impact of an impending holocaust of humanity. This forthcoming human holocaust will be far more horrific than any form of Draconian birth control imaginable. It will be environmental selection in a most brutal form. It will ultimately control irresponsible overpopulation, in a Malthusian manner. When it happens, the human suffering and death will be on an unprecedented scale. This worst example of an unsustainable human population growth tragedy is the downside of global warming. .

Global warming is inevitable; and it is not just due to greenhouse gas emissions, which are spiraling increasingly out of control as presently overpopulated countries raise their standard of living. Conservation of the few in the developed nations cannot possibly offset the increasing per capita consumption and directly-related greenhouse gas emissions of the many in overpopulated countries. There should be a sincere effort to conserve any and all possible resources worldwide. Nevertheless, the Milankovitch Cycle is going to markedly increase global warming, perhaps even further than any man-made greenhouse gas effect on increasing temperatures. Recall that the earth is now approaching the closest extreme of the 100,000 year cycle, which now occurs mainly in the southern hemisphere, due to the seasonal position in the orbit; however, the difference in distance is only 3%, and this distance is steadily decreasing. The calculated outcome of this is an inevitable global warming in the Northern hemisphere. As an analogy, this situation is parallel to mid-June of a seasonal year; there could be a long, hot summer ahead, so to speak. Also recall that the 23,00 year precession portion of the cycle is giving the northern lands more summer sun exposure today; i.e., the North Pole is aimed closer to the sun in the present direction of wobble. The calculated outcome of continued orbit rounding and precession is an inevitable increased intensity of sun exposure to the northern hemisphere; land masses are to the North. Global warming and global cooling primarily depend on sun exposure in the Northern hemisphere (and greenhouse gases).

The northern hemisphere is now warmer than it has been for 3000 years. It will only get warmer, because further rounding of the ellipse and precession can only result in a prolonged increase in the intensity of global warming. In 3000 years the orbit will be very round, and precession will still be in a northern direction, although somewhat less so. Both the summers and winters will be hotter than they are today (more sun intensity). What this Milankovitch Cycle prediction means to mankind is that even if there is no further increase in greenhouse gases, increased global warming environmental impact will certainly occur, and that this increased global warming outcome is inevitable. Recall that even the Milankovitch Cycle itself is increasing greenhouse gas levels in the atmosphere. For mankind, any reduction in the generation of greenhouse gases will help to lessen the severity of the global warming; no one will say that further increases in greenhouse gases are inconsequential. Unfortunately; greenhouse gases (e.g., CO_2) are geometrically increasing with expanding populations and production-related improved human development conditions (standard of living); this only worsens the outlook for a looming environmental tragedy.

No area will be environmentally harsher on humanity than sub-Saharan Africa; it is unavoidable. In this location, the only possible outcome of further global warming is increased drought, thirst, crop failure and starvation, and on an unprecedented scale. Sub-Saharan Africans already know about drought, thirst and starvation; it is already here. The drought of today is only significant in comparison to recent decades; the current rapidly-growing population is already well beyond environmental sustainability. As global temperatures increase, soil moisture will decrease. The great drought to come will be far worse than the droughts of recent history; it will be a drought, the likes of which have never been seen for over a thousand years. Imagine the heart-wrenching images of starving humanity frequently seen on television; then imagine the images of an environmental crisis that is far more severe than that of today. Today the suffering and death in African refugee camps is caused by the growing imbalance of population and resources, accompanied by the predictable anarchy and war, with people further ravaged by disease from overcrowded conditions. And the heart-wrenching images seen on television today are only a hint of what is to come. This upcoming carrying capacity adjustment will be most severe; suffering and death will surpass anything seen in human history. There are four times as many people here as there were 50 years ago; numbers are increasing exponentially. This is the fastest growing population on the planet. The topsoil is well on its way to becoming the worst topsoil on the planet. There cannot be a happy ending here. If population continues to increase at the present rate, it is possible that unless all these people can relocate, nearly all of the people here could die. Where will they go? Worldwide, it appears as if the areas most heavily populated will be most affected by global warming, embracing the inevitable future, severe drought. India has tripled its population in the last 50 years; overpopulated rural western India's drought is only beginning. Even presently-productive areas in the US could lose most of the soil moisture, decreasing production and increasing wildfires. Worldwide agriculture and worldwide economy will take a turn for the worse. BBC news estimates that there will be 50% loss of world crop production due to global warming alone in this century. They also estimated a 50% increase in (population-related) food demand by 2020. The numbers are there; the numbers do not lie. As global warming involves the entire planet, there will be winners and losers. Greenland (glaciers gone) will (temporarily) become green again; global warming will be even more dramatic at higher latitudes.

Sea level will rise with global warming. With global warming, rains will shift to higher latitudes and melt landlocked glaciers. Sea water expands as it warms. As global warming progresses, flooding of coastlines will progressively worsen. Populations everywhere are concentrated on coastlines. Sea level today is rising at 2mm (possibly 3mm) per year; tropical mangrove wetlands are matching this rate by forming new land. But tropical mangrove wetlands are rapidly disappearing to make room for cities, housing and fish farms. Estuary wetlands, like mangrove swamps, will also form new land because of decreasing river water velocity. But estuaries are favored locations for cities and are often dredged for transportation, as in the Mississippi delta. Dredging for shipping channels increases flow velocity; increased velocity increases erosion. Wave action, from traffic wakes, also increases erosion. As world populations increase, further development is increasingly concentrated on pricy beach-front property, supposedly protected by seawalls. Barriers, like seawalls, increase the intensity of storm erosion in their periphery. Eroding seaside beaches are increasingly developed for recreation, and then have to be continually replenished with more sand. Winter storms move the sand deeper and can form longshore bars in shallows; however, this sand will relocate to the beach in the summer, providing that barriers, like seawalls, aren't constructed to increase the intensity of erosion. People unwisely build beach-front homes on barrier islands; this island is the transient longshore bar, and a seawall here will make the island disappear. It's a matter of understanding both geological processes and the physics of wave action. There is an inadequate understanding of estuary and coastline hydrodynamics, to the population's peril. What was once land-forming mangrove shoreline, land-forming estuary, and stable beach, will be the areas the sea will first invade the land. This environmental Russian roulette is similar to inviting tragedy from the inevitable flooding of a population that was foolish enough to build commercial property and housing in a river floodplain (floodplains will eventually flood with increasing severity since levees and dams will raise the river-bed). So, as land-locked glaciers melt and warmer seawater expands, global warming will certainly flood the most populated seaside areas with salt water, just as earlier global warming (following the last ice period) flooded the Mediterranean Sea and Black Sea in ancient historical times. Cleopatra's palace, in old Alexandria, is completely underwater. As with Cleopatra's palace, future visitors to Anchorage, Seattle, San Francisco, Los Angeles, Houston, New Orleans, Mobile, Tampa, Miami, Jacksonville, Savannah, Charleston, Newport News, Boston and New York City will need underwater diving equipment. Florida will cease to exist, as will Bangladesh. If Pleistocene sea levels mean anything, and they do, the entire East Coast of the United States will be flooded under water extending inland for hundreds of miles. This happened 6000ya (Milankovitch Cycle) and it will happen again; unfortunately, since that time greenhouse gas levels have doubled. Today, two out of three people live less than 100km from the coast; nearly all will be displaced. This will destabilize society.

Global warming increases El Nino Pacific warming, which increases incidence and/or intensity of Atlantic basin hurricanes. Record-breaking numbers and intensities of 2004 hurricanes were surpassed by the 27 hurricanes in 2005. At one time thought impossible, a South Atlantic hurricane, Catarina, struck Brazil in March of 2004. El Nino Pacific warming will also increase the frequency and intensity of typhoons (Pacific hurricanes). In 2004 (20 typhoons—7 severe), Japan and Australia experienced hurricane extremes never before seen (2005: 16 typhoons—7 severe). As global warming increases in degree, tropical storms will increase in number and intensity (2 category 5 storms hit the East coast of Central America in 2007; there was another hurricane on the West Coast when the 1st one struck.). Further inland, similar increases in the number and intensity of storms should be expected. In 2004, an all-time record high for tornadoes in the U.S. occurred. This will further destabilize society.

Only a climate change like that seen following Europe's Medieval Warm Period (700–1200 AD) could be a worse scenario for humanity. That climate change was "The Little Ice Age" (1200–1900 AD). Recall that a century ago, world population was less than one billion people; this may have been the world's maximum carrying capacity throughout The Little Ice Age. If this type of climate change were to occur again, productive countries, presently helping to offset food shortages in irresponsibly overpopulated countries, would suddenly be unable to feed their own populations. There will be little hope of reciprocal aid from today's irresponsibly overpopulated countries, already enduring a prolonged, severe population-resource crisis. As their overpopulation increased, limited availability of life-saving resources for individual population members would have greatly decreased, limiting further population growth in the form of a horrific population crash. It is likely that they will still be overwhelmed by the crisis from the great (global warming) drought. Today, for example, most of these irresponsibly overpopulated countries still cannot adequately manage their own population sustainability; how could they possibly achieve self-reliance and help others? In spite of massive support from developed countries, they only continue to increase in population and continue to request more aid. This could never be a stable situation.

The Medieval Warm Period (700–1200AD), a small global warming, ended with the melting of northern glaciers, breaking down the Gulfstream circulation and the introduction of less saline water in the northern end of the stream. The northern end less-saline water then no longer sinks to supply cold water for this part of The Deep Ocean Conveyor (thermohaline circulation); this then decreases Gulfstream momentum. An even more severe episode of Gulfstream breakdown occurred when a glacial freshwater dam broke and flooded the North Atlantic 8200 years ago, resulting in a sharp, and long, global temperature decrease; it was far colder and lasted far longer than The Little Ice Age. At this time long ago, mankind suffered a great deal, but still survived in small numbers, if reliable food sources were available. So, it is likely that the Medieval Warm Period (700–1200 AD) caused the Little Ice Age (1200–1900 AD), due to glacial melting, which caused decreased salinity in the Gulfstream, which then decreased the feed of warm water to the west coast of Europe, and caused the failure of agriculture (Berlin is at the same latitude as Edmonton). It is interesting to note that the (cooler) Dark Ages, or Great Retreat of European Agriculture (500–700 AD), preceded the Medieval Warm Period; there was even a Roman Warm Period before the cooler Dark Ages. There seems to be warming and cooling cycles within our present interglacial period. Recall that we are still living in a warm interglacial period during an ongoing two and one-half million year ice age (ice age caused by continually decreasing atmospheric CO_2 levels). Recall that the Gulfstream has only recently slowed and directed 30% of its flow towards Africa, warming the Atlantic basin and increasing hurricane activity. Recall that if global warming continues, and it will, for every degree Fahrenheit the Earth warms, the earth will experience an additional one percent of increased rainfall. This rainfall will not be evenly distributed; previously noted, there will be marked changes from previous patterns. In general, recall rainfall will increase more with increasing latitude, diluting Gulfstrean salinity and melting northern glacier ice, further diluting the salty ocean water with the previously land-locked glacial meltwater. Other factors may have also contributed to the Little Ice Age; different factors can have an additive effect. The Maunder Minimum of sun (and sunspot) activity from 1645 to 1715 AD during the Little Ice Age was particularly cold. In Iceland, the environmental impacts of a very small flood basalt eruption in 1783 were notable; recall that Franklin described 1784 Europe as a "year without a summer." There were many Krakatoa-level volcanic eruptions during throughout the "Little Ice Age;" recall that Tambora's eruption led to another "year without a summer" in 1815. Climate is almost never as constant and optimal as it has been recently. Climate dominates the environment.

As Europe went into the deep freeze of the Little Ice Age, European agriculture could not support the existing human population levels of that time. Agricultural production was strongly constrained, due to limiting climatic conditions during The Little Ice age. The favored staple, cereal, would not grow with any reliability. There was no bread; there was no beer. Crop failures and starvation were commonplace, worldwide. Diseases like the Black Death (plague), smallpox, tuberculosis and typhus were enabled by malnutrition and crowded conditions, which were necessary to escape the cold. Nothing promotes the spread or the increased virulence of a disease quite like the culture medium of crowded conditions. Resource readjustment did not happen quickly; starvation was not uncommon, even after the plague subsided. Only after most Europeans died and survivors began migrating to America did European recovery occur (early 1600's), placing severely limited regional environmental resources under less population pressure. European immigration to the Americas was at great expense to Native American populations; it looks as if the European immigrant population only slowly replaced Native American numbers that had died from European diseases (Mann 1491).

World population only began to grow exponentially at the end of the Little Ice Age (1900). World population levels of today represent greater than a tenfold average of any time greater than about a century ago. **Reflecting on the high probability of worsening global warming, followed by another Little Ice Age, Great Retreat of Agriculture, or whatever it will be called, dooms any large populations in the not-that-distant future.** In *Collapse*, Jared Diamond listed five factors that caused the fall of past societies: ecological collapse (ecocide-damage that people inadvertently inflict on their environment); climate change; hostile neighbors; decreased support by friendly neighbors; and finally, society's response to the above environmental issue (or combinations of the above environmental issues). All five society collapse factors are population related and/or resource related. The logical consequence of poor population management or poor resource management is anarchy and social breakdown. Resource management without population management is not possible; that is, resource management and irresponsible overpopulation cannot possibly be sustainable. If resources do not increase, any population increase decreases the resources available to each population member. If resources were to increase, any corresponding increase of population cancels any advantage provided by the increased resources. Malthus noted that populations will increase faster than any increased resource availability, due to the geometric growth patterns of the former contrasted to arithmetic growth patterns of the latter. These are conditions of life. Overpopulation damages the environment; the damaged environment cannot support the overpopulation. And that destabilizes society, causing war, famine and disease.

Society's response to the present conditions of life will be the critical issue today. The most appropriate response is clear; there is no mystery.

The solution lies in controlling irresponsible overpopulation. Some do not seem to do this. In Africa, polygamy (males possessing more than one wife) is so common that many states allow the first wife to inherit the greatest percentage of family wealth upon the patriarch's passing. It is not uncommon for such men to father hundreds of children in this polygamous society. In monogamous societies, the overpopulation issue is all about parental irresponsibility; poor parents procreate perennially, producing pressing population problems. In monogamous societies, men, particularly in some cultures, seem to be far less responsible than women (problem partners—demonstrated by patterns of HIV-AIDS infection). This needs to change, and for more than one reason (disease prevention as well as birth control). Any sexually-transmitted disease (or any other disease) compromises the immune response, and, with exposure, enables HIV-AIDS infections. Even in the United States, AIDS is the leading cause of death for African-American women between 25 and 40 years of age; the death toll in sub-Saharan Africa is staggering. Women deserve better; their own life is at great risk. Who controls the birth control? Hopefully, it is the mother; mothers seem to be more responsible in matters of children. Educating and empowering women is the single most effective means of population control. Studies show that the education and empowerment of women increases employment opportunity, reduces fertility, and reduces mother and child mortality. An empowered, educated mother is the best health-care, for both herself and her children. Educating and empowering women also positively affects political interactions by giving women a greater role in government. This can often moderate violent irrational emotional extremes, as in initiating a war. Even though most human behavior is determined by emotions, emotions do not do very well with problem-solving. If parents (it really takes two) cannot be responsible, extreme government control (like male-dominated authoritian China) is the next best hope to effectively control extreme overpopulation. Excessive populations become oppressive, and their governments must also become oppressive. Otherwise, the default population controls of starvation, war, famine, and disease will once again become the limiting factors. The theories of Malthus, Darwin, and Diamond are all relevant in this upcoming crisis of overpopulation. Mankind would like to make the rules, but mankind does not make the rules; the environment makes the rules.

Where food is scarce and violence common (note—violence is linked to chronic hunger), parents may try to cope by marrying their daughters off, usually to much older men, as soon as the girls enter puberty. But the marriages themselves can harm the young brides. Such unions often end girls' educations and trigger significant health risks. Pregnancy is the number one cause of death worldwide among girls between the ages of 15 and 19. A child is statistically likely to be born within the first 2 years of marriage. The number of these girls facing these risks is staggering. In the Democratic Republic of the Congo, for instance, 74% of girls ages 15–19 are married. In Nepal, 7% of girls are wed before they turn 10. Throughout sub-Saharan Africa, girls' and women's chances of contracting HIV substantially increase after they marry. International human rights standards set the minimum age for marriage at 18; however, many countries permit marriage under 18 to marry with parental consent. Often, child marriages occur without regard to statutory law (Roth—National Geographic, May, 2006).

Deadly disease is not limiting the fastest population growth on the planet. Sub-Saharan Africa has the highest birth rate, the highest rate of population increase and the lowest use of contraceptives of any major region in the world. Africa, particularly sub-Saharan Africa (India next in line) has the highest HIV-AIDS infection rate today; yet, they still have the highest growth of population, in spite of the deaths from AIDS. Recall that women in these countries are (for the most part) uneducated and are definitely not empowered. They marry very young and have no power in the family or society. Any women in these societies that can be educated first will tend to marry later; this decreases HIV infection rates and limits family size. Education improves the conditions of life for these women and their families. Sub-Saharan Africa really needs help; they truly need food and medicine, but they must give consideration in exchange. There must be followed-up accountability on the education and empowerment of women as well as accountability to make sure corrupt government officials do not seize the food and medicine, and sell them on the black market. Amnesty International estimates that 25% of Africa's gross economy is lost to corruption. Africa is not really that poor; it has large reserves of oil and 20% of the world's mineral wealth. Africa is just overpopulated.

Survival tomorrow primarily depends upon the education and empowerment of women, today. This will never work unless it is implemented in presently overpopulated countries, where women have the fewest rights, least education and no control of their lives. Improvement in the conditions of life (through increased resource availability) cannot occur if a population increase cancels any potential benefit that could have been provided to a certain number of individuals. Said otherwise, increased resource availability can be negated by further irresponsible increase in population, and this only increases the magnitude of the original limited resource problem. Irresponsible overpopulation increase becomes a negative limiting environmental factor and inevitably causes a net reduction of any increased resources (for all individuals in a population). That is an unstable situation and it only allows a tragic outcome. The greatest threat to a stable society is another unstable society.

When conditions deteriorate, anyone that can relocate to find a better environment will relocate to find a better environment. There must be limits on something, be it resources, space, or population. Everyone in the overpopulated areas will try to relocate to areas of better-managed populations with better environments. This is perfectly understandable; everyone wants a better life. For some population members, it will be a matter of life or death. Irregardless, life in irresponsibly overpopulated areas is not good; if it is at all possible, people must leave and go somewhere else (to a better environment). According to BBC News, there were more than 191 million migrants in 2005, more than ever before. If all these people belonged to the same country it would have the fifth largest population on earth; this is the figure for just one year. Migrant numbers have more than doubled since the 1960s; this 191m number nearly matches the US population of the 1960's. About half of all migrants are bound for North America and Europe. The country that attracts the most people from abroad is the United States, with 35 million in 2005, followed by Russia and Germany with 13.3M and 7.3M people respectively, in 2005. China sees the most people leave, followed by India and the Philippines. Uncontrolled immigration will understandably cause problems. In 2005, around 30 million migrants entered the global workforce illegally; by definition, these people are not afraid to break the law. In almost every world capital, population movements are being viewed with alarm. In Amsterdam, Bern, Bonn, Brussels, Paris and Vienna, the large refugee/illegal migrant influx from Eastern Europe and the Third World is cause for great concern. Each year, about a quarter million illegal African immigrants enter Europe, bringing with them crime, riots, terrorism, polygamy and AIDS at an alarmingly disproportionate level. In Washington, DC, there is increasing disquiet caused by Central American, Chinese, Middle Eastern and Cuban/Haitian boat people. In Japan, the key concerns revolve around the illegal influx from China and Southeast Asia. As populations increase, immigration problems will increase. Immigration to these less-populated countries will change a potentially tragic outcome, but this change will only become more tragic, not be a change for the better. While immigration may buy some time, it increasingly worsens the magnitude and intensity, as well as the location of the inevitable population crash when it does occur (negative environmental selection). The new location of the larger and more intense population crash should be of particular concern. Population management without immigration management is not possible. Resource management without population management is not possible. The logical consequence of poor population management or poor resource management is anarchy and social breakdown. The logical consequence of poor immigration management is anarchy and social breakdown. Anarchy and social breakdown lead to widespread suffering and death. This is already happening in Africa and it can happen anywhere; Africa is not exclusive on widespread suffering and death.

In the *Collapse* chapter on Ultimate Causes of the End, Diamond observes that "illegal immigrants from poor countries pour into the overcrowded lifeboats represented by rich countries." Increasing the numbers of an otherwise stable population through uncontrolled immigration compromises a stabilized population by eventually causing carrying capacity problems. At best, it is a decreased support by "friendly" neighbors. The source country of the neighboring immigrants, even if friendly, is compromising neighborhood support of the country that has a stable population. If a pattern of uncontrolled immigration from neighboring overpopulated countries is allowed to continue, responsible populations will eventually experience internal collapse, due to this overpopulation-related immigration into the responsible countries. Uncontrolled immigration is like a non-hostile invasion; it may change to become increasingly hostile as immigrant numbers increase. Responsible populations will experience increasingly militant and hostile internal activities from the uncontrolled immigrant populations, associated with an increasing loss of control in the responsible population's destiny. There would be an eventual reverse-assimilation caused by increasing numbers of uncontrolled immigrants; the immigrant-burdened country would then become just like the overpopulated country that was the source of the immigrants. Almost assuredly, there was an imbalance between the irresponsible overpopulation and adequate resources in the source country that caused the residents to emigrate in the first place. People do not change if there is no reason to change. As the power base of the immigrants in the new country increases, the cultural attitudes and behavior of the immigrants from the overpopulated country will not change, and they will then only continue to increase overpopulation and deplete any limited resources in the new country. This outcome becomes an external source of stable population destruction from the desperate neighboring overpopulations. Good fences make good neighbors.

Jared Diamond indicated that everyone was in the collapse scenario together, and this is so true. It will be very important for everyone to help their neighbor, because everyone will be affected by a carrying capacity collapse. However, this does not mean to surrender to past practices of business as usual. Ignoring the problem will not make it go away; it will only make the problem a bigger problem. This is a growing tragedy of the commons; the outcome is not in question. Resolving the problem of irresponsible overpopulation growth will not be easy, but it will be far easier than managing the alternative. The only real solution to preventing a future collapse is to assist in the development of responsibility in population management. Survival strongly depends upon the education and empowerment of women, today—and it must happen today in the areas of irresponsible overpopulation. This is where the help must be focused.

Otherwise, any developed, responsible country (that does control overpopulation and protect the rights of women) must be on guard to deter an inevitable uncontrolled immigration through implementation of an effective border security protocol, and to deter the inevitable hostile armed invasion through an effective armed protocol. Diplomacy simply does not always work, particularly while under attack. An inadequate self-defense is fatal to any society. A hostile invasion is a poor alternative compared to encouraging timely widespread population control. A non-hostile invasion is a poor alternative compared to maintaining border security. Recall that uncontrolled immigration carries a significantly greater risk of a continually-increasing spread of anarchy and violence, and the far larger social breakdown that predictably results from exceeding the environmental carrying capacity (the greater the numbers, the greater will be the anarchy and violence). Jared Diamond's *Guns, Germs and Steel* chapter—A Natural Experiment of History, contains an interesting story about a peaceful island society that actually did live within its available resources and managed their population size. When outside invaders arrived, the enlightened islanders preferred to settle the confrontation peacefully, with diplomacy, offering to share their resources with the newcomers. The invaders killed and ate the diplomats, and hunted down the survivors, enslaving them. Over time, they killed and ate the slaves. The islanders could have successfully repelled the invaders by resisting, meeting the invading force with equal or superior force; they did have the ability to do this but failed to do so (big mistake).

Ecological collapse (damage that people inadvertently inflict on their environment) in irresponsibly overpopulated countries will happen, as will climate change; both processes are presently in progress. Environmental breakdown breaks down societies. Security is essential to freedom in any civilization; anarchy destroys civilization from within, enabling the rise of tyranny in governments. Even tyranny may not choose to manage overpopulation; tyranny often initiates war (need soldiers) so as to gain increased resources elsewhere, reducing and exploiting other populations in the process. War is often the tragic outcome of poor population-resource management by governments. Wars are caused by greed. Wars are caused by desperation. War is not the appropriate answer to population control problems. No one fights a war without suffering also; and, beware the side that has little to lose. Most of humanity's decision making process is emotionally motivated, often exacerbated by desperate circumstances. When it comes to problem-solving, emotional decisions made under duress do not work out very well. It is best to address overpopulation problems as early as possible. It is also important to understand that wars are a significant historical pattern of mankind's behavior; wars will never go away. A highly capable self-defense protocol is the best deterrent against an outside decision in favor of war. If a violent attack does occur, self- defense requires a decisive response. To not retaliate against a violent invasion would be the worst possible choice; however, over-reaction should be avoided. No invasion must be allowed to succeed by sheer numbers. Surrender to any neighboring hostile or non-hostile overpopulation invasion is not an option for anyone's long-term survival. It only postpones and worsens the degree of suffering and death in the certain collapse of any overpopulated civilization. Global warming-related drought and famine are going to occur, war or no war. Worldwide, economies will be devastatingly dismantled; chaos will follow. The cold will then return; present population patterns will not be sustainable. The overpopulation levels of today will be reduced, one way or another.

The main reason for the goal of a sustainable population is to minimize human pain and suffering; this cannot be attained unless all populations can reach this goal. Educating and empowering women is the appropriate answer, but to date, this has occurred only in the developed world; that makes it only a partial answer. Outsourcing is supposed to raise the Human Development Index; it has not done this effectively in the areas of irresponsible overpopulation growth. It fails to do so in the 3rd and 4th world when higher production improves resource availability, but because of a corresponding great population increase, any improvement in conditions for life is more than offset by these higher population increases. These higher population increases create a highly unstable situation. When higher population increases happen, outsourcing benefits no one; it only contributes to a greater potential problem. Population control must occur worldwide and it must happen very soon. The risk of anarchy and war predictably increases as irresponsible overpopulation increases.

Survival tomorrow primarily depends upon increased consideration of the environment, today, as well as the education and empowerment of all women everywhere, today. Without this life-saving society response, irresponsible overpopulation will eventually destroy the environment and excessively limit the resources necessary for life. The certain, looming climate changes will devastate humanity, leading to widespread, decreased support by friendly neighbors, creating desperate people that will then likely become hostile neighbors. There are environmental limits to any population. Whether these numerical limits of overpopulation are perceived, or not perceived, makes no difference; the environmental limits exist, whether they are acknowledged or not. Unless society adapts to the upcoming challenge to make a world-wide difference in lessening a frequently-ignored pattern of irresponsible overpopulation (by adequate performance in educating and empowering women), environmental limits will impose a certain and severe population crash in the not-too-distant future. There is serious trouble ahead if irresponsible overpopulation does not become responsible. Otherwise, declining environmental conditions and increasing overpopulation numbers add up to an upcoming carrying capacity-related environmental, and population, collapse.

Even today, before the full impact of global warming strikes, the overpopulation impact on the environment, and the resulting environmental impact on mankind generates a sobering preview. The human environmental impact on an area varies directly with the human population of the area; it is an additive effect. Overpopulation-related overly-intensive planting causes an acute shortage of the two most valuable resources—water and topsoil. Overpopulation necessitates the elimination of native vegetation for widespread agriculture, exposing bare ground for overly-intensive planting. Eliminating native vegetation breaks down the transpiration portion of the local water cycle, decreasing rainfall. Loss of native vegetation combined with bare-ground agriculture causes topsoil loss, which then causes increased susceptibility to drought, as well as flooding. Topsoil holds moisture best; plants need topsoil to grow. Resource management without population management isn't possible. The logical consequence of poor population management or poor resource management is anarchy and social breakdown. Environmental breakdown breaks down societies. The great African drought has only just begun; the breakdown of African society is already underway. Increasingly intolerable hunger and thirst create desperation, and lead to extreme responses. People that have never been truly hungry just cannot relate. No one in desperate circumstances cares about an endangered species of toad, or any other endangered species. Irresponsible overpopulation led to starvation and genocide in (the Sub-Saharan region of Africa known as) Rwanda a decade ago; it is once again becoming a problem. Overpopulation was temporarily reduced; therefore, fewer people had more land and more food per person. Many survivors freely acknowledge that the real problem was that there were just too many people and not food enough to eat. Just over a decade ago, drought precipitated a major crop failure and famine, but the real cause of the drought-related breakdown was overpopulation, which inevitably led to desertification and topsoil loss, which then led to the drought, crop failure and starvation. There was not enough water for the crops and not enough food to feed all of the increasing population. The same conditions that precipitated the crop failure, which ultimately caused the starvation in the last crisis, are returning. There will not be enough food to feed all of the people, again. If the choice is to make no change, there will be no change; the (tribal-related) genocide will also return. Rwanda was a (tribal-related) class war; the starving lower-class resented the well-fed upper class and began killing them. The upper class retaliated. This (tribal-related) conflict between classes over a shortage of resources has happened elsewhere. Rwanda has no exclusive on genocide.

Overpopulation numbers will always add up to a carrying capacity-related environmental collapse. If irresponsible overpopulation is not addressed worldwide, water and food shortages, and shortages of other essential resources will increasingly become more of a widespread problem. Environmental degradation (eventually) causes a predictable society degradation; society degradation (= war, famine and disease) is the ultimate population check. Overpopulation devalues humanity; "outsiders" are devalued of their humanity and treated like vermin. In a resource conflict, any group of "outsiders" is at risk of mistreatment. Overpopulation exaggerates tribal-thinking. Racial issues are caused by any race that believes they deserve special treatment. Some races think that they are entitled to additional advantages today because they didn't have the advantages they deserved in the past. Wasn't this the type of tribal-thinking that caused a racial problem in the first place? Tribal-thinking is decidedly uncivilized, inconsistent with any equal opportunity for all; it blatantly promotes racial privilege. Emotionally-linked religious bias has also caused mankind to devalue "outsiders" since early beginnings (Diamond—*The Third Chimpanzee + Guns, Germs and Steel*). Religious differences in the Middle East have been the cause of wars for thousands of years; religious killings have been a daily Middle Eastern occurrence for decades. Does anyone really believe God chose to support only one particular Middle Eastern religious leader with a particularly nasty agenda—one that horribly compromises "outsiders" (as in the unrestricted slaughter of "other" human beings), made a " preferred provider" arrangement with this religious leader, and offered "exclusive membership" as well as an "afterlife" to those that follow (the homicidal agenda of) this "preferred provider?" (Think about it—this is intelligent-albeit malevolent-design to decrease the population of "outsiders" and gain resources +/or more power for exclusive members!) And since followers of the "preferred provider's" malevolent design are "exclusive members" of the one and only "true faith," as long as members follow the "preferred provider doctrine" (which was God-given only to him and his followers), all descendents of the followers also become "exclusive members" of the one and only "true faith." And, they can do anything they want to "outsiders," because abusing "outsiders" (as in exploiting or killing "outsiders") is never counted by God as a transgression against any members of the one and only "true faith." Isn't it about time to finally understand this type of tribal-thinking? Actually, thinking is the wrong choice of word—following is a better word. Following doesn't require thinking, and following the "will of God" can justify any nasty agenda! More than one Middle Eastern religion believes they are "God's chosen people" and more than one religion devalues, and often mistreats, "outsiders." How many have already died in the name of religion? The ongoing genocide, fueled by religious differences, refuses to go away. But the problem isn't really caused by religion (as the problem isn't really caused by race)! The problem is caused by tribal-thinking and not-thinking; tribal-thinking is destructive and not-thinking is negligent. Modern nuclear warfare is a potential fast-track to another massive extinction event; humanity is at risk. Nuclear club membership is growing steadily; efforts to stop nuclear proliferation are ineffective (no surprise). Unless learning and logic prevail, the education and empowerment of women worldwide won't occur, and appropriate consideration won't be given to environmental protection; then the calculated outcome of overpopulation will be different path to population control—the hard way.

42

The Ongoing Losses

Human overpopulation not only threatens the future of humanity, it threatens the conditions of too many other lives. Irresponsible human overpopulation is causing a significant increase in environmental destruction, an ongoing loss of biodiversity and an increase of environmental extinctions today; humans will suffer as a result of these losses. BBC news notes that over a third of the world's plants and animals are now threatened with extinction. As a direct result of irresponsible overpopulation's expansion in developing countries, complete extinctions of innumerable species happen far too often in the tropics. When tropical forests are cleared and burned, local extirpations of resident populations as well as extinctions of entire populations are daily occurrences. Even at higher latitudes, local losses of resident populations happen far too often; this occurs with ongoing wetlands loss. Local extirpations that do not result in immediate species loss still cause a loss of biodiversity. Intra-species loss of biodiversity has harmful effects that limit genetic adaptation, reproduction and species survival. In other words, local environmental extirpations have the potential of eliminating traits that are useful to survival. Even though some species members may survive elsewhere, a genetic bottleneck effect may occur as population numbers and diversity decline. This means the expression of potentially beneficial rare genes is less likely. As with inbreeding, any recessive deleterious genes present will be over-represented in the survivors. Wildlife is suffering an unacceptable ongoing loss; should it be protected? The difference between past significant extinctions and mass extinctions may have only been a difference in the latitude of the environments affected. The destruction of tropical habitats is the most predictable cause of mass extinction; 90% of all species are tropical. The next significant extinction may not have to wait for the return of the ice; it may be underway today in the human-overpopulated tropics. Setting aside a few parks will not be enough; governments of the world must encourage more environmental responsibility from people throughout the land. The land is more than property to be used or abused; it is a community to which we all belong. "The land," a term coined by wildlife ecologist Aldo Leopold, includes the soil, water, air, plants and animals. "The land" is the ecosystem; it encompasses the conditions of our life, and all other conditions of life everywhere on the planet. We all have an obligation to care for it; we live here. The ongoing losses represent an ongoing tragedy; the ongoing losses exemplify another "tragedy of the commons."

Global warming will cause changing environmental conditions and this will impact nearly all life. Greenhouse gases, notably CO_2, are on the increase, and this intensifies the present global warming trend. Mankind has already eliminated half of the planet's forests. The burning and destruction of the rainforest greatly contributes to elevated levels of atmospheric CO_2 and CH_4; it also eliminates a main source for short-term photosynthetic control of CO_2 reduction. Even more importantly, the burning eliminates the habitats of most terrestrial species on earth. And while grass wetlands and tropical mangrove shoreline swamps may produce some atmospheric methane as global warming intensifies, they bury far more carbon in the black organic sediments below. The ongoing destruction of grass wetlands and mangrove shorelines eliminates the ultimate long-term terrestrial management of elevated atmospheric CO_2 levels. Grass wetlands at higher latitude bury huge amounts of carbon with seasonal cold vegetation die-back. In spite of some token effort of wetland protection, wetlands are disappearing worldwide; tallgrass prairie wetlands and carbon-rich peat bogs are still being drained and burned. In the same manner as burning the rain forest, this wetland drainage and burning contributes to elevated levels of atmospheric CO_2 and CH_4, and it also eliminates a major source of long-term photosynthetic control of CO_2 reduction. Not only does this draining and burning of wetlands significantly contribute to both short-term and long-term global warming, coastal and inland wetlands are the habitats of most forms of life on land that do not live in the rain forest. Coastal mangrove swamps and estuary wetlands are the nurseries for a very high percentage of ocean fish. Recall that increased atmospheric CO_2 is primarily managed through ocean limestone deposition, oceanic dead zone expansions significantly interfere with this ultimate management of global warming (discussed in upcoming chapter). Dead zones are toxic environments for aerobic life, and are, well, dead. Wetlands detoxify the nutrient pollution that causes the dead zones.

A temperature increase of one degree C (Centigrade) in a Pacific coast tide-pool can stop the hearts of the resident crab population. It would not take much of a further increase in tide-pool temperature to cause a local extinction of these important community members. This would profoundly affect the coastal tide-pool community food chain. Will it affect us? Global temperature has risen 0.6 degree C in the last 200 years. This warming has had significant repercussions in coral reef ecosystems, which comprise the most significant aquatic animal base on the planet, and provides the major habitat of near-shore tropical marine diversity. The Millennium Assessment of 2005 revealed that 20% of tropical corals have died and another 20% are in danger; thermal pollution (ocean water warmer than 30° Centigrade) is the leading cause. Nutrient pollution and damage from inappropriate contact are two other significant contributing causes. Indonesians are using dynamite to harvest fish on local coral reefs. One-quarter of all sea-creatures call the coral reef climax community home. Coral reefs and shoreline wetlands are the nurseries of tomorrow's fish populations. Their losses will unquestionably affect all of mankind's future seafood sources. The marine wetlands of Louisiana are rapidly disappearing; damming upstream (traps sediment in slow water areas upstream), channeling stream meanders (ditching increases erosion by increasing flow velocity), channel-dredging (for adequate shipping depth), and building levees and dikes (act like dams and raise the stream bed) on the Mississippi River prevent normal seasonal flooding, which otherwise replenishes the wetlands of the area, keeping these lands from sinking below sea level. Any sediment that could stabilize the loss is diverted to the Gulf of Mexico. Saltwater intrusion continues to move further inland, and combines with ship transportation bank erosion, in a growing pattern of wetland destruction, initiated by this breakdown of saline-compromised, wake-damaged channel bank vegetation. Even an anti-trapping mentality has let to nutria overpopulation (an aquatic rodent introduced for trapping revenue), and subsequent overgrazing of wetland plant life, further degrading the wetland ecosystem. Wetlands are nurseries for 75–90% of the fish and shellfish harvested in America. The result of this wetland loss is a disappearance of the major southern seafood source and a loss of storm surge protection for New Orleans and adjacent coastal areas. Regional wetlands are disappearing at the rate of 25 to 30 square miles (an area the size of Manhattan) a year, bringing the sea ever closer to the city. For every mile of coastal wetlands, storm surge height can be reduced by one foot. It was wetland loss that led to a Gulf of Mexico encroachment inland, in the direction of New Orleans, which then set the stage for the worst natural disaster in U.S. history. Even though hurricane Katrina (in 2005—Category 4 storm) spared New Orleans from its worst winds by passing slightly east, a devastating 18 foot surge of water from the Gulf of Mexico flooded Lake Pontchartrain, which then spilled over its dikes into the city, flooding New Orleans. It could have even been worse for the residents of New Orleans. It could have been better for the residents of New Orleans. Protecting against wetland loss would have protected New Orleans, as it had done for centuries prior to the wetland loss. Farther east, around Gulfport and Biloxi in Mississippi, the ocean storm surge exceeded hurricane Camille's reach (of 1969), with reports of water flooding structures up to 30 feet above sea level. More than 1,900 square miles of wetlands have been lost from the Mississippi delta in the past 75 years; many of the barrier islands ringing it have become little more than tidal sandbars. Scientists project, at this pace, that the delta will be virtually gone by the end of this century, leaving a sinking New Orleans even deeper below sea level and right at the edge of the gulf. Ongoing losses like these are unacceptable; it is much too costly to ignore environmental preservation. All biologic and environmental systems are interconnected. Our Biosphere (planet) is a very large ecosystem. Damaging the environment only damages us. What will it take to make humanity improve respect for the environment?

43

A Choice

The conditions of life, for humanity, and for life's biodiversity, need better protection. **At issue is a choice between the quantity of human life and the quality of human life. A quality human life requires a quality environment.** Without change, the future view for all is that of an overpopulated resource-poor (topsoil-poor) country of today, like Haiti. The trees are gone; the topsoil is gone. Haiti is plagued by severe environmental problems, food shortages and social unrest. Life in a bad environment is not only unpleasant, it is seriously endangered. This future could easily parallel the society breakdown of Rwanda. Jared Diamond's *Collapse* contains many stories of environmental breakdown and death. There are numerous instances of environmental damage and/or climate change that did not allow society to maintain formerly high population numbers. What were the people of Easter Island thinking as they cut down the last trees on their island? By that time, it was already too late; instead of building statues, they should have been building boats to get out of there. Prior to that time, they could have implemented conservation and/or population control in order to prevent the looming crisis. Too late, they realized that there was nowhere to go; there was no escape. The island's tree loss led to topsoil degradation and failed food production. The large population could no longer be supported; there was no happy ending. The same problem can happen anywhere, on a much larger scale. Large-scale environmental breakdowns and widespread death have happened in the past and they will occur again in the future. History does tend to repeat itself and those that do not learn the lessons of history do seem to be doomed to repeat those lessons. In this situation, the ability to adapt to environmental stress truly depends upon changing behavior as a result of learning.

It is prudent to not encourage a large-scale environmental breakdown. The rich farmland of the Mississippi drainage feeds one of twelve people in the world; it has not had a significant drought for 700 years

(Fremling—*Immortal River*). The "Little Ice Age" of the last 700 years is over. Will the area's rainfall shift further north with global warming? Is the drought in the Southeastern US a hint of what lies ahead? Major droughts, such as those that have occurred before, will return and make the dust bowl of the 1930's minor in comparison. Continued topsoil loss in this drainage area will not support adequate agricultural production; it is the topsoil that holds the moisture and nutrients, affording better drought protection. Continuing wetland drainage lowers the water table; the drainage-tiles of agriculture short-circuits natural processes and dump toxic excess fertilizer into waterways, creating "dead zones" in streams and oceans (Fremling—*Immortal River*). If global warming is followed by a cooling, recall that North Atlantic cold once shifted rainfall further south and crippled a Mayan civilization. When the World Millennium report states that present resources cannot support present consumption by present populations, doesn't that mean something?

Overpopulation and environmental degradation create an environment that selects against humanity. There was a time when the survival of the human population was threatened by low numbers; that is not the case today. The inability to adapt to change leads to extinction in the biological world; this principle can also apply to societies. There are dead-end societies that did not adapt and ceased to exist, just as there are dead-end species that did not adapt and ceased to exist. Population and resource problems qualify as an underlying factor for nearly all failed societies, assuming that the adjacent societies warrant consideration. Emotional decisions at the time did not solve existing problems; learning and logic are better able to solve population and resource problems. In the *Collapse* chapter on Ultimate Causes of the End, Diamond notes that "the values to which people cling most stubbornly under inappropriate conditions are those values that were previously the source of their greatest triumphs over adversity." Large families do not necessarily promote family survival. Adaptation needs will change with time; society's adaptation response is key to its survival in the future. The environment played the primary role in the fate of mankind's past societies. No environment is permanent; change is inevitable. Society will choose how it will impact the environment; inaction is a choice. There are no survivors when adaptation to change is too slow to meet the demands of rapid environmental changes. The events of the past have left a large imprint on the present. Society could continue to disregard the environment and suffer the consequences, or society could learn from past mistakes in order to avoid repeating them.

Learning is demonstrated by a change in behavior. This change in behavior must include everyone (all societies); a partial effort will only result in partial success, which will not be enough. A too little, too late response will be too little, and too late. Two somatic (learned behavior) adaptation responses are necessary for survival. The first necessary adaptation response is managing population, through the education and empowerment of women worldwide—today. The second necessary adaptation response is environmental protection; this is linked to population management, but it is not the same as population management. The environment will continue to play a primary role in the fate of mankind's societies. Negative impacts on the environment will result in negative impacts on the future of mankind. Our treatment of our environment today is the key to the fate of our society; this treatment may include correcting past environmental abuse.

The environment cannot provide for humanity if humanity does not provide for the environment; it is in the best interest of all to become more environmentally-aware. It is time for conservationists and preservationists to join forces to pursue a common goal of caring for the environment, and not focus upon their differences. There is a place for both conservation and preservation; one complements another. Promoting conservation of the rain forest, for a financial return on eco-tourism, will meet with better success, for a hungry resident population, than any altruistic goal of preservation. Preservation of an endangered species is important, but extinction is mainly caused by habitat loss. The aquatic environment (both freshwater and oceanic) is the most fragile. Habitat management may require different combinations of preserves having restricted human access, conservation areas that allow limited human access, and areas of habitat restoration that even allow shared access with humans. It is important to choose a priority system for the most important goals; i.e., some goals are more important than others. Areas with inconclusive differences of opinion belong on the back burner; they are a waste of time and energy that are badly needed elsewhere. Wildlife needs support based upon sound scientific principles and management experience, not illogical emotional extremism. Environmental extremists and eco-terrorists have hijacked a mainstream concern for the environment and compromised the goals of both conservation and preservation. Extremism has no place in environmental protection; it is counter-productive and undermines popular support.

It is time for governments to update farm subsidies and mainly subsidize conservation tillage practice that does not fill streams with precious topsoil; topsoil loss is by far the most important resource loss to manage and protect. This subsidy priority not only protects an increasingly-endangered fresh water resource, it also protects civilization's most significant common asset, food supply. Historically, it is topsoil loss (from bare ground agriculture) that inevitably leads to inadequate food production. Inadequate food supply leads to society breakdown; ecological suicide carries a dear price. The richest topsoil in the world (20% of world total) was in Iowa; fall plowing, historically practiced by the European ancestors of Iowa farmers, has caused over half of this topsoil to wash into the Mississippi, polluting the water. Bare soil only absorbs 20% of the rainfall before sheet, rill, or gully erosion washes it into the waterways; this lowers the water table and short-circuits the local water cycle. This negligent, topsoil-eroding, drought-enhancing, field tillage is still practiced by most of the Iowa farmers today, in spite of a quarter-century of education effort; there is an incomprehensible resistance to change (learn). Another downside of fall-plowing, exposure of the soil to the atmosphere, even creates a significant source of atmospheric CO_2. Mass production for mass populations is the way of today; government subsidies presently favor large cooperate farms. Large cooperate farms have larger machines that are less able to practice conservation tillage (Fremling—*Immortal River*). Is subsidizing large corporations for destructive land use a good use of taxpayers' money? Smaller family farms, with smaller fields that are surrounded by prairie-grass perimeters, and have diversified agriculture under reduced tillage, have a place in the future, if government will do the right thing (with more environmentally-friendly subsidies).

It is time to expand protection of fresh water supplies (and its wildlife) worldwide. Deforestation and negligent agricultural practices lead to stream headwater degradation and topsoil erosion, which enters the streams as sediment; sediment adds to fertilizer-induced oxygen depletion and blocks photosynthesis to create even larger dead zones. Surface waters flowing through residential, industrial, agricultural and mining areas have become contaminated chemical catastrophes. In the long run, draining wetlands, building levees and damming streams to prevent seasonal floodplain wetland cycles do not expand agriculture or control flooding. Respectively, they only lower the overall water table, raise flood crests and raise the stream bed to encourage future flooding (Fremling—*Immortal River*). Wetlands, the most effective stream filtration and chemical detoxification system, are still on a yearly decline in every state in the U.S. Wetlands can also decontaminate the biological pollution of freshwater, such as the presence of disease-causing bacterial pathogens. Subsidized restoration of flood-plain wetlands is needed for raising the water table, stream detoxification and flood control; both wildlife and humanity will benefit from this environmental support. Periodic flooding of the floodplain is important to the ecology of the region. The flood provides a "reset mechanism" for the environment that both renews and stops plant succession, supporting the habitat diversity necessary maintain cover and nesting habitat necessary for many native fish and wildlife. If this periodic flooding did not occur, the further progression of ecological succession would decrease essential habitat diversity (Fremling—*Immortal River*). In times of flood, wetlands serve as a "sponge" to store and slowly release floodwaters, raising the water table and supporting later times of low stream flow; wetlands also trap large amounts of suspended sediments in floodwaters (Fremling—*Immortal River*). This need for expanded fresh water protection is even more immediate than an enhanced food supply; there is enough food, but not enough fresh water. If the Millennium Report is correct (it is), fresh water will soon become a significant source of serious contention in future society. A difficult choice must be made between water for residents in urban centers with growing populations, and water for agriculture to feed the growing populations in urban centers. In this dilemma of a limited resource problem with no real choice, the only correct choice is population management. Otherwise, continued over-utilization of limited freshwater supplies will only decrease future freshwater supplies and result in increasing contamination of the remaining inadequate freshwater resource. Fresh water shortage is a ticking time bomb for overpopulation.

It is time to pay more attention to ocean environmental damage. The aquatic environment is more fragile than the terrestrial environment. This damage isn't often seen, as the damage is hidden from view; however, the underwater destruction is shocking and distressing to those few that have seen it. Eighty percent of oceanic pollution originates on land. Dead zones caused by excess agricultural nitrate and phosphate fertilizer runoff adversely affect the entire food chain. There are around 150 coastal dead zones in the ocean today; this number is increasing as developing countries desperately try to feed their exploding populations by adding more and more fertilizer to their topsoil-depleted agricultural lands. The ocean contains 95% of the living space for complex life on the earth. Primary production is mostly at sea. Primary production (photosynthesis) is the primary regulation of world atmospheric carbon dioxide (CO_2) and oxygen (O_2) production. Photosynthetic *Prochlorococcus marinus* is the single largest producer of organic matter (and oxygen) on earth. *Prochlorococcus* cells dominate the temperate and tropical oceans; they are the most abundant organism in the oceans. *Prochlorococcus marinus* accounts for up to 80% of oceanic primary production. **Prochlorococcus marinus is the primary control of global warming (CO_2 reduction), even though the water's surface delays the process of ultimate limestone lock-up in ocean sediments.** *Prochlorococcus marinus* evolved in an environment of very low nitrogen availability. When cultured, high light-adapted strains of top world producer *Prochlorococcus marinus* can grow in a medium with ammonium ions, but cannot not grow in a medium where nitrates (NO_3)- were the only nitrogen sources (Kenyon College Website—*Prochlorococcus marinus*). As dead zones increase in size, pollution-damaged *Prochlorococcus* cannot do their job as well. This means that the primary global warming management of the planet is being increasingly compromised. Said otherwise, the inability of most phytoplankton to manage increasing levels of atmospheric CO_2 is playing a major role in the rising atmospheric CO_2 levels! And, as a result of ozone depletion in the stratosphere, there is even possible phytoplankton damage from increased ultraviolet ray exposure. The tropical coral reef, mangrove swamps and grassy estuary marine wetlands are all under assault and losing ground today, to a greater degree than ever imagined; recall that these are the nurseries for marine life (read this as our seafood supply). Over-fishing seriously compromises the future food supply of mankind, and inappropriate harvesting methods (that destroy habitat and cause collateral damage to non-targeted marine life) further degrade this marine environment on which we are food-dependent. Deep-water trawls destroy deep-water coral reefs, the most abundant corals in the ocean. This destruction of habitat has the same outcome as the use of dynamite while fishing shallow tropical reefs; there is only a one-time harvest before the area becomes no longer sustainable for food production. Problems will also occur as the deep ocean circulation returns discarded wastes to the surface and into the (our) food chain. The oceans are used as dump sites in countries with irresponsible population growth. The ocean dumping of nuclear waste, solid waste and raw sewage has been discontinued by developed countries.

It is time to require more responsible population control from countries requesting aid for their irresponsible populations. This is not an unreasonable demand; developed countries have already controlled overpopulation. Importantly, it is the underdeveloped countries with irresponsible overpopulations that are requesting aid from developed countries with responsible populations. Educating and empowering women is the single most effective means of population control. Population responsibility is far better than society collapse, the logical result of combined overpopulation and inadequate resources. It is the overpopulated countries that degenerate into anarchy when the inevitable next environmental mismanagement or climate crisis occurs. Supporting the growth of even larger populations only magnifies and spreads the risk of an even greater collapse.

It is time to worry about depleting any non-renewable resource. Petroleum is becoming increasingly expensive to find, extract, refine and deliver to markets. Production of this non-renewable resource has already peaked and started to decline. As the petroleum supply decreases, increased production costs will combine with the inevitable supply and demand price increase, to raise a market price already spiraling out of control. The price of oil/barrel went from $57 in April 2007, to $117 in April 2008, largely due to the increasing per capita demand of geometrically increasing populations. As fuel for the tractors of agricultural mass production becomes less and less available and more and more expensive, compromised mass production (supporting mass populations) will increasingly move the overpopulated masses in undeveloped areas further into poverty and starvation. During the same yearly time span noted above, corn increased 31%, rice increased 74%, soy increased 87% and wheat increased 130% in price (BBC news). With foreseeable future estimates of world dependence on fossil fuel remaining at a realistic 90% of all energy sources, conservation efforts of a few individuals will mean little against an irresponsible overpopulation's demand by the many. The petroleum example further illustrates that the cost of a limited resource in high demand is not limited to dollars, as evidenced by the discovery of oil-field funding for Middle East sources of world terrorism. Why do religious fundamentalists in this area, who aspire to reject all western technology, now claim to need nuclear technology? If this area already has the majority of the world's oil; there is little need for a petroleum alternative here. As irresponsible overpopulation continues to increase, energy-dependent future food and water shortages will become far more serious than high gasoline prices.

It is time for all businesses to understand that environmental responsibility is good for business; it is time for the public to clearly communicate that they will not financially support environmental irresponsibility from any business community. Some members of the business community have taken a role in strongly supporting environmental preservation. For example, the Dupont Chemical Company has taken a pro-active role in environmental protection. Dupont gifted 16,000 acres adjacent to the Okefenokee Swamp to the Conservation Fund (August 27, 2003). The Okefenokee is one of the oldest and most well-preserved freshwater areas in the U.S., and the world. Originally planned as a titanium strip mine, Dupont donated both the land and mineral rights following environmental impact studies. Dupont had previously gifted 18,000 acres in smaller parcels (new total 34,000 acres). It was one of the most environmentally significant corporate land donations in the history of the United States. Additionally, Dupont refuses to do business with any company that is not sensitive and responsible to the preservation of the environment. Dupont sets a fine example for other business entities to follow. Other companies must market their environmental responsibility. British Petroleum is beginning to do this, and this approach to marketing has public support. The public, including other business, must expand this support of environmental responsibility. If this hurts outsourcing and irresponsible developing countries, so be it.

Environmental education can be a learning process. Learning too late is learning the hard way. Protected environments protect the conditions of life, as well as all life within the protected environment. Protection pays.

Recall that as long as an environmentally-integrated genetic adaptation of the phenotype (e.g., large brain) is maintained, and as long as this structure enables learning (behavior change—a somatic adaptation), a viable adaptation to changing environmental conditions is possible. The continuing selective forces of any moderate environmental stress will continue to eliminate less-adapted individuals; they will not pass through this time into the future. In this situation, the ability to adapt depends upon somatic performance, changing behavior (learning). Without this change, history will repeat itself, on a scale matching that of the overpopulation.

44

Epilogue

Comprehensive understanding of the environment requires a scientific approach. Scientists have collectively created our most efficient tool for analyzing nature; it's an inborn system of self-correction. They seek to disprove other scientists' theories, even as they support their earnest attempts to advance human knowledge. Political considerations are short-lived; errors cannot endure for long. Those conclusions that survive the attacks of other scientists eventually become scientific "Laws." These laws eventually become accepted as valid descriptions of reality (Tyson). Will the status of the environmental role in biology and human destiny be advanced? Only time will tell.

"It is not so much that I have confidence in scientists being right, but that I have so much in nonscientists being wrong.... It is those who support ideas for emotional reasons only, who can't change."—*Isaac Asimov*

45

Environment and Evolution Questions

There will be detractors. There is the possibility that I am completely wrong, about everything. There is also the possibility that I am completely right, about everything. Neither case will be true. The only certainty is that some of the concepts contained herein may be correct. If greater environmental consideration and improved world population responsibility are enhanced by a controversy of the contents, this is not a bad legacy for my efforts.

The text at times seems overly complicated; why?

It is difficult to avoid biology and chemistry in any discussion of the environment (or evolution). An effort was made to organize and clearly communicate concepts in the least possible complicated manner; some repetition was utilized (perhaps over-utilized) to support this effort. Hopefully, this text is written at a level so anyone that has completed high school biology and chemistry can comprehend a good deal of the material presented. Other material requires a higher level of educational background.

How much effect does the atmospheric CO_2 level have on climate today?

Atmospheric CO_2 level (380ppm) and Milankovitch cycle position (earth orbit is getting closer to sun) seem to be the two major players in determining world climate today, but other issues (such as variations in sunspot activity, oceanic currents, volcanism, and extra-terrestrial impacts) play some role in world climate. Present cycle position and atmospheric CO_2 levels have been "just right" for humanity, but neither one will last for long. Increasing atmospheric CO_2 levels and sun proximity will cause global warming and carrying capacity issues for humanity in the near future. Yet even today, glaciers of the ice age are only a few degrees away. If atmospheric CO_2 levels were to drop to early ice age levels (160ppm of 2½mya), mankind would try to do anything possible to elevate atmospheric CO_2 and escape the horrific icy onslaught, which is as inevitable in the distant future of humanity as the imminent global warming.

What is a species?

A species is a biological concept that does not apply to most life on earth (simple life); and, with plants, it can be, at best, unclear. In simple life forms, there is no sexual reproduction, and speciation may be irrelevant (if it even exists); however, adaptations are always relevant. Speciation is less significant to simple life than adaptation to a changing, isolated environment. Angiosperms and insect pollinators have been in a protracted co-evolutionary dance, due to the plant's first appearance of flowers and fruits; this interrelationship has lead to their (flowering plants and insects) dominance on earth. Polyploidy, aneuploidy, and hybridization are important determinants of the sheer numbers of plant species (genetically-determined perspective). Virtually every plant on earth has one or more of these phenomena in its history or direct origin; hybridization followed by polyploidy is a dominant mode of speciation in plants. Oaks and dandelions exemplify how unclear the species concept may be in plants. All oak species can interbreed with all other species (although some require an "intermediate," as in A can breed with B, and B can breed with C; therefore, the genes of A can get into C, via B). Dandelions are apomictic—their seeds form without fertilization; therefore, they are asexual but seed-producing.

It is helpful to visualize a species as a collection of very similar genotypes and phenotypes, constantly influenced by environmental selection. In animals, speciation is itself a basic adaptive phenotype, expressed as a sexual behavior. An animal species is a pool of individual organisms that selectively breed with one another; speciation is a stabilizing behavioral constraint that maintains genetic compatibility in sexual reproduction. If the unlikely other-species fertilization of an egg does occur, it simply does not work to integrate the genotype of an individual with a given norm-of-reaction, to the genotype of another organism with a very different norm-of-reaction, even in the relatively stable internal environmental conditions of development. Environmental selection in the internal environment of disharmonious embryologic development often results in abortion. If a hybrid animal survives to birth, genetic disharmony between an impossible integration of different adaptive phenotype norms-of-reaction only worsens; environmental selection increasingly limits the genetic disharmony. Invariably, under the best of conditions, animal hybrid viability is below that of individuals having a normal genotype with an environmentally-proven norm-of-reaction. Genetic disharmony in the genotype of even the most viable hybrids almost assures sterility issues in the germ cells of individuals that survive to reproductive age.

What is the origin of (a) species?

In simple life forms, speciation loses relevance; however, the origin of a simple life "species" may be due to a genetic change in expression of the phenotype in a changing, isolated environment. If this has survival advantage, it may become fixed and predominate in the vegetative-reproducing population. In complex life, the origin of a species may also involve genetic change in expression of the phenotype in a changing, isolated environment. A complex life breeding population could include different phenotypes, including any new phenotype change with survival advantage (adaptation); this change may become fixed and predominate in the sexually-reproducing population.

In David Mindell's book, *The Evolving World*, in the chapter, (Plant) Domestication: Evolution in Human Hands (in a discussion of wheat species), he states that "The process of hybridization among disparate forms and of increase in the number of chromosomes (polyploidy is the norm) via duplication (sometimes of entire genomes) is a common method of speciation in plants, and the increase in genetic material allows for the evolution of novel traits and protein functions." Speciation in complex animal life is increasingly the result of sexual selection, a behavior phenotype, which may itself be constrained by isolating factors in the environment. The origin of a species in complex animal life is due to differences in sexual selection preferences among individuals in a population. Darwin correctly understood that isolation of a group with selective preference in breeding behavior results in genetic changes over time. As the isolation continues, genetic changes continue. As far as separated populations rejoining and reestablishing genetic exchange, the changes that accrue over time eventually result in future incompatibility between the ancestor species phenotypes, and the offspring species phenotypes.

How does natural selection affect the origin of species?

If the definition of natural selection is not limited to the process of fatal competition within a species, there are many other processes of natural selection, such as sexual selection to consider. Sexual selection plays a major role in the reproduction of complex animal life; sexual selection creates genetic bias in the offspring (affects genetic patterns in the progeny). Sexual selection is probably responsible for the origin of species in sexually-reproducing animal populations. Sexual selection is a natural process; it is logical to include it as part of any natural selection process. So far, this text has demonstrated that natural selection is not one process (mechanism), but a combination of several processes (mechanisms) that lead to the preservation of favorable variations and the rejection of injurious variations (result). If natural selection is ever discussed as a mechanism or process, chances are that environmental selection is the process involved. The case has been made for the primary role of environment in natural selection (a change in gene frequencies in a population, owing to fitness of phenotypes' reproduction or survival among the variants—Levinton). While life's genetic variation from mutation over time is random (chance variation), environmental selection is not; it is directional, and this direction is determined by the existing environmental conditions. The selective forces of the environment play a major selective role in natural selection (preservation) of more-fit individuals; environmental selection determines the pathway of survival. Sexual selection is a part of ecosystem dynamics; sexual selection and resultant speciation is a self-stabilizing genetic (phenotypic) constraint in sexually-reproducing animals. Adaptation to ongoing environmental change is a necessary mechanism for evolution to enable biologic survival in a variably-adverse environment. Preserving past adaptations is an essential part of creating new adaptations; the new are built upon the old. Nearly all adaptations are environmental adaptations. Evolution is genetic success in response to environmental conditions. The environmentally-integrated adaptive process results in the natural selection (preservation) of environmentally and genetically designed more-fit individuals. Natural selection is the genotype preservation of successful phenotypic adaptations to previously-experienced environments. The genetic inheritance of environment-proven phenotypes (natural selection) is an essential biologic mechanism of evolution; it preserves phenotypic adaptations to previously-encountered environments. Adaptations may be species-limited.

What evidence exists to support the hypothesis that speciation is the result of sexual selection in animals?

This topic is beyond the scope of this text; the topic is another text in its own right. This is not even the first time something like this was suggested; it is the first time it is seen in its environmental context. Darwin's *The Origin of Species by Means of Natural Selection,* and *The Descent of Man and **Selection in***

Relation to Sex both contain excellent discussions on sexual selection, as does Douglas Futuyma's *Evolutionary Biology*. Levinton, out of several other possibilities, acknowledges sexual selection as a possibility that leads to speciation.

In an effort to minimize extensive discussions and supporting evidence, I omitted some very interesting, but voluminous, possibly marginally-relevant material related to the conditions of life, regarding the origin of species. I was tempted to present the case for, and the evidence supporting, the idea that species generation and maintenance was due to sexual selection (where it exists), or that a population of a species is limited by an environmental barrier, and that speciation origin is actually not accomplished by the means of a natural selection (fatal competition within a species) process, as Darwin suggested in his first edition. Natural selection is the genetic bias of environmentally more-fit types; it supports maintenance of the species' gene pool. Had Darwin included his sexual selection as one process of a combination of many other environmental processes, comprising an overall process of natural selection, it may have been more correct. This is not a criticism of Darwin; he just did not have genetic or ecological information of today available to him, nor did he have access to sexual selection experiments in animal behavior as discussed in the Futuyma text. Darwin did all of the pioneering groundwork; he essentially got it right 150 years ago—he is my hero. If a reader wants to follow this further and arrive at their own conclusions on this matter, the included references should suffice. A topic can be "Googled" on the internet. I tried to limit the topics related to the conditions of life; it is such a comprehensive subject, yet, sexual selection and speciation are certainly important processes in the conditions of life.

Can I provide an example, within a species, where fatal competition and other forms of environmental selection are interacting with genetic bias?

Africanized honey bees (known colloquially as "killer bees") are hybrids of the African honey bee (*Apis mellifera scutellata*) and the European honey bee (*Apis mellifera linguistica* or *Apis mellifera iberiensis*). In areas of a suitable temperate climate, the more aggressive survival traits of Africanized queens and colonies outperform less aggressive European honey bee colonies. With interbreeding, queen "killer bees" (subspecies) hatch earlier than queen European bees (subspecies); African queens kill all unhatched European queen competitors in the hive. Their competitive edge leads to the dominance of African traits. In Brazil, the Africanized hybrids are known as assassin bees, for their supposed habit of taking over an existing colony of European bees. According to this lore, their queen waits outside while several worker bees infiltrate the hive by bringing in food, where they will then locate and kill the queen. The new queen will then enter and take over the hive.

The chief difference between the European races or subspecies of bees kept by American beekeepers and the Africanized stock is attributable to selective breeding. The most common race used in North America today is the Italian bee, *Apis mellifera linguistica,* which has been used for several thousand years. Beekeepers have tended to eliminate the fierce strains, and the entire race of bees has thus been gentled by selective breeding. In central and southern Africa, bees have had to defend themselves against other aggressive insects, as well as honey badgers, an animal that also will destroy hives if the bees are not sufficiently defensive. In addition, there was formerly no human cultural tradition of beekeeping in Africa, only bee robbing. When anyone wanted honey, they would seek out a bee tree and kill the colony, or at least steal its honey. The colony most likely to survive either animal or human attacks was the fiercest one. Thus, the African bee has been environmentally-selected for ferocity (Wikipedia).

Is species selection possible?

It is. Species differences often include morphological differences that may be of adaptive value in the environment. This does not necessarily mean that environmental survival advantage is species-specific. Environmental survival is adaptation-specific; this adaptation must be inheritable. Because speciation represents breeding populations, environmental influences on natural selection among species has a long-term result, seen as a morphological trend in the fossil record.

How do I define macroevolution?

Macroevolution is the big picture of historical and ecological types of life. It is totally integrated with the small-scale processes of microevolution, such as mutation at the molecular level. Neither is irrelevant.

Is mass extinction an example of microevolution?

Mass extinction is really about the predominance of death from massive habitat destruction e.g., most all types of life on earth die. This would be a "macroevolution" event. Microevolution enters the picture when environmental stress conditions generate a genome-related adaptive survival response, which is beneficial to the developing organism and passed to offspring. This response requires a small amount of genetic variation that was originally the result of rare mutations (in a population), which can provide genetic expressions that may become favorable adaptations in a new environment; this often involves epigenome-related change. Existing mutations (in the genotype) often have neutral value in the existing environment, but will sometimes have survival advantage when expressed as a new phenotype under the stress of new environmental conditions. If an adaptation provides environmental advantage, it becomes incorporated into a new genetic bias of the phenotype. Natural selection genetic bias is the genetic bias of survivors (perhaps confined to fringe areas) that had the ability to adapt to the stressful environmental conditions of isolated areas, and were able to genetically transfer this adaptation of the phenotype to offspring. Following this microevolution event, the appearance of new types becomes a macroevolution event.

Do I believe that natural selection is the only mechanism of evolution?

To answer the question, no; it is not the only mechanism of evolution; natural selection genetic bias is a current result of evolution that has been influenced by mechanisms involving both heredity and environment (usually epigenome expression modification through environmental signals). The limiting factors of environmental change create an ongoing selective mechanism that never ends; successful phenotypes are preserved through inheritance. Recall that the Foundation of Modern Evolutionary Biology chapter established that natural selection is more of a result than it is a process. If it is a process, it is a combination of processes, primarily involving environmental conditions, whereby successful phenotypic adaptation components are continued in the genome as the genetic bias of natural selection. So, natural selection is the genetic bias of environmentally more-fit types (the preservation of favorable variations and the rejection of injurious variations); this is a result, more than it is a mechanism of evolution. Natural selection genetic bias provides environmentally time-tested genetic raw material (pathways) for environmental adaptation, the driving force of evolution. Extinction is primarily due to habitat destruction, and in the case of massive habitat destruction, natural selection is a genetic bias of survivors (perhaps confined to fringe areas) that had the ability to adapt to the stressful environmental conditions of isolated areas. Survivor offspring had the genetic bias of more-fit types, and diversified/disbursed into new environments, exploiting new opportunities. These environments were often increasingly moderating, but were still limited by environmental selection. Evolution is a change in types of life over time. Both reduced biodiversity and extinction play a major role in the pattern of evolution; even a lack of biologic change influences the overall pattern of types of life over time. Evolution has major environmental mechanisms. Any limiting mechanism in the environment, such as climate, predation, competition, even disease, will affect the genetic diversity, response to stress, and distribution.

Why do I think nearly everyone else believes the environment plays more of a passive role in evolution?

I greatly respect Darwin and his theory of natural selection, as the explanation for the origin of species (different types of life). His work has been regarded by many as the greatest scientific discovery of all time. In trying to support the importance of inherited variation, he assumed a stable environment. The environment is not stable, even when some consider it to be stable. It is only right that his views are strongly supported today; they deserve to be strongly supported.

It is not wrong to supplement the overall perspective on the process of natural selection. Many people, including myself, believe that the order in the natural world seems to come from more than chance genetic mutation (accident) and chance survival (accident), modified by an inherited advantage in competition for limited resources. Selective forces in the environment are variable, and are as important as genetic variation. Environmental selection is the gatekeeper of genetic success. Environmental selection plays the major role in extinction. Extinction and environmental change play the major roles in environmental creation; environmental creation may be subtle. Something is very special about environmental creation; it enables the selective forces of the environment to interact with life's genetic potential. The creation of an isolated changing, environment must occur to initiate environmental adaptation, the most important life-changing phenomenon. Heredity and environment are co-dependent; adaptation is an indirectly-integrated environmental and genetic process. Environmentally AND genetically determined evolution (change in types of life over time) results when life comes under the selective forces of an isolated, changing environment; life will adapt or not adapt to that unique environment. Natural selection is the result of an environmental selection process; natural selection is the resulting genetic bias. Natural selection is the genotype preservation of favorable environmentally-related phenotypes. The phenotypes preserved in the genotype are expressed by the epigenome, and it is environmentally-sensitive.

Extinction can also be due to inbreeding, genetic drift, and other biological processes that weaken the gene pool; how can I say extinction is primarily due to habitat destruction?

I will not argue these possible causes of extinction; I simply do not believe that these are very significant as causes of extinction. There are many causes of extinction, and they seldom act alone. I tried to substantiate habitat destruction as the primary cause of extinction in the chapters on Major Environmental Extinction Events, Extinction Events Initiate Change, Chicken Little Was Right, Fire Down Below, Constant Extinction, Ecological Succession, Extinction Event Significance, Environmental Extinction and Environmental Creation. In turn, this environmental destruction plays a significant role (triggers the creation of special environments with selective influence) in evolution.

Can you have evolution in a stable environment?

Environmental creation includes and follows environmental extinction events. Environmental creation can also occur in a more steady-state environment. Soot created a special environment in Darwin's England that favored the survival of dark-colored moths over the survival of light-colored moths. Today the soot is gone, and light-colored moths have the environmental advantage of better camouflage to escape predation. Evolution occurs in any isolated, changing environment; it may be subtle.

What about non-adaptive traits?

Non-adaptive phenotypes are eliminated by environmental selection. Assuming the non-adaptive phenotype experiments in the chapter on Phenotype and Environment, Difficulties on Theory, and Genetically-Determined Evolution have been reviewed, recall that under severe environmental stress, only adaptive phenotypes survive environments lethal to non-adaptive phenotypes. Non-adaptive traits may persist for a time in environments of lower stress. Neutral-value phenotypes may persist for an even longer time. To be a neutral phenotype, any slight disadvantage must be offset by an equal value of advantage in environmental adaptation. Non-adaptive phenotypic traits in wild populations are likely to be present as genotype norm-of-reaction possible, random neutral phenotypes, produced by some degree of environmental stress; some of the random neutral phenotypes may even lie outside the genotype norm-of-reaction. If there are genotype expressions of phenotypes that appear to be non-adaptive, these non-adaptive phenotypic traits may have been adaptive to an ancestor at an earlier time. Non-adaptive phenotypic traits may be related to (i.e., another aspect of) a beneficial adaptation. All traits are not all adaptations; a given trait may be just another feature of an adaptive phenotypic trait. The red color of hemoglobin and the green color of hemocyanin are both traits of color for oxygen-carrying blood pigments; the traits of color are only another manifestation of the oxygen-transporting molecule adaptations. Non-adaptive traits may be genetically-linked to adaptive traits, either as true linkage on a chromosome, or, due to genetic constraint, inseparable from a beneficial adaptation; the latter may involve developmental constraints. Pleiotropy, a single gene expression of multiple traits, has been discussed previously in the text.

Is stabilizing selection adaptive?

Stabilizing selection is environmental selection for the conditions of life in an isolated, unchanging (stable) environment. By using environmental feedback to display the most appropriate phenotype, developing organisms maximize inheritable adaptation options with the plastic expression of the best-match preserved phenotype adaptation from the norm-of-reaction; the environmental feedback may even allow small adjustments in epigenome expression (genetic accommodation). "(In a population), stabilizing selection operates by culling-out extremes from the phenotypic distribution; a modal phenotype is thus favored" (Levinton). Organisms continually adapt to the environment over time, assuming natural selection's genetic potential to do so. Stabilizing selection is highly adaptive in a stable environment; but, it may reduce biodiversity. Under environmental extremes, a special commitment to the previous environment is not necessarily adaptive.

What do I think about cladogenesis?

The generation of shared character states is a genetic-related adaptive response to environmental challenge. In animals, a species is a sex club that propagates existing genetic bias. An isolated environment under unusual stress will produce novel phenotypes in developing individuals. If the novel phenotype has an environmental advantage (an adaptation), its survival and reproductive ability will be improved compared to other members of the species. An offspring with significant advantage (by historical proof), let's say a jaw muscle mutation (autosomal dominant loss of myosin) in a robustus hominid, will not develop a crest in the top of the skull for strong muscle attachment. The environmental selection will really be for more cooperative behavior between compromised early human ancestors. Loss of the skull crest likely enabled the formation of a larger brain and this further enabled cooperative behavior and learning. Thus a mutation leads to a suite of adaptations in a stressful environment. The unique mutation was a cut away from the ancestor population, forming a clade that was originally nested within the ancestor population. A similar pattern works equally well in plants; however, flowering plants are married to insect pollinators, and either or both may be species-specific.

What do I think about genetic constraints on phenotypic variation? What do I think about developmental constraints?

Genetic and developmental constraints throughout ontogeny and phylogeny are "hard-wired" biological expressions of genome action or functional limitation; I consider genetic constraint expressions as a form of past environmental "fail-safe patterns."

What percentage of beneficial alleles is lost before fixation? What do I call that process?

Assuming the experiments in the chapter on Phenotype and Environment, Difficulties on Theory and Environmentally-Determined Evolution have been reviewed; an expressed gene's fixation into 100% of a population comes from inheritability of an environmentally-supported parent's transfer of preserved traits to offspring. In detail, any newly expressed gene's contribution to the phenotype may or may not be adaptive; the latter is more likely. Non-adaptive phenotypes are costly non-efficient burdens, and cannot be long supported in a challenging environment. However, an expressed gene that contributes to an adaptive phenotype will be carried onward, preserved within offspring as preserved genome bias of natural selection, and displayed when the newly-inheritable environmentally-integrated phenotype offers environmental advantage.

In the preceding clade mutation example, "fixation" of an inheritable autosomal dominant trait favored rapid establishment through favorable selection of offspring. It is also possible to fix a gene expression of neutral value (non-adaptive) into 100% of the population, due to environmental stress alone, providing the stress is continued over some time and that the stress is below the lethal limit of the phenotype expression (Waddington stress studies previously discussed). It is even statistically possible to have a non-expressed neutral gene fixed into 100% of the population (Futuyma). In his chapter on Genetics, Speciation, and Trans-specific Evolution, Levinton provides argument against genetically-determined evolution when he notes that: "The rate of chromosomal fixation is inversely proportional to body size (Benson).... Despite the widespread occurrence of chromosomal races, the evidence does not support any concomitant morphological differentiation.... Phenotypic shifts, when they occur, are liable to be rapid and of short duration, relative to longer periods of stabilizing selection.... Climatic shifts are often equally sudden and will select for rapid evolutionary change.... Chromosomal change, therefore, was likely not the cause, but was more likely the effect of evolutionary radiation and speciation.... Rapid speciation of mammals, which seems to have occurred over time spans of thousands of years, is inconsistent with the average fixation rate of new chromosomal arrangements, which is maximally on the order of one lineage per million years." The percentage of beneficial alleles lost before fixation is variable (forget fixation). Environmental conditions are variable; environmental conditions determine what is beneficial and what is not beneficial. Evolution is genetic success in response to environmental conditions. In mass extinctions, beneficial alleles already fixed in populations are lost, due to bad luck. I call this "chance." What would you call it?

What do I think about peak shifts?

Common environmental variation of the phenotype is formed by expressing existing adaptations, already preserved in the norm-of-reaction (as natural selection genetic bias). The combined interaction of existing natural selection genetic bias (the time-proven, best possible genotype norm-of-phenotype-reactions to previously-encountered environments) AND environmental selection (gatekeeper- setting direction +/or ground rules) helps insure maximum "adaptation" to the existing environment. (This "adaptation" relates to environmental sorting for selectable expressions of inherited adaptations from natural selection genetic bias—not to the origin of adaptations.) The epigenome is the environmental interface. The expression of preserved adaptations to common environmental variation is managed by plastic (epigenetic) change during organism development. Peak shifts will mainly involve this plastic range of common phenotype expression during development of resident organisms. Peaks were introduced in this text's Foundation of Modern Evolutionary Biology chapter. And while shifts could involve differences between only two alleles, because it is related to environmental variability, it almost certainly depends upon inheritable epigenetic expression differences, and almost certainly involves expression of more than one locus. But for a discussion example, assume there is only one genetically-determined peak of "optimum conditions" for a population of best-fit phenotypes, and maximum "adaptation" to this "environment" results in increasing the numbers of individuals, having the particular gene expression (say heterozygous) at one locus. Assuming a very stable environment (unchanging environmental selection), increased "specialization" (stabilizing selection—a phenotype-sorting for this expressed genotype; and/or epigenetic-genetic accommodation) to this particular environment causes the "adaptive" peak to increase in magnitude (increasing the number of individuals in the population) with the genetic expression advantage ("beneficial alleles"), and to narrow genetic diversity (as the number of other individuals without this expression decline). Environments are never stable for long. Environmental change will occur and then environmentally-select for alternate phenotypes with alternate genotypes (environmental selection for expression advantage in a different, but not uncommon, environment). The environment will select for plastic (epigenome) phenotype change in developing offspring (from natural selection's preserved norm-of-reaction) that involves plastic expression of other alleles (in pathways to preserved phenotypes). The "beneficial alleles" become "alleles non-grata." Levinton notes adaptive peaks undergoing major shifts with sudden landscape change are associated with a pattern of "constancy, sudden change, and then constancy; morphological change is to be expected." Levinton also discussed environmental correlations strongly associated with inversions, but this seems to be limited to fruit-flies. Darwin stated that natural selection (the preservation of favorable variations— rejection of injurious variations) does not infer perfect adaptation, only selection of the best-qualified competitive candidate.

The frequency of alleles in a population can change in the absence of "natural selection." What do I call that process?

I would call this random genetic variation in a population.

What do I think about genetic drift? What do I think about the founder effect?

Genetic drift is the fundamental tendency of any allele to vary randomly in frequency over time due to statistical variation alone (Wikipedia). According to Futuyma, "Adaptations do not result from genetic drift, so this is not responsible for many of the most interesting features of organisms... genetic drift constitutes evolution by chance alone" (a genetically-determined view). If adverse environmental selection leads to a sharp reduction in population size, environmental selection is involved and genetic drift is excluded by definition. A small population of survivors could pass through a "bottleneck" and present with a non-typical allele frequency, and these individuals could create a "founder effect," but it is a stretch to label such a "founder effect" as pure genetic drift. Futuyma continues, "Moderately common alleles, which contribute the most, are likely to be carried by the founders; if the colony increases very rapidly in numbers, further loss of genetic variation will be slow, but if it remains small, genetic drift continues to erode genetic variation very rapidly."

Can genes be reverse-transcribed?

Brosius, in the first essay of Vrba's *Macroevolution, Diversity, Disparity, Contigency* suggested retroposition (reverse transcribed RNA to DNA) as a mechanism to reintegrate retronuons into genomes. In many eukaryotic lineages, the process of retroposition is still very active. All types of RNA's can be reverse transcribed and their reverse-transcribed DNA reintegrated into genomes as retronuons. A nuon is any discrete segment of nucleic acid (Brosius and Gould). Respectively, about 38% or 42% of the mouse or human genomes, are retronuons (excluding mRNA), compared to the 1.5% of the human genome that are exons, coding for proteins. Brosius and Gould suggested that it is conceivable that much of the mammalian genomes are probably derived ancient retronuons, formed by retroposition. Retroposition predominantly leads to "junk DNA," but this may be coopted to form new genes.

What do I think of evolution as a process of intelligent design?

I think intelligent design is basically a variation of creationism. A design does exist; it is natural selection genetic bias and it is indirectly environmentally-integrated. Others will see this as proof of intelligent design. The issue is inconclusive. Because of religious conditioning, it is better to separate science, whose laws are demonstrable, from personal belief, which is not demonstrable, but based on faith in the truth of the belief itself.

In the text, I mention limiting factors and their affect on wild populations; how do they work?

One of the earliest examples of a limiting factor and its dynamics involved predation. Hudson Bay company records of purchased Canada lynx pelts and purchased varying hare pelts reflected population fluctuations of each from 1850 to 1930. In peak productive years, a female hare can produce 16 to 18 young; predation by the lynx is a predominant form of hare mortality. When the population of hares increased, it provided an abundant food source for the lynx; the lynx population then increased. When the population of hares decreased from increased lynx predation, the lynx population was soon decreased. There was a time lag which placed the lynx population peaks and lows slightly behind the hare peaks and lows; these cycles repeated on a 10-year frequency. The cold of the Canada winter was not the primary limiting factor in this instance; the limiting factor of cold suppressed the number of different species, so the two abundant populations revealed the primary limiting factor of predation.

The study of limiting factors is often more complicated, even in high latitudes where reduced biodiversity simplifies the number of biological variables. The University of Alaska and The US Fish and Wildlife Service have developed sound and efficient caribou management over the last half century. The state and federal agencies have spent millions of dollars on research, combined with extensive experience. The findings reveal that caribou population welfare is mainly dependent upon herd carrying capacity, weather, and predation. The carrying capacity of a herd is the maximum number of animals sustained by the herd's habitat quality and range. When the carrying capacity is surpassed, the over-crowded caribou herd must decrease or the ecosystem (environment) will rapidly deteriorate, as will the herd. While both can occur, usually the caribou suffer most. "Nature" does not reward overpopulation. The herd's habitat and range aren't chosen at random. Caribou prefer to eat mosses (lichen), grasses, sedges, mushrooms, willows, and flowers. Caribou are browsers; they usually do not overindulge in any one food source. Food diversity protects this sensitive habitat from excessive environmental impact due to overbrowsing. In addition, most caribou migrate twice yearly in habitual patterns; this range extension protects this sensitive habitat from excessive impact due to overbrowsing. This instinctive behavior of utilizing diverse food sources and migration over a large range helps them secure the future and quality of their forage. Habitat and range carrying capacity varies yearly, due to variations in the weather. It also appears that stability in nature is required for nutritional fitness and endurance. Weather has the power to limit the growth of a particular herd. The environment's carrying capacity is only as high as that which is sustainable by the worst part of the winter (the major limiting factor in many parts of Alaska). Deep snows hinder access to valuable food supplies and prevent effective escape from predators. Drought can damage sensitive vegetation on which caribou thrive; heavy rainfall has the same damaging effect. Limiting factors help balance the population. Limiting factors, like the rainfall, can also be excessive. When range quality is less than ideal, when weather is un-favorable, or when predation is excessively high, it is common- place to see increased herd mortality, decreased calf production, disease and/or famine. Range quality problems can be serious. Recall that "nature" does not reward overpopulation. Weather-related population crashes usually recover.

Predator-prey cycles usually even things out, unless man interferes (as in predator overprotection). Any one factor (caribou herd habitat quality/range, weather, or predation) can cause a population decline effect. The above limiting factors can combine for devastating effects on the herd; combinations of limiting factors frequently occur. Under-predation and over-predation extremes are both a problem (like the rainfall limiting factor). Too many predators have proven to limit the growth of big game (herbivore) densities; 80% herd calf kills by wolves are documented. Sound predator control, especially of wolves, has reversed severe herbivore herd declines in almost every study. In extinction events, top predators get hit the hardest and are among the first to die; this principle also applies here. It is the wolf that is more threatened by over-predation. There are political preservationist and environmental protectionist problems with predator control. Unsophisticated, emotional wolf-lovers don't understand; "nature" does not reward over-predation. Hunters are predators; management regulates hunting and this has worked very well; populations managed for hunting fare better than many "protected" populations. Hunters pay for conservation and most wildlife management activities they were the first conservationists.

If Iowa has lost more habitats, more landscape areas of habitats, and the most wildlife (number and biodiversity), can anything be done to protect what is left?

Currently, 238 species of native Iowa plants and animals cling to an uncertain future, while listed as threatened or endangered species. Any endangered species' protection is primarily dependent upon its relationship to the ecosystem's food web and other interacting environmental conditions; any effort to save an endangered species will be more successful if the entire habitat is protected. Preserves of all kinds can really help, protected areas can add-up. Abandoned homesteads can be wildlife sanctuaries, where wildlife is seldom disturbed. By eliminating mowing and/or spraying these areas, insect survival is increased, and so is the survival of birds and mammals that feed on them. Flowering shrubs and plants sustain pollinators, serving as a hummingbird, butterfly and bee food source. Insects that are attracted to the flowers will feed other wildlife. Dogwood and serviceberry provide berries (food). Rock piles, brush piles and dead trees provide shelter for wrens to turkeys or for chipmunks to deer.

Reversing an ongoing habitat loss through a humanity-compatible habitat expansion offers the very best hope for both wildlife and mankind. One way to improve non-crop farmland habitat is to manipulate those areas where existing environments or conservation techniques designed for other benefits (e.g. soil erosion control) are found. There are numerous current-use agricultural practices suitable for enhancing wildlife habitat. In general, these areas tend to be long and narrow. Research indicates these linear areas may not be suitable for ecological specialist species (specific habitat requirement); however, if managed carefully, many ecological generalists (general habitat requirements) can benefit from such areas. The Conservation Reserve Program (CRP) is the federal government's single largest environmental improvement program and one of its most productive and cost-efficient. The Conservation Reserve Program provides buffer strip assistance by subsidizing landowners to not plant crops on steep-sided erodable land near streams. The grasses and trees in this buffer area reduce soil transport into waterways and the resulting habitat benefits riparian and aquatic wildlife. The CRP is administered by the U.S, Department of Agriculture's Commodity Credit Corporation (CCC) through the Farm Service Agency (FSA). CRP may be the most important subsidy of today. The CRP-related farm bill saves highly erodable topsoil and creates huge areas of wildlife habitat. If it is converted to prairie grasses, it becomes the very best habitat for Iowa wildlife. In 2005, national conservation payments stemming from the farm bill totaled $7 billion. When it comes to the protection of wildlife, nothing else comes close to this. In 2005, the combined budget for Bureau of Land Management and Forest Service (both manage 16% US land mass) was $5.4 billion, including the budget to manage fish and wildlife. In 1996, wildlife resources became a farm bill priority. The main significance of this is that when wildlife is tied to other programs, farmers are more likely to return these highly-erodable lands to bare-ground crop production (corn and beans). But corn prices are increasing, subsidies must compete.

In the past, helping endangered species often meant acquiring lands to protect crucial or unique habitats. Today, the Landowner Incentive Program (LIP) grant from the US Fish and Wildlife Service provides Iowa with more than $137m for a 75% cost-share with private landowners that wish to voluntarily protect or enhance habitats used by threatened or endangered species. There are other subsidies that generally support habitat statewide. WHIP, the USDA's Wildlife Habitat Incentive Program also helps landowners enhance, protect and develop wildlife habitat on their property. Through WHIP, the Natural Resources Conservation Service (NRCS) works with landowners to develop wildlife habitat plans and provide cost-sharing assistance for implementing wildlife habitat management practices. The Environmental Quality Incentive Program (EQIP) is a USDA conservation cost-share program designed to encourage and support conservation of natural resources on private lands on a voluntary basis. EQUIP subsidizes formation of grass waterways to minimize erosion in cropped areas; this provides valuable habitat, but they can be a problem for nesting birds during spring rains. The Iowa Financial Incentive Program (IFIP Cost Share) and Resource Enhancement and Protection (REAP) also subsidize grassed waterways to reduce erosion and provide habitat. In addition, the Environmental Quality Financial Incentive Program (EQIP) subsidizes the formation of earthen structures designed to minimize erosion by slowing runoff water to non-erosive speed on moderate to steep slopes (slowing water slows erosion). Properly managed terraces provide valuable nesting cover and foraging for some birds. EQIP also subsidizes tree planting and shelterbelts. Windbreaks are permanent plantings of trees, most often located on the north and west sides of farmsteads. These are designed to lessen impact of prevailing winds, protect livestock, protect young crops, and control blowing snow.

In addition to an economic benefit, properly designed windbreaks are valuable wildlife habitat. Several species of breeding birds use windbreaks as nesting sites; small mammals use them for food and nesting habitat.

The tallgrass prairie is the most endangered ecosystem in the nation. The restoration of Iowa Prairie habitat is occurring at the federal, state and private levels. Restoration is a process of reestablishing, or bringing it back to the original condition. It involves identifying a prairie remnant (or other suitable restoration area), evaluating the site quality, and planning appropriate long term management. Understanding plant tolerances and preferences helps match the restoration site with plant characteristics. Successful prairie restorations mimic native plant affinities, creating diverse prairie patches, similar to naturally occurring communities. Knowing the differences between types of prairie communities is important to management plan. Wet prairie is found where clay, silt, loam or organic soils predominate; wet prairie is typically found in low-lying areas, but may be in uplands where soil holds water. Dry prairie is found on sandy well-drained soils, high above the water table; dry prairie is usually found in areas where rainwater rapidly runs off or soaks in, or where rainfall is less than 10 inches. Mesic prairie is found in areas between wet and dry, where rainwater doesn't collect; drainage isn't as rapid as in the dry prairie. The Iowa State Department of Transportation is realizing that there is a great cost savings in maintenance if roadsides are restored with native grasses. In other words, replanting the roadside ditches in native prairie grasses saves mowing and herbicide expenses; when blizzards whip across the prairie, prairie grass also acts as a snow fence, keeping ice and snow off of the roadway. Prairie grasses return every year and do not interfere with adjacent agriculture. Prairie grass roadsides are an ideal wildlife refuge; prairie grass roadsides have the potential to provide habitat for what wildlife populations still remain, thereby utilizing already existing public land. In Iowa, lack of winter habitat (that provides shelter from storms) is usually the primary limiting factor for all wildlife struggling to survive in a sometimes-brutal, snowy, icy landscape of plowed fields and mowed ditches. Native prairie grass provides winter shelter from the storms, along with a much-needed food supply; high-quality food is needed to combat heat loss from the extreme cold. Deer feel secure in prairie grass cover, and are less likely to dart in front of cars, as they tend to hold in cover as danger approaches. Deer invariably run, straight ahead, when exposed to danger in an open ditch with short mowed grass. Deer are the most dangerous animals in Iowa (even if cougars are around). In addition to state and federal highways, county secondary roads divide nearly the entire state in square mile sections. The largest tract of public land in the state is in the roadside ditches. The restoration of the tallgrass prairie habitat along roadsides offers the greatest hope for a timely widespread protection of what remains. Some county engineers are adopting the lower maintenance prairie grass replacement of the maintenance-intensive short brome-grass, which provides no winter food or shelter. Other county engineers are reluctant to change; they appear to be insensitive to the increased energy costs of mowing, insensitive to wildlife destruction from mowing, insensitive to herbicide costs, insensitive to herbicide damage to wildlife, and insensitive to protecting what remaining wildlife struggles on, in marginal existence, primarily due to an increasing loss of habitat. The public owns this land that is often poorly-managed for wildlife. The public is in very good position to exert pressure on public employees to improve habitat, through prairie roadside restoration, at no additional cost to anyone. Many environmental problems result in frustration, due to the fact that there is little that a single individual can do. In the case of roadside prairie restoration, the individual can make a real difference. Prairie flowers are beautiful in the summer.

Europeans, unfortunately, realized the damage that extensive wind and water erosion produced when they increased field sizes to that of those seen in America. European agricultural production is restricted to the limited topsoil that still remains after centuries of misuse. The same erosion occurs in America; but topsoil loss is less obvious, due to greater topsoil thickness. Half of the Iowa topsoil is gone; it took just over a century for the loss. It's time to change to better conservation methods before the other half is lost. One way to minimize the topsoil loss to wind (and water) erosion is to reduce field size. The small family farm can be the best steward of the land. If a 2½% border of prairie grass frames each small field, wind and water erosion of that field will be greatly minimized; restoring (4 sides = 10%) the field perimeter to native grasses protects the entire field. Periodic Spring burning prevents inappropriate tree growth. Periodic partial harvesting is a cost-effective, less-polluting source of electricity or ethanol. Topsoil belongs in agricultural fields, not in streams. If prairie perimeters are combined with no-till or minimum-till agriculture, there is even greater promise of saving valuable topsoil. This field-framing prairie grass planting will supplement wildlife preservation to a significant degree, providing travel corridors, food, shelter and less-disturbed areas away from roadsides. In that farm subsidies are basically government assistance for the benefit of all society, it makes sense to help the small family farm preserve topsoil and wildlife, at least as much as present subsidies have been displacing small family farm society (at a great cost to the rest of society) by preferentially subsidizing environmentally-unfriendly big corporation farms? (Market-based price guarantees for crops unfairly favor large corporation farms better able to operate on a small margin.) These federal deficiency payments comprise the greatest subsidies for agriculture; they involve land under intensive bare-ground agriculture in corn and soybeans. Bare ground agriculture suffers the most extreme erosion. By comparison, today's subsidies for soil conservation, habitat, and wildlife are quite small. The World Trade Organization is progressively eliminating market-based subsidies. In the long-term, subsidies based only on the conservation of civilization's most important resource (topsoil) have a much better chance of survival. And, they have a better chance of doing the most good.

Wetlands support favorable levels in the water table, minimize flooding and detoxify pollution from fields and streams. Restoring Iowa wetlands habitat is crucial; most prairie wildlife is dependent upon wetlands for some part of their life cycle. Ducks Unlimited and the U.S. Fish and Wildlife Service have joined in efforts to restore Iowa wetlands across the state. Individual landowners, with government assistance, are also creating and restoring the lost wetland community. In diversity and form, created wetlands should imitate the ecological processes of those occurring naturally. The restoration process may be as simple as removing a drain, or more complicated and costly. Interested individuals should contact local professionals to discuss objectives, site selection, wetland design, evaluation (documenting plants and animals over time change), management and permits. Some unplanned Iowa wetlands are forming rapidly. In cases where a large shallow lake is fed by a feeder stream, wetlands can form in a half-century, filling the entire lake with silt deposition so extensively that the former lake can be crossed at certain times of year without getting wet feet. The silt comes from topsoil loss, related to negligent bare-ground (corn and bean) agriculture.

Nutrient pollution exists in every lake in Iowa. As fertilizer runs-off bare-soil agricultural fields, this fertilizer influx causes over-fertilization of the water. An algae bloom (eutropification) decreases photosynthesis as turbidity increases, and the bloom's increased respiration removes dissolved oxygen in the surface water, making the conditions difficult for all aerobic life. Installing sewer lines on shoreline residences, instead of allowing septic tank field drainage into the lake, helps reduce this eutropification; however, if any agricultural runoff drains into the lake basin, eutropification is rampant. Most Iowa lake waters are silted into extreme turbidity by feeder streams; this silt suspension results in a brown glow at the surface and total darkness not far below, as depth increases. The darkness (even more-so than turbidity) eliminates any photosynthesis. Silt pollution, Iowa's number one water pollutant, depletes oxygen at all depths. This limits aerobic life; dissolved oxygen is depleted by any aerobic respiration. Particularly in deeper water, both silt suspension and eutropification sharply reduce oxygen levels in all Iowa lakes, degrading the habitat of aerobic life. Late summer thermal stratification prevents any surface aeration of deeper waters. Water clarity (Secchi disk readings) in lakes not silted-full by feeder streams is less than half that of a half-century ago. The clarity situation improves somewhat under the winter ice, when streams no longer drain the frozen fields, but snow cover on the ice will increase the vegetation "die-back" (and decrease photosynthesis), as the darkness increases with depth of the snow cover.

Iowa rivers and streams must be included in any plan for wildlife habitat improvement. Iowa has six major drainage basins and over 27,000 miles of perennial streams. Exchanging agriculture, cities and transportation corridors for prairies, wetlands, and forest has degraded Iowa's riverine systems. Mussels in streams have been compared to canaries in coal-mines; both are highly sensitive to environmental danger. At one time mussels lined Iowa streams, bank-to-bank. They are absent from around half of Iowa streams today; silt from topsoil loss is the main, but not the only, cause of their decline.

Bare-ground row crops in rural areas and concrete surface cover in urban areas have increased water runoff to streams, bypassing groundwater or water table replenishment. The loss of grass and forest bank areas has caused entrance of soil, nutrients, insecticides, herbicides, and greatly increased amounts of water into waterways. Pesticides for weeds and insects, and nutrient pollution threaten the water supply for all life. Combined with levees and damming waterways (levees and dams raise the stream bed, promoting flooding), and channelization (channelization eliminates habitat, accelerates flow and promotes erosion), stream silt burden, along with nutrients and pesticides, join to create pollution that has degraded 99% of Iowa streams. Iowa streams turn clear when agricultural runoff decreases. Under the ice, they are crystal-clear every winter. Before the arrival of bare-ground agriculture, they were crystal-clear and unpolluted year-round (floods excepted). Presently, nutrient and topsoil pollution is beyond excessive in Iowa streams, especially after rainstorms or snowmelt. The Des Moines River is unsurpassed as a nutrient pollution producer in the U.S. (Fremling). The pollution effect carries to the Gulf of Mexico, damaging coastal areas by forming increasingly-expanding dead zones. Two-thirds of the water carried in streams passes during seasonal floods and is lost to the water table, unless held in floodplain wetlands. (Partly due to the increased amounts of water entering the streams and partly due to the interruption of natural flow patterns, there has been a long-term increase, not decrease, of flooding intensity, and flooding frequency.) At these times of greatest stream flow, silt and eutropification reduce oxygen levels in streams, degrading the habitat of aerobic life. Previously mentioned, streamside wetlands filter-out silt, nutrient pollution, toxic pesticides and pathogens. Biotic diversity in Iowa waterways has been damaged from loss of habitat; this damage must be better managed (through habitat restoration).

Restoring Iowa rivers and streams requires understanding of stream ecology, including land-water interrelationships. Moving water aerates the unpolluted stream. Some plants and/or animals are adapted to the swift current of small stream; others favor the soft, slow flow of a river's backwater. All plants, including aquatic, require nutrients, oxygen and sunlight to flourish (primary production). Many aquatic plants and animals have specific habitat requirements such as water temperature, O_2 levels, water depth and velocity. The high velocity of channelized streams is not life-friendly to most aquatic life in Iowa. Accompanying the "in-stream" plant community is the "streamside" (riparian) vegetation; both in-stream and riparian vegetation influence water temperature and productivity along the stream flow. A variety of stream invertebrates assists in breaking down large particulate material into small pieces and dissolved substances. Also important to stream productivity are the decomposers (aquatic bacteria and fungi); they change organic debris into detritus. Others feed on detritus, and others feed on them; nutrients recycle. In-stream treatments can improve habitat. The damming of low gradient streams slows water flow and increases sedimentation; damming is a problem in Iowa. Wing-dams used on Mississippi provide valuable habitat, until silt accumulates and eliminates this habitat. Consideration of erodability of adjacent stream bank is critical to any alteration. Water current deflectors protect deep water habitat and reduce bank erosion. Boulders (> 2' in diameter) create pools and eddies important to invertebrate and fish habitat; in smaller streams, smaller boulders serve the same purpose. Underwater structure is underwater shelter for wildlife. Stream-side modifications also improve habitat. Levees not protecting heavily populated areas should be removed; levees act like dams alongside streams and prevent wetland flooding, decrease drought stream-flow, lower the water table and raise the stream bed to increase future flooding. Livestock exclusion from the stream protects the stream-bank and this creates less erosion and less pollution. Fencing (crossings too) and off-stream watering protects the stream integrity and livestock water source. Stream-bank stabilization is critical for steep or poorly vegetated banks. Resloping the bank may be necessary, which may include installing a layer of various sized rocks (rip-rap) or tire and rock. For esthetic reasons, planting willow cuttings or grasses are often preferred (riparian buffer strip).

Properly managed woodlands meet the basic needs (food, water, shelter and space) of wildlife year-round. Management plans may focus on attracting one specific type of wildlife; others may be more general and manage for variety. Regardless, management requires knowledge about target species' habitat requirements. For example, deer prefer edges and dense under-story; however, pleated woodpeckers prefer deep, mature woodlands. Wildlife needs that dead tree; forest birds nest in dead trees. Woodpeckers especially control tree insects; they also make a primary cavity inside a tree. Chickadees, bluebirds and wood ducks utilize this as a secondary cavity. Squirrels, raccoons and possums make dens in dead tree cavities. Downed logs are used by chipmunks, rabbits, weasels, gray and red fox. Logs are also valuable to frogs, toads, lizards, and snakes; they use them for winter hibernation. Maintaining three hard and two soft snags/acre of woodlot helps improve woodland habitat shortage. The primary goal of any woodland restoration should include providing suitable habitat for the establishment of over-story, under-story and non-woody plants. Woodland restoration should define management goals and objectives. Small woodlots will be mostly an edge environment. Large woodlands are needed for deep-woods species. Restoration should consider woodland age and proximity to water. Restoration should develop a specific plan: site selection, planting methods, plant or seedling sources, and maintenance require consideration. Woodland habitat restoration is most suitable on once-forested sites (soil texture and color are lighter than prairie). Soil resources (pH, texture, drainage) influence the decision. The Natural Resource Conservation Service has soil locations and descriptions. Starting from scratch (bare soil) takes decades; there should be an adjacent woodland for a seed source. An inventory of existing plant (and animal) community provides natural history clues and influences the process. Introduced species should match the site; restoration should consider winter hardiness, insect and/or disease resistance, ultimate size and form, and the purpose in the landscape. Further details may be found by consulting the Iowa State University Extension Office Publications: *Managing Iowa Habitats* (sourced extensively in the above answer).

46

Human Population and Resource Questions

Is it fair for developed countries and the people within them to have all the wealth?

Karl Marx, Vladimir Lenin, Joseph Stalin and Mao Tse-tung didn't think so; they had other plans, in which the government took care of everyone, the wealth and everything, except for the environment. For the members of society, communism does not encourage enough production incentive to generate significant wealth from the sale of surplus goods; furthermore, a compromised communal environment does not allow enough production to even generate wealth. Communism's environmental concerns are closely related to the "tragedy of the commons" experience; recall that the opportunity to gain wealth was compromised by environmental destruction (ecological suicide). The environmental destruction seen in former socialist bloc countries is as horrific as their consideration of human rights. Fairness did not exist there.

Wealth is generated as the result of trade, created from surplus production and its sale at a profit. This requires hard work, beyond meeting daily subsistence needs. Wealth must involve responsible population control. Is the North American family, with an average of two children, earning an average income, rich, or poor (by world standards)? Increase the family size to ten children, without increasing the income; are they (still) rich or (now) poor? Do it again, generation after generation; what happens to the amount of valuable resources available for each child? What happens to the educational, health care and employment opportunities? Population irresponsibility has consequences; it deprives everyone, especially the children, from quality of life. Previously noted, overpopulation concerns are also closely related to the "tragedy of the commons" experience.

Doesn't a person from an underdeveloped country have the same rights to the world's resources as a person from a developed country?

There is little purpose in discussing rights in societies where even the most basic rights do not even exist; human rights are definitely not a priority in communistic societies. The 2nd world (communism) failed to improve the quality of human life. The poor overall quality of life itself, as well as communism's extremely compromised, unimaginably polluted, toxic environment, led many members of the society to a premature death. There is even more of a conspicuous absence of human rights in overpopulated, underdeveloped countries. That is, rights and freedom fare far worse here than under communism. Today, in sub-Saharan Africa, there are more than 800,000 slaves in Niger; this is 8% of the population! The same (or very similar) traditional institutions of slavery exist in Benin, Chad, Ghana, Mali, Mauritania, Nigeria, Somalia and Togo! Increasingly-more-overpopulated India also has slavery—today (BBC news)!

Overpopulation severely devaluates human rights. Overpopulation severely damages the environment. And, similar to, but even worse than communism, where basic human rights are a luxury, the rights to the most basic environmental resources are an even more of a luxury in overpopulated underdeveloped countries! There is a conspicuous absence in the quality and quantity of even "free" resources, such as clean air and fresh drinking water, in nearly all 2nd, 3rd and 4th world countries. The total lack of human rights and horrific lack of environmental quality in overpopulated undeveloped countries certainly isn't right! Overpopulation devalues human conditions of life.

Access to resources of value, and wealth, do not appear to be a right, but the result of trade, created from surplus production and its sale at a profit. This requires hard work, beyond meeting daily subsistence needs. It also requires responsible population control. There must be a balance between population numbers and resource consumption. The 1st world has achieved this balance. Ownership promotes responsibility. Problems inevitably occur if the balance is lost. Surplus populations are populations that are out-of-balance. Surplus production is not possible when irresponsible populations over-utilize a finite quantity of non-renewable resources; even over-utilized renewable resources can quickly become non-renewable resources (e.g. fresh water and fish). Surplus populations over-utilize resources and an increasing overpopulation increasingly decreases the possibility of surplus production. Surplus production and its sale at a profit is necessary to transition from a subsistence economy to a productive economy. Irresponsible overpopulation will not allow the transition to a sustainable society. The inevitable shortage of essential resources from overpopulation ultimately leads to an exploitation of others; the shortage of essential resources ultimately leads to anarchy and war. Does an irresponsible surplus population have a right to forcibly take whatever they want from the hard-working production efforts of a responsible population? The exploitation or destruction of "outsiders" is the very nature of war. I have one grand-child; this is not uncommon in developed countries. In undeveloped countries with irresponsible population growth, it is not uncommon for people of my age to have 100 grand-children. Ask me if I believe my single grand-child has no more right to resources of value than each of the 100 other grand-children (in this example).

Where is the compassion for humanity?

The answer to compassion concerns requires a visit to overpopulated sub-Saharan Africa today; in addition to the area's unthinkable slavery problem, half of the countries included in this area are already suffering badly from increasing drought. The inevitable conflict has already begun. Overpopulation of both pastoral Arabs and black farmers created the resource shortage of water and land that underlies the conflict. Darfur has faced many years of tension over land and grazing rights between the mostly nomadic (pastoral) Arabs, and black farmers from the Fur, Zagawa and Massaleet communities. Black farmer refugees from Darfur say that following air raids by government aircraft, the Janjaweed (Arabs) ride into villages on horses and camels, slaughtering men, raping women and stealing whatever they can find. Millions have fled their destroyed villages, with many heading for camps near Darfur's main towns. In the camps there is not enough food, water or medicine. Janjaweed (Arabs) also patrol outside the camps; Darfurians say the men are killed and the women raped if they venture too far in search of firewood or water. Many women report being abducted by the Janjaweed and held as sex slaves for more than a week before being released. Over 200,000 have perished as a result of this conflict; it is an "ethnic cleansing." Overpopulation exaggerates tribal-thinking; tribal-thinking is destructive. Compassion cannot accept having it continue another day; yet, it continues and all efforts of peacekeepers have failed to date.

This overpopulation-drought-related genocide can be compared to the earlier-discussed overpopulation-drought-related genocide in Rwanda a decade ago. Overpopulation and ecocide (ecological suicide) are inseparable; population pressure problems in the form of conflict are inevitable with this combination. Rwanda was already densely populated when Europeans arrived, due to moderate rainfall and an altitude too high for malaria or the tsetse fly. The new world crops expanded food supplies. Public health and medicine increased life expectancy. The population of Rwanda grew rapidly (3% yearly), and it continues to do so. To feed the expanding overpopulation, every bit of land was devoted to intensive agriculture. Conservation practices were ignored; topsoil was lost (ecocide). Predictably, food production failed (with drought from desertification). Overpopulation had already divided land into parcels so small that famine could not be avoided. The famine caused serious social problems (population pressure); families turned on each other (land disputes), at first in the courts. Crime rates rose sharply. Ordinary rank and file peasants believed that there were too many people on too little land, and that a reduction in numbers would make more available for the survivors. The Rwanda "ethnic cleansing" was really a (tribal-related) class war of have-nots killing the wealthier members of same community; retaliations followed. Nearly a million were murdered, with government sanctions. Overpopulation exacerbates tribal thinking; tribal-thinking is destructive. The killing stopped over 10 years ago, but the population continues to grow; visitors today see "a sea of children." The Rwanda genocide could happen again. Hunger changes behavior. Compassion cannot allow this to repeat itself.

In other areas of sub-Saharan Africa, overpopulation and (La Nina-related) drought had been causing increasingly severe crop failure and starvation. The drought in eastern parts of Africa is not over, in spite of a recent (El Nino-related) flooding, affecting countries from the Horn of Africa and south to Mozambique. Zimbabwe, under anti-apartheid leader Robert Mugabe, is currently facing a terrible food crisis, economic crisis and political crisis. (In a reverse of apartheid, tribal-thinking blacks displaced all white farmers, quietly murdering large numbers of them.) The country urgently needs to import 1.2 million tons of food to avoid famine. The African continent is susceptible to droughts, partly because of geography, but partly due to poor agricultural practices. Topsoil loss and desertification continue to worsen, due to the rapidly expanding population's rapidly expanding agricultural efforts, further increasing the topsoil loss and drought intensity (ecological suicide).

This area has the fastest-growing population on the planet. It has the greatest rate of infection and highest number of AIDS infections on the planet. This is a destabilizing situation. Safe sex and overpopulation control are not practiced by the irresponsible population. What are they thinking? What kind of person intentionally creates a certain-to-happen legacy of starvation, disease and suffering, to their children? Picture the heart-wrenching images seen on television; then imagine it getting worse, much worse. The education and empowerment of women is needed more here than any other place on the planet. An empowered, educated mother is the best health-care, for herself and for her children. Studies show that the education and empowerment of women increases employment opportunity, reduces fertility, and reduces child mortality. It also positively affects political interactions by giving women a greater role in government; it can moderate hostile extremes. A tribal-thinking Muslim cleric in India recently asked for volunteer fighters in Iraq and Afghanistan. An educated, empowered woman shouted him down and told him to go himself; they have better things to do. India is becoming a player in the global market. Both India and Africa must focus on controlling overpopulation. In the 1960's India's population was 350 million; to date, this number has nearly tripled. India has made recent progress in the education and empowerment of women; India has also made recent progress in controlling the AIDS rate of infection.

Worldwide, the carrying capacity of the land, particularly in some areas, has been exceeded by human overpopulation. This is a destabilizing situation; it's ecological suicide. If there is no change, there will be no change; it will only worsen. The only compassionate option is to help to make a change to sustainable population levels, before environmental selection further intensifies the suffering and death. Compassion does not enable increased suffering and certain suicide. Compassion demands a change in irresponsible behavior that leads to disaster. Compassion demands helping people to help themselves. History abounds with many short-sighted short-term solutions that resulted in long-term suffering for humanity; there is nothing at all compassionate about enabling learning the hard way. Something must be done, and it must be done now. Compassion demands it. There is no feel-good easy solution to irresponsible overpopulation dangers. Something bad will happen to these people, no matter what. How bad should it be? Of all choices, short-term solutions that endanger the long-term survival of all humanity are the worst possible choices. There is no compassion in supporting the ever-increasing (ecocidal) overpopulation and the predictable social meltdown. Isn't the best solution for people to gain control over their lives by becoming self-reliant and responsible? China is doing this with state-imposed mandatory abortions after one child; it is already beginning to pay dividends in their quality of life. While this form of population control may seem to be excessive, excessive populations warrant a better alternative than the Rwanda scenario. Neighboring Burundi and Congo are presently on the brink of social (Rawanda-related) meltdown. Educating and empowering women is a far better alternative in population management and improving the quality of life than either the China or Rwanda extremes. In general, choices based only on emotion are ill-advised. Emotion has a place in life, but not in making decisions that affect the future of all of mankind, as well as the future of other life. These decisions must be based on objective evaluation of scientific data; the decisions must support both humanity and the environment (so crucial to the future of humanity). Population adherence to the decisions is vital to their own self-interest.

Some believe that the African drought is not the fault of the resident population; what if it is not their fault?

In general terms, world weather is temperature and moisture-dependent. Ocean-sourced rains fall year-round in equatorial areas worldwide; 90% of all rainfall occurs at sea. Land areas are mainly to the North. Moving to higher latitudes, there are wet and dry seasons, with summer rains (e.g. monsoons) near the Tropics of Cancer and Capricorn. Proceeding poleward, dry summers persist up to latitudes of around 45 degrees, but westerly winds can bring winter rains. Recall that the African continent is susceptible to droughts partly because of geography. The Sahara desert area was once fertile, and a major center of human population within historical time (e.g. 6000ya). Precession changes in the earth's orbit, and the Himalayan uplift have created great environmental changes. The Himalayan uplift, caused by the Indian continent colliding with the Asian continent, is still ongoing. It has blocked much of the westward flow of moisture north of the equator, causing the very wet monsoon in southeastern Asia, as well as the rain shadow of desert in the Middle East and Africa. As the uplift continues, the African and Middle-Eastern drought worsens. In addition to this, 7mya, another geological uplift gained significance along the entire east coast of Africa (from Ethiopia to South Africa), forming what is known as "the wall of Africa." It may have played an even bigger role in blocking ocean-sourced moisture to Africa. Both the Sahara Desert and Kalahari Desert are located in predominantly dry high-pressure areas, caused by downdraft Hadley Cell air circulation. Hadley Cell air circulation begins as heat-generated equatorial uplift and upper atmospheric cooling (= rain) of the moisture-laden air. The Indian Ocean is rapidly warming, seemingly from global warming; however, cold water upwelling from the Deep Ocean Conveyor keeps precipitation mainly in the eastern areas.

At different times of the year there is a reversal in the direction of flow in warm surface water and associated atmospheric moisture for the (Indian Ocean) North Equatorial Current. In August and September, during the wet southwest monsoon, the North Equatorial Current flows from west-to-east as the Monsoon Current, pulling the warm surface water (and moisture) towards India, away from the African coast. During the dry monsoon season (February and March), cooler winds with far less moisture (north-easterly) blow from the Himalayas to the Indian continent and then towards Africa; this aids in the development of an east-to-west flow of the North Equatorial Current. Upon reaching the east coast of Africa, the east-to-west flow of the North Equatorial Current surface water turns southward, crosses the equator, and either becomes an Equatorial Countercurrent or the Mozambique Current. The Equatorial Counter-current results from the need to balance the movement of water to the west in the northern and southern

Indian Ocean, caused by the west-flowing North Equatorial Current and South Equatorial Current. In El Nino years (western Pacific low pressure area collapse), this current continues and intensifies in the Pacific Ocean. This strong Equatorial Countercurrent exists to the south of the east-to-west flowing North Equatorial Current; and to the north of the east-to-west flowing South Equatorial Current. At the equator, this west-to-east flowing countercurrent pulls warm equatorial surface water and atmospheric moisture away from Africa, directing it back eastward, just as the Monsoon Current did six months previously.

A good portion of the remaining warm surface water (and atmospheric moisture) flows southward along the African coast as the Mozambique Current. The Mozambique Current flows south along the east coast of Africa from the vicinity of the equator to about 35°S, where it becomes known as the Agulhas Current. The warm Agulhas Current runs south and west along the East coast of Africa, further supported by the east-to-west flowing South Equatorial Current; both combine just below Madagascar, flowing towards the near-freezing waters of Antarctica, before returning eastward (West Wind Drift Current) towards Australia. Because of the discussed moisture-robbing ocean currents, Africa gets only a small portion of equatorial moisture compared to Southeast Asia, India and South America. Continuing this discussion, off the eastern African coast, covering nearly half of the distance from the Tropic of Capricorn to the equator is the Island of Madagascar. Madagascar also robs the east coast of Africa from moisture. It deflects the majority of the warm water and moisture in the east-to-west flowing South Equatorial Current to the south. In addition to this, another counter-current, the Madagascar Current, flows to the north along the African coast, east of the Agulhas Current, along the west coast of Madagascar, carrying colder, drier southern waters that are further cooled by deep coastal upwelling; this also results in a loss of rain to Africa's interior. The seasonal warm surface water flow towards Africa and associated moisture along the coast should still provide eastern Africa with moderate, if not large amounts of rainfall; it does rain adequate amounts on a seasonal basis in areas very near the equator. Equatorial rain is supported continent-wide all along the areas of rain forest.

What should be, is not to be; significant desertification is occurring as a result of rain forest destruction in both Africa's eastern and western forested areas. When this desertification is combined with the desertification of poor agricultural practices, the result is ecological suicide. Overpopulation has led to desertification from loss of the rain forest, desertification from loss of all other native vegetation, and desertification from the loss of topsoil (from erosion). More than 80% of African soil is seriously degraded and is on the verge of permanent failure.

In earlier times, farmers survived by clearing new land for each season's plantings while allowing old fields to lie fallow and replenish their nutrients; with overpopulation, this no longer happens. Africa's four-fold increase in population during the past 50 years has forced farmers to grow crop after crop on the same fields; this has depleted the soil of all nutrients. The physical nature of the soil has changed. The soil has become so compacted (like a brick) that it can no longer hold water. Roots cannot penetrate; even weeds cannot grow. Fertilizer cannot help. The soil needs to be replaced. How do you replace over 80% of Africa's soil? The loss of productive land has driven farmers to clear even more marginal land for agriculture; this marginal land rapidly fails.

This loss of productive topsoil is the fault of overpopulation and this large population must bear some responsibility for this. And, the desertification from loss of native vegetation is likewise caused by an overpopulation impact on the environment; the overpopulation must bear some responsibility for this too. All the above tends to rob Sub-Saharan Africa of moisture; pre-existing lower humidity and higher interior African daytime temperatures further decrease the relative humidity over land that is already in a predominantly dry high-pressure Hadley Cell. So, except for desertification from the overpopulation-related loss of rain forest and other native vegetation, and from overpopulation-related overly-intensive agriculture, combined with overpopulation-related loss of topsoil, the drought is really no one's fault. The problem is that Africa is already prone to drought and African overpopulation leads to environmental suicide (ecocide). Overpopulation is not the fault of the weather, but unlike the weather, something can be done about it, and, like the weather, the problem can change for the better. Overpopulation leads to environmental breakdown and anarchy; Africa is defined by anarchy.

Present aid efforts only provide emergency food relief (necessary) and emphasize further education of men for planting even more higher-yielding food (increasing intensity of agriculture); this increased intensity in agriculture is self-defeating. Proper educational effort must be focused on the education and empowerment of women. Continuing the irresponsible population increase and environmental degradation can only lead to increased suffering for all. It is more important to fix the problem than to fix the blame.

Didn't the African drought end with the summer floods of 2007?

Paradoxically, overly-intensive agriculture promotes both drought and flooding. Floods occur because bare ground can only absorb 20% of the rain compared to native vegetation. The summer floods extended from coast to coast, from Senegal and Liberia to Ethiopia, Uganda and Kenya; these floods temporarily stopped the inevitable worsening drought. Changes in climate can be very unclear when viewed on the short term; the pattern of increasing drought is not necessarily ended with one year of floods, especially when between 65–80% of the water bypasses the water table and races to the sea, taking even more precious topsoil in the process. While the loss of native vegetation certainly predisposes the immediate region and downstream areas to flooding, these floods are most definitely not the fault of overpopulation (well—perhaps some of it is). The floods are just a different source for a crop failure crisis. Crop failure from either flooding or drought leads to starvation in overpopulated areas that have exceeded the land's carrying capacity. Unlike a prolonged drought, flooding from interior rains is more of a short-term problem and can be weathered by short-term aid. The flooding that will be a long-term problem is the one when most of the world's population is permanently displaced from a global warming rise in sea level.

Is it fair for developed countries with responsible population growth to have disproportionate population growth from illegal immigration?

What do you think? The logical consequence of poor population management or poor resource management is anarchy and social breakdown. Resource management without population management is not possible. Population management without immigration management is not possible. The logical consequence of poor immigration management is anarchy and social breakdown. Excessive immigration is a looming problem for all 1st world countries. The United States population numbered around 200 million in the 1960's. Baby boomers of that time made a sincere effort to limit family size (Ehrlich published *The Population Bomb* in 1968) and be more earth-friendly. Immigration is the cause of a disproportionate population growth. The present population increase to 300 million is mainly due to immigration, and there has been little indication of any environmental responsibility from the recent influx of immigrants, many (>30%) of which entered illegally.

Disrespect for the law is a bad start; while I am not willing to say that immigrants are bad people, they can be bad people. According to Iowa Rep. Steve King, illegal aliens commit 12 murders and kill another 13 daily through drunken driving incidents every day in the U.S. While the latter figure isn't premeditated murder, it is a form of vehicular homicide; people (famlies) are still being killed. The 25 homicides/day total more than 9,000 U.S. residents killed every year by illegal aliens! As World Net Daily's Joseph Farah points out, that means far more people are killed by illegal aliens in one year than have been killed in the Iraq and Afghanistan wars since their inception! Rep. King also notes that on average, 8 children are sexually abused by illegal aliens—every day; that's nearly 3000 sexual abuse cases annually. And all that's just a small portion of the illegal alien crime wave. Some 325,000 illegal alien criminals will be incarcerated in state and federal prisons this fiscal year. A GAO study found that illegal aliens (wanting amnesty) commit, on average, 12.6 criminal offenses per person. Incarcerated illegal aliens have committed over 4.1 million crimes. Prison system numbers in the U.S. have experienced an eightfold increase in the past half-century; this is at taxpayer's expense. Is this "fair?"

I lived in South Florida during the Muriel immigration (1978) when Fidel Castro purged his prisons of more than 125,000 people: mostly murderers, thieves and drug dealers, and put them in an armada of boats headed for the nearby Florida Keys. Immigration officials were overwhelmed. While the Cuban immigrants of 20 years earlier were certainly productive members of society, the Murielitos were a source of many problems. The resulting crime wave overworked police, flooded jails, overwhelmed community budgets, and made living in the Miami region almost unbearable for a time. Nearly 30 years later, it still inflicts problems in the community's culture. In addition, the waves of Haitian immigrants into Florida are definitely not enhancing South Florida's present quality of life. Is this "fair?"

In another situation, similar in many ways to the Muriel immigration to Florida, a parallel occurrence is happening along American states that border Mexico; but, it is on a far, far larger scale. Immigrant minorities that become the majority change, and tend to be more abusive to the new minority (the former majority) than any treatment they received when they were a minority. At the same time, the former-minority, now new-majority wants affirmative action continued, to their advantage and to the disadvantage of the now new minority (the former majority). Is this double-standard "fair"? Excelsior, the national newspaper of Mexico, stated: "The American Southwest seems to be slowly returning to the jurisdiction of Mexico without firing a single shot." Anyone that crosses the border anywhere in the Southwestern United States will have very little difficulty seeing a sharp contrast between Mexico and U.S. states like California, Arizona, New Mexico or Texas. There is a reason that these areas are so different and this reason seems to be more related to the culture of the country than to the geography. Think of this difference, and then picture what these states would be like if they had always remained as part of Mexico (hint—no difference). Why does everyone (especially Mexicans) want to live in the U.S., rather than Mexico? Yet, Gloria Molina, Los Angeles County Supervisor, has declared: "We are politicizing every single one of these new citizens that are becoming citizens of this country.... I gotta tell you that a lot of people are saying 'I'm going to go out there and vote because I want to pay them back.'" For what—allowing them to come here? This is a non-hostile invasion that is becoming increasingly hostile. Recent widespread demonstrations have clearly shown that the immigrants are not limiting their influence to the Southwest. Is this "fair?" Immigration is the cause of a disproportionate population growth in the U.S. University of Texas professor Jose Angel Gutierrez, stated: "We have an aging white America. They are not making babies. They are dying. The explosion is in our population.... I love it." Further comments indicated widespread contempt of non-immigrant population control; there is clearly another agenda in progress and it is not population control in the environmental context previously discussed. Uncontrolled immigration can be a real problem, especially as the minority becomes the majority. Clearly, setting an example in population control for others, by itself, does not work. Is it "fair" for responsible citizens to manage population numbers when immigrants refuse to do so?

Mexico has a long-standing drug culture; drugs and crime are strongly correlated. "La Cucaracha" is a famous song from the Mexican revolution, describing a marijuana-stoned cockroach's inability to run in a straight line. Marijuana is usually the first introduction to drug abuse. At the present time, according to the DEA, 80% of the highly-addictive drug, ice methamphetamine, enters the U.S. through Mexico; illegal aliens are the main distribution source! Because of omnipresent drugs, Mexico is literally a war zone; police corruption is rampant and only the outlaws have guns. Benjamin Arellano Felix, the former head of Mexico's most powerful drugs cartel (based in Tijuana), led this cartel for 20 years (an insight into Mexican corruption), developing a maze of contacts with suppliers of cocaine and heroin in Colombia. The organization he set up is believed to still turnover billions of dollars a year, supplying up to half of the cocaine presently entering the U.S. Any positive increase in cultural diversity from immigration is more often a disproportionate increase in negative cultural displacement, strongly associated with increased crime-related drug abuse. People just don't seem to change behavior; areas of the U.S. are changing into another Mexico. As immigration increases, crime has risen and the quality of life has diminished. Is this "fair?"

A half-century ago, apart from areas of cultural Latin-American influence, there was a relatively nonexistent drug problem in the United States. The Vietnam War changed all that, as drug-addicted American soldiers returned home. According to a New York Times article dated May 19, 1994, 20 years after the great influx of legal immigrants from Southeast Asia, 30% are still on welfare compared to 8% of households nationwide. There has been little indication of either social or environmental responsibility from this recent influx of immigrants. Asian-American gangs in America are heavily involved in drug trafficking and are extremely violent. The violence, like home invasions and innumerable homicides, extends beyond urban areas. Asian-American gangs are primarily made up of second-generation Laotian and Vietnamese immigrants whose parents came over from the least-developed mountain regions. Their core members come from hill tribes, such as the Khmu, the Mien and the Hmong. Even if they live in a city, the Hmong exercise a strong cultural tradition in hunting—but it is in their tribal way. Their cultural hunting traditions certainly don't respect wildlife or wildlife management laws; in fact, it doesn't respect much of anything. Hmong hunters commonly shoot anything they see, flagrantly violate game management laws, drive local residents from public hunting areas, boldly trespass on private lands and intimidate the regional countryside population. When confronted by landowners, life threats have often been made; landowner lives have been taken. A Wall Street Journal editorial dated December 5, 1994 quotes law enforcement officials as stating that Asian mobsters are the "greatest criminal challenge the country faces." Well, perhaps so, perhaps not.

That particular perspective changed September 11th, 2001, when a group of (illegal) immigrants took America for suicidal-homicidal airplane rides. Many believe the U.S. over-reacted with the invasion of Iraq (perhaps so). Tribal-thinking religious extremists, specifically Islamic fundamentalists, pose a terrible immigration threat to all of Western societies; they have declared war on Western civilization. Can it be true that their religious leaders now assure a place in heaven for any Islamist that kills a non-believer? If this is the case, continued conflict with others is assured. Islamic extremist behavior brings disgrace upon all Islam; it compromises and will eventually damage all Islam. While any religion can have a problem with extremists, the Islamic extremists of today appear to have an agenda of world domination through the use of terrorism. Religious conflicts always lead to the worst extremes in war (e.g. tribal-thinking holocaust ethnic cleansing in WW 2). These extremes eventually generate reactionary extremes; moderate Islamists need look no further than recent events in Yugoslavia to see their likely future.

Palestinians admittedly had their own problems with what they perceived as an illegal immigration; but, did that justify such extreme behavior towards the U.S.? Worldwide sympathy for the Palestinian's plight faded on that September 11th. No human being should have to live within a walled ghetto, managed by an oppressive police state. (Extremes generate reactionary extremes—former Nazi ghetto residents became Israeli ghetto-builders). Palestinian extremists have assured that a ghetto life for all Palestinians will be the case for the foreseeable future. This goes far beyond unfairness. Palestinian desperation continues to spiral into a widening conflict of death and destruction, with increasingly greater and even still greater extremes. The logical consequence of poor immigration management is social breakdown and anarchy. Ask the Palestinians if they think that the Zionist immigration was "fair." Additionally, an amazingly high birth rate and a high child survival rate will lead to overpopulation and widespread social problems; this is not the fault of infidels. An ever-increasing Palestinian overpopulation, which was made possible by improved public health, medicine and nutritional improvements in the Palestinian conditions of life, ultimately did not allow an improvement in their conditions of life. The recent pandemonium in Gaza exemplifies that the logical consequence of poor population management is social breakdown and anarchy. This is tragic documentation of a predictably destabilizing situation.

The Asian gangs didn't go away; they were just upstaged by an even more dangerous group of immigrants. Returning to the "less dangerous" immigrants, the L.A. Times stated that Orange County, California is home to 275 gangs with 17,000 members; 98% of which are Mexican and Asian. Think about this; this is the undisputed outcome of recent immigration (DRUGS, CRIME + ANARCHY). Afghanistan's major crop is heroin; the heroin is more dangerous to the U.S. population than the Taliban are. Colombia's major export over past years seems to have been cocaine. MS-13 gang members from Central America clearly have no respect for U.S. immigration laws, or for that matter, any other laws. They are predominantly illegal immigrants, consistently violent and are forming nationwide alliances with other Latin gangs. Violent crime in the U.S. had been steadily decreasing for decades; it is now on the increase and this increase definitely has no connection to legal ownership of firearms by responsible citizens. Organized crime has a new face. This goes far beyond unfairness. Social breakdown and anarchy are the logical consequences of poor immigration management.

It would not be "fair" to ignore the fact that the U.S. needs workers. But cheap labor is short-lived; immigrants quickly unionize and the price goes up. Does the U.S. really need that many unskilled workers? The Heritage Foundation Study is an economic analysis of low-skilled workers authored by Robert Rector. It examines how much they make, spend, and receive in government services. The study finds that the average household led by a person without a high school diploma received $32,138 in U.S. government benefits and services in 2004, but paid only $9,689 in taxes. The result is that these households received $22,449 more in benefits than they paid in taxes. Multiply that by 17.7 million low-skilled households and this is an enormous fiscal problem. The labor department states that immigrants make up about 15% of the U.S. work force; it is estimated that one-third of these workers in the U.S. are undocumented (illegal). Of the entire U.S. workforce, 1 out of 20 workers are undocumented (illegal). The study focused on low-skilled workers because it is estimated that nearly two-thirds of illegal immigrants fall into that category. That means that amnesty would cost U.S. taxpayers billions of dollars. Social systems have been unfairly exploited by illegal immigrants. In reality, the cheap labor would have been taking the same money to subsidize workers in the U.S. (bring the jobs back home?). Illegal immigration has a too dear a price. For this discussion to be "fair," it must be emphasized that not all immigrants are problems. Properly screened immigrants that do follow immigration protocols tend to make good citizens. In this group, most all have a good work ethic and have good family values. If law-abiding immigrants wish to live here and become productive members of society, they should be welcomed into the U.S. Still, even if immigration is a "good thing," there can always be too much of a "good thing." The number of immigrants into any country must be managed solely to benefit its own citizens. Every country has rules for immigration. Every person in the Americas is an immigrant or a descendant of an immigrant, and these immigrants had to follow protocols and learn the language. U.S. citizens cannot just decide to move to Mexico if they want; to do this they must follow a demanding protocol, which includes financial requirements and learning the language. It is only "fair" to ask Mexicans that want to move to the U.S. to do the same. This "fair" approach should apply to all immigrants.

What can be learned from historical overpopulation?

Compared to the relatively colder climate (Great Retreat of Agriculture) of The Dark Ages, which began around 400 AD, The Medieval Warm Period—700–1200 AD, was a warm climate optimum. It was nearly as warm as the climate of the 20th century, which permitted a tenfold increase in world population. The Medieval Warm Period was a mild global warming; cereal crops flourished. More food meant more people could be fed. It was even warm enough for grapes to grow and for wine to be produced in England. The European population reached the greatest numbers in its history; these numbers would never be surpassed until the great tenfold leap of overpopulation in recent time. There were so many people in warmer Scandinavia that the regional overpopulation caused serious resource shortages (800 AD). This forced a foraging for new resources; these foragers were known as the Viking raiders. And they did acquire a lot of resources. European countries ran out of currency as they paid tribute to the raiders for not raiding. After a couple hundred years of terrorizing Europe, soldiers from the terrorized areas finally repelled any new Viking raiders that hadn't already invaded/immigrated and integrated. But when the Vikings finally stopped raiding, what were all these local soldiers to do? There were now a lot of people and a lot of soldiers. Why, they could do some raiding and terrorizing of their own, and call it a crusade! And since it was all done in the name of the one "true faith," God was certain to favor their fortunes! Because of tribal-thinking, they could commit any atrocity in the name of God, and harming heathen "outsiders" was not even going to be a problem! Even before the Viking raids, the furious conquest of Spain by the tribal-thinking armies of Islam had involved raiding incursions into France (in the name of Allah). The Muslims were the handiest foreign invader still around. It was payback time, and payback hits where it hurts the most. It was time to liberate the Holy Land from these Muslim conquerors. Many were killed. After the crusades wound down, The Little Ice Age then began to re-balance the overpopulation problem with repeated crop failures. The effects of the plague that accompanied The Little Ice Age are enumerated immediately following this discussion. The tribal-thinking Holy Wars really never stopped; Holy Wars raged on and on, both in the East and the West. Eventually tribal-thinking Christian soldiers, with equal ferocity, drove the tribal-thinking Moor invaders from their lands in Spain. After the victory, what were all these Christian soldiers to do now (sound familiar)? To seek their fortunes, the conquistadors (and other Europeans) then turned Westward, to the newly-discovered Americas. There were riches awaiting them and a population of non-believers ("heathens" or "outsiders") to reduce.

What is the status of present and future food production for present (+ future) populations?

Although there is enough food production today, reserves are becoming conspicuously low. Australia, a major producer (#2 worldwide) of the world's wheat supply, is experiencing the worst drought in 100 years. Prices are on the increase. From March 2007–March 2008, wheat has risen in price 130%. In the same time period, soybean products have increased in price 87%, rice has increased in price 74% and corn has increased in price 31% (BBC news). The U.S. is using corn to produce ethanol to reduce dependence on foreign oil (only get 30% return). In the same time period, oil went from $57 a barrel to $117 a barrel; this is the main reason for the food price increase. Petroleum is the energy source of agricultural mass production; it is the raw material for agricultural chemicals. Looming shortages of petroleum and food, combined with an ever-increasing population and ever-increasing standard of living, are creating a supply and demand problem. BBC news said there was only a 40 day food reserve in early 2008. They also estimated a 50% increase in food demand by 2020. While the Green Revolution and Gene Revolution may have provided for the current tenfold population increase, there is no assurance that food production can surpass current production and meet the increasing demand. Cropland and fisheries are on the decline. Even without climate change, food production may have already hit the wall.

Famine and overcrowding are bedfellows; this connection can be seen in the famine-related refugee camps of today (also see with wars and natural disasters). Overcrowding provides the optimal conditions for the transfer of epidemic disease from direct or indirect contact, passage of lice or fleas, human waste contamination of food or water, and even airborne droplet infection. "It is often disease, secondarily connected to a resource failure, natural disaster or war, which brings about the conditions of population check, or collapse; the ability to separate disease from resource failure has been imperfect in the past." (Bray—*Armies of Pestilence*)

What can be learned from a historical population decline?

It has been said that: "The black death was the greatest calamity that has befallen the human race" (MacArthur). (In that this same disease killed a comparable number of Europeans in the Plague of Justinian, there is a real basis for this view.) Humans were first infected by the rodent populations of central Asia in 1346 (perhaps from around modern Iraq and Kurdistan). Infected rats and humans arrived in the Crimea in 1346 or 1347. Genoese ships fled the outbreak in the Crimea, and arrived in Constantinople, Messina, Sardinia, Genoa, and Marseilles in 1348. And while it did not surpass the percentage loss of humanity seen with humanity's near-extinction in the Mount Toba eruption 74,000ya (where an estimated 99% died, only 1% survived—about 10,000 of 1,000,000), the sheer numbers lost to the plague over a limited time were staggering. Europe (with ±100 million people) may have lost nearly 50% of its population (50,000,000) in as few as 5 years. It would continue to kill Europeans for another 300 years. *The Pied Piper* of Hamlyn may be a memory of the 1362 plague outbreak… as it carried off mostly children. A familiar custom that still exists today originated from a Black Death outbreak in Avignon France; if you sneezed, you were suspect of pneumonic plague, and the response was to wish you good health (today, its gesundheit!). (Bray—*Armies of Pestilence*).

"The period between 1348 and 1420 witnessed the heaviest losses (in Europe). England had a population of 5–7M in 1348; half this number had died from the plague by 1350 (England still only had around 1M people in 1806). In Germany, about 40% of the named inhabitants disappeared. The population of Provence was reduced by 50% and in some regions of Tuscany, 70% were lost during this period. Furthermore, this plague continued to devastate Europe for over another 200 years. Historians have struggled to explain how so many could have died. There are problems with the long-standing theory that it was just caused by a medical illness; so, social factors are also examined. A classic Malthusian argument has been put forward that says that Europe was overcrowded with people; even in good times it was barely able to feed its population. A gradual malnutrition developed over decades, lowering resistance to disease. And competition for resources in short supply meant more warfare; warfare further compromised the harvest, and increased the malnutrition…. The conditions of the poor became so bad that they achieved a net zero population growth. The economic conditions of the poor also aggravated the calamities of the plague because they had no recourse, such as fleeing to a villa in the country…. The poor lived in crowded conditions and could not isolate the sick…. After the plague and other exogenous causes of population decline, wage increases, because of a lower labor supply, led to a redistribution of wealth. However, this did not happen right away because property owners resisted change through wage freezes and price controls. The wage freezes and price controls were partly responsible for popular uprisings, such as the Peasants Revolt of 1381(many landowners had to harvest their own crops). By 1500 the total population of Europe was substantially below that of 200 years earlier, but all classes overall had a higher standard if living." (Wikipedia) There are less painful ways to manage population and resources.

If disease is a major limiting factor of mankind, what is the environmental connection?

Infectious disease that is significant enough to impact large populations is endemic or epidemic in nature. Endemic disease has a constant environmental exposure; over time, the population may evolve some genetic resistance. For example, malaria is endemic, transmitted by mosquitoes from regional wetland areas in tropical climates. Malaria may be the prime candidate for the #1 historical killer of mankind throughout history; mosquitoes constantly infect and re-infect people. Malaria parasites injected into the bloodstream may encounter a limited response from the host's immune system. Natural killer cells and conditioned antibody-producing plasma cells in the circulatory system, and Kupfer cells in the liver kill many, but not all, invaders. Without immune system conditioning, infants (and even young children) have almost no ability to survive a high incidence of exposures. And because the parasite rapidly mutates, all attempts to form an effective vaccine and immune response have failed. There are 300 million cases yearly, even today; 90% are in Africa. Even today, there are over 1 million deaths yearly. In addition, tropical Africans commonly have a recessive sickle-cell anemia allele that is an adaptive trait evolved to combat mortality from malaria; it affects red blood cell shape and turnover, which subsequently provides resistance to malaria infection. Severe anemia problems may arise when both recessive alleles are present (homozygous); in this case, the sickle-cell red blood corpuscle shape and their rapid turnover can be life-threatening. Nevertheless, heterozygote sickle-cell alleles are effective in providing the great majority of Africans with significant resistance to malaria. Children with the heterozygous sickle-cell trait have a greater chance of surviving an initial infection. Statistically, the risk from endemic malaria is greater than the risk from the homozygous recessive trait. Independent of the sickle cell trait, Sean Carroll, in *Regulating Evolution*, notes: "Nearly 100% of West Africans lack Duffy proteins on their red blood cells, which also makes them more resistant to malarial infection. Duffy protein, which usually appears on the surface of red blood cells, is used by the malaria parasite to form part of a receptor for the invader. Duffy protein is also found in the brain, spleen, and kidney; each is regulated by a separate enhancer sequence. The Duffy gene's red cell enhancer in these individuals is disabled by a mutation of a DNA enhancer sequence, but there is no effect on Duffy enhancers elsewhere in the body. Still, malaria is getting worse and worse. When populations were much lower and individual groups were sparsely scattered throughout the entire area, malaria did not have such an attractive culture medium in which to spread; today the opposite is true. Large populations in cities like Nairobi were once at an altitude that was high and dry, too cold for malaria-transmitting mosquitoes. With recent global warming, mosquitoes from lower elevations have invaded the city and spread the disease while utilizing their rich human food supply. Outsiders experience a malaria epidemic if they enter endemic malaria areas; there is no genetic resistance and death is common. There is evidence that malaria was spread along trade routes from tropical Africa and temporarily supported (warm livestock barns) in coastal lowland areas near Rome. (All cities with good harbors became silt-filled inland swamps, formed by topsoil erosion and river delta deposition). A devastating malaria epidemic may have been a significant cause for the fall of the Roman Empire (along with Justinian's Plague, climate change, productive topsoil loss, loss of sea trade, dependence on foreign troops for defense and various other causes).

Epidemic disease runs through human population as infection cycles; the exposed population that survives often develops resistance. Originally contracted from animals like cattle, the grim reaper is still with us. Tuberculosis, once eradicated in the U.S. infects two billion people and kills two million/yr worldwide. If the criteria used is number of the dead, it is a strong contender for the worst killer disease in mankind's history. It is now presently teaming-up with AIDS (especially in rotating prison populations) as a major threat, and it is gaining momentum to sharply impact mankind, again. There is particular concern about drug-resistant strains of TB. Epidemic diseases need large populations to spread, or they will die out. Nothing really intensifies death from epidemic disease quite like overcrowding. Overpopulation overcrowding exposes far more individuals to infection. In 1918, a flu virus killed up to 50 million in the overcrowded conditions during The Great War; the number killed by the flu was greater than the number killed in combat during WW1. With people packed into the trenches, trains, trucks, and hospitals, the 1918 Spanish (Kansas?) flu virus turned more lethal, just as the avian flu does on crowded chicken farms today. Avian flu is seldom lethal in wild populations. There is a reason the flu of the winter season spreads so quickly; it always seems to come from Southeast Asia, where existing overpopulation has people, pigs (intermediate host for virus genetic recombination yields variations that bypass lasting immunity in humans) and poultry crowded closely together. The worst killer-plague epidemic reached the Americas around 1500, and it seems to have killed even more people over a shorter time period than the Black Death killed in Europe. After Columbus arrived in the new world, an estimated 95 million (95%) of the Native American population soon died of epidemic disease (mainly smallpox), also originally introduced from animals (Diamond—*Guns, Germs & Steel*); the New World was more crowded than originally thought (100M+). The Native American populations and cities paralleled Europe or Asia in size, but these populations were nearly exterminated by disease. Recall that at an earlier time, the Black Death may have reduced 100 million Europeans to 50 million in as few as 5 years (1347–1352AD). Although plague likely originated from rodents in western Asia, the crowded conditions created by "The Little Ice Age," which suddenly occurred following a "Medieval Warm Period," no doubt played an additional environmental role; brutal cold created crowded conditions inside the shelters. The Plague of Justinian (452 AD) was an earlier episode of the same disease that was later called the Black Death. It too was associated with a cooler climate. The Plague of Justinian, The Great Retreat of Agriculture (colder climate) and The Fall of Rome introduced the Dark Ages. Two out of three people died; some estimates of the death toll from Justinian's plague do reach as high as 100M.

As human overpopulation encroaches further and further into tropical ecosystems, the environmental disruption creates change and instability in the ecosystems. Human incursions cause exposure to new tropical diseases. The overcrowded humans and their numerous domestic animals are increasingly exposed to an environmental opportunistic virus threat; this is how viral epidemics begin. Many virus infections come from the tattered edges of the tropical rain forest, or they come from tropical savanna that is being rapidly settled by people. The tropical rain forests are the deep reservoirs of life on the planet, containing the great majority of the world's plant and animal species. The rain forests are also its largest reservoir of viruses, since all living things carry viruses. When viruses come out of an ecosystem, they tend to spread in waves through the human population (epidemics). Emerging viruses include: Triple E; West Nile; SARS; H5N1 Bird Flu; Lassa; Rift Valley; Oropouche; Ricio; Q; Guanarito; VEE; Monkeypox; Dengue; Chikungunya; Hantaviruses; Machupo; Junin; Rabies-like Mokola; Rabies-like Duvenhage; LaDantec; Kyananur Forest brain virus; Semliki Forest Agent; Crimean Congo; Sindbus; O'nyongnyong; Nameless Sao Paulo; Marburg; Ebola Sudan; Ebola Zaire; Ebola Reston; and HIV-AIDS (Preston—*The Hot Zone*).

AIDS is arguably the worst environmental disaster of the 20th Century. Any existing infection (like malaria) increases the risk of AIDS transmission. Humanity's overpopulation-expansion and environmental encroachment in Sub-Saharan Africa's rain forest, and nearby savannah, is the underlying reason that HIV-AIDS species-jumped, probably in the region of Lake Victoria (in 1950's). The environmental source is unknown; any existing disease, like malaria, facilitates the development of the virus in the population. The AIDS virus may well have jumped into the human race from African primates, from monkeys or anthropoid apes. For example, HIV-2 (1 of 2 major strains of HIV) may be a mutant virus that jumped into us from an African monkey known as the sooty mangeby, perhaps when monkey hunters or trappers touched bloody tissue. HIV-1 (the other strain) may have jumped into us from chimpanzees, perhaps when hunters butchered chimpanzees. A strain of simian AIDS virus was isolated from a chimpanzee in Gabon, West Africa. So far, this is the closest thing to HIV-1 that anyone has yet found in the animal kingdom. From the area around Lake Victoria, HIV-AIDS easily spread down the Kinshasa Highway to the west coast (in late 1950's–early 60's); once the road was paved, it quickly spread in the rapidly-growing population. Lake Victoria is near Mount Elgon, the source area of the deadly Marburg virus in the 1970's. It's a form of a nightmarish hemorrhagic fever (25% fatal) in which the virus can rapidly liquefy infected individuals. An environmental source has been identified—green monkeys from Uganda. Not far away, something even worse, Ebola Sudan, species-jumped to mankind. The environmental source is unknown. This virus also rapidly liquefies infected individuals; it has a higher fatality rate (50% fatal) and seems to be more contagious. At nearly the same time, the worst-possible-nightmare surfaced, an outbreak of Ebola Zaire, killing so many so quickly it died out before it could spread from a remote location. The environmental source is unknown. Ebola Zaire virus is airborne (highly contagious), rapidly liquefies infected individuals, and basically kills everyone, leaving few (if any) survivors. Ebola Zaire is known as a "slate-cleaner;" i.e. if it were to escape into the world population and become widespread, it is so contagious and so lethal that it could conceivably eradicate nearly every infected human being. Only the Russian genetically-modified smallpox (developed for germ warfare—theoretically kills everyone), can kill like Ebola Zaire kills. Ebola Zaire, Ebola Sudan and the Marburg virus are all from the same general rain forest/savannah region of Africa. Ebola Reston, a near-exact copy of Ebola Zaire broke out in a monkey quarantine area next to Washington DC on two occasions in the early 1990's. Five hundred individually caged monkeys died in the airborne infection that spread from room to room. In both cases, there were no survivors. There is no cure for any of these hemorrhagic virus infections; they represent the highest threat level of disease known to mankind today (Preston—*The Hot Zone*). Still another new ebola strain appeared in Uganda, late in 2007 (BBC news). BBC News also reports that researchers found a massive die-off for gorillas in (West Africa's) Congo's Lossi Sanctuary between 2002 and 2004. Ebola was passing from group to group of the endangered animals, and appeared to be spreading even faster than in humans. The apes were infected by fruit bats, but were also transmitting the virus among themselves. The Lossi outbreak killed an estimated 5000 gorillas, about as many gorillas as survive in the entire eastern gorilla species. In 2002 and 2003, several outbreaks of Ebola flared up in human populations in Gabon and Congo. Outbreaks of the disease in humans have sometimes been traced to the bush-meat trade.

Africa has no exclusive on dangerous environmental viral diseases. The Asian bird flu H5N1 strain has species-jumped to man; most infected individuals do not survive. It hasn't yet transferred human-to-human, is presently not highly contagious, but deadly nevertheless when infection does occur. A recent outbreak in China early 2008 may yet prove to be a realization of the worst fears, a highly contagious human to human adaptation. Birds are the natural host of influenza. Rural families living in close proximity with animals in overcrowded Asia allows the viral "predator" a wonderful opportunity to reproduce itself; the pig is an intermediate host in many flu virus species-jumps to humanity. SARS is a corona virus that also species-jumped to man; the environmental source may be a civet cat. It is highly contagious (airborne) and most infected individuals do not survive. Invading the tropical environment carries a dear price; the cost only continues to increase.

Much of the above discussion focused on viral infectious disease. There are few defenses against viral diseases. There are many forms of disease; nearly all have an environmental connection. Examples include: fungal (infectious), bacterial (infectious), viral (infectious), parasite (infectious), prion disease (infectious—perhaps physical chemistry precipitation of protease-resistant beta sheets), genetic disease, nutritional disease, metabolic disease, neoplastic disease (cancer is discussed in Biochemistry of Genetic Mechanisms), toxic chemical disease, toxic radiation disease, behavioral disease, and advanced age-systems failure. Voluminous books have been written on each topic; comprehensive coverage of all is beyond the scope of this brief discussion.

Although the management of bacterial disease has been greatly enhanced by antibiotics, any discussion of disease must at least include the growing resistance of bacteria to these antibiotics. Bacterial killers are making a comeback. Resistance occurs when a bacterial mutation adapts to an antibiotic. Inadequate low dosage and short-term patient dosage facilitate the bacterial adaptation; bacteria survivors and their offspring are thereafter immune to the antibiotic. Strong warnings to health professionals in 1960's were not followed, due to an even stronger public demand for antibiotics in cases where the administration of an antibiotic was really not necessary. Additionally, the antibiotic regimen prescribed was not followed by many patients. Developing countries bought antibiotics over-the-counter, proper dosage and completion of regimen time almost never occurred. Veterinarians used the same antibiotics for pets and livestock, leading to widespread environmental presence, which further enhanced bacterial resistance. Bacteria from pets and found in meat will be antibiotic-resistant. Now, antibiotics are everywhere in the environment, enhancing the development of enhanced bacterial resistance. Life-threatening infections are increasingly unresponsive to antibiotic therapy; the Vancomycin intravenous treatment of last resort is failing to manage an increasing number of fatal infections. Over 75% of the 150 most commonly prescribed drugs in the U.S. came from organisms. For most of the world's population, there may be no better reason than this to support biodiversity; our very lives depend upon protecting our medical resources for our future survival.

What is the relationship of the AIDS epidemic to overpopulation?

Irresponsible overpopulation decreases the quality of life. Infectious disease always does best with large populations of overcrowded hosts. Irresponsible overpopulation and infectious disease (AIDS) are bedfellows; both increase suffering and death. Sub-Saharan Africa has the highest birth rate, the highest rate of population increase and the lowest use of contraceptives of any major region in the world. According to BBC news, nearly 2 out of 3, or 64% (25.8 million), of all people with HIV (> 40m worldwide) live in Sub-Saharan Africa. Worldwide, there are 5m new cases and 3m deaths yearly. In 2005, 2.4m people in Sub-Saharan Africa died of an HIV-related illness, and a further 3.2m were infected with the virus. Despite falls in adult HIV prevalence apparently under way in Kenya, Uganda and Zimbabwe, there is little evidence of declining epidemics in this region as a whole. In fact, prevalence levels remain "exceptionally" high and might not yet have reached their peak in several countries. Between 20–25% of adults are already infected in Zimbabwe, Botswana, Namibia, Zambia and Swaziland; South Africa has a high number of people (now 800,000) needing terminal stage AIDS treatment. Twelve of the eighteen countries that have suffered HDI development declines between 1990 and 2003 were in sub-Saharan Africa, with Southern Africa "hit hardest." South Africa has plunged by 35 places, to 120 on the global Human Development Index (HDI), Zimbabwe dropped 23 places, and Botswana fell 21 places. It does not help that Robert Mugabe's version of murderous apartheid against Caucasian farmers led Zimbabwe to starvation, failed economy, social breakdown and anarchy. A mass exodus of the Zimbabwe population into South Africa and other neighboring countries is presently spreading Zimbabwe's extreme lack of sustainability into the surrounding areas. HDI declines were also noted for Lesotho, Swaziland and Zambia. Africa and India have the highest AIDS infection rates today; surprisingly, they also have the highest growth of overpopulation, in spite of the deaths from AIDS. Not long ago India recorded the number infected as 4 million; India had 1 in 7 of the world's HIV+ infections at that time (BBC news). Although India has recently noted a decrease in the rate of HIV infection (now has 1 in 10 of world's HIV+), India appears as if it will overtake China as the world's most populous nation, surpassing the 2 billion in 2025.

What are my thoughts on the overpopulation solution in Tom Clancy's *Rainbow Six*?

This book is a discussion topic for every environmental science or environmental issues class I have taught. Although somewhat extreme itself in many ways, it provides insight into environmental extremism and eco-terrorism; there is more than one type of eco-terrorism. Eco-terrorists are simply terrorists, environmental extremist terrorists; they should be treated accordingly. Environmental extremists alienate society from environmental support; environmental extremism is counter-productive.

How does the public support or not support environmentally-friendly business?

The first thing the public must do is to move away from any support of emotional environmental extremism. Environmental extremists alienate mainstream society from environmental support; environmental extremism is counter-productive. Organizations like Greenpeace and PETA have a place, as systems to watch for potential environmental threats and unethical treatment of animals; yet, because of extreme behavior, they have managed to alienate most of society. Both have even supported eco-terrorism; this is neither appropriate nor acceptable.

The University of Alaska and The U.S. Fish and Wildlife Service have developed sound and efficient caribou and moose management over last ½ century. State and federal agencies combine extensive experience with information from millions of dollars on research. Larger animal population evaluation is based on aerial photo census. The Alaska Game Board has the daunting task of studying proposals, hearing testimonies from game biologists and other professionals, and then passing judgments into policy based on sound management theories. Alaska fish and game studies indicated the need to relocate bears and reduce the wolf population around the village of McGrath. PETA has interfered with this; PETA has been campaigning for an Alaska tourism boycott, because of the planned wolf aerial kill. They successfully recruit members every time Alaska does something like this. If PETA really cared about animal welfare, they'd be equally concerned for the threatened moose population. If PETA really cared about wolves, why are they willing to see even greater numbers of wolves starve to death in their inevitable population crash? Does PETA care if the Eskimos had nearly nothing to eat during the winter? There is no relief planned; there is no supermarket in McGrath, nor is there the financial base for building one. Regrettably, it does look like PETA doesn't really care about the wolves (or moose or people), but all those uninformed, usually urban, new members do (care). Are the PETA folks just in it for the money? What really happens to the money when someone "adopts a wild wolf?" Has it ever been used to purchase and feed a single wolf's daily need of five pounds of meat per day? The point is, predators have a place, but must be managed like any other wildlife, with policies based on scientific information, not emotion. Solutions to complex problems are seldom solved by emotional approaches. All facts must be considered in order to find the best possible solution to any problem; sometimes the facts require readjustment in solution implementation. Emotionally-based solutions to problems just do not work, unless there is only an agenda to collect money ($28M in 2005 PETA donations) from an uninformed, usually urban, idealist. The city mouse and country mouse fable comes to mind.

Public support (or lack of public support) should be based on supporting the business community that has taken a role in strongly supporting environmental preservation. Business must market its environmental support. The public, including other business, must expand this support of environmental responsibility. Although it may be a result of criticism in the book, *Fast Food Nation* (Eric Schlosser), McDonald's restaurants are now more supportive of environmental considerations. It's good for business; this business is becoming environmentally responsible. Partnering with the agency, Conservation International, they are moving toward a "socially responsible food supply." Any business that considers the triple bottom line of environmental responsibility, social responsibility and economic viability, deserves preferential public support. The public is beginning to increasingly support organic produce; this diversified aspect of agriculture is not only more profitable to the small farmer, it is also friendlier to the environment. Local farmer's markets selling diversified agricultural produce help everyone.

The fine example set by Dupont Chemical Company (mentioned previously), where they not only practice environmental responsibility, insist on doing business only with others who also practice environmental responsibility, and "put their money where their mouth is," by giving to, not taking (money for a cause) from, environmental welfare. At General Electric their "green is good for business" motto is no secret; they are definitely making more money in green business. Other companies, such as Chevron and British Petroleum have taken pro-active environmental roles; it pays. Even Bank of America claims to be "green." A Subaru plant in Indiana claims to generate zero landfill waste. Power companies, like Florida Power and Light, strongly support alternative energy sources and demonstrate environmental responsibility; it pays.

The Marine Stewardship Council certifies companies using environmentally-responsible harvesting methods in fishing; boycotting non-member companies that irresponsibly damage the marine environment makes far more sense than boycotting Alaskan tourism, as advocated by PETA. In fact, Diamond's *Collapse*, reveals that The Alaska Department of Fish and Game represents the largest U.S. fishery certified to date by the Marine Stewardship Council; environmental responsibility certification just doesn't get any better than this. This should be supported. Specifically, this relates to the wild salmon fishery; pending accreditation is the Alaskan Pollock fishery, which accounts for half of the U.S. catch. Also pending accreditation is U.S. West Coast halibut, Dungeness crab, and spotted prawn, as well as U.S. East Coast striped bass and Baja California lobster. Certified overseas companies include MSC founder Unilever, now Gorton's Seafood in the U.S., Western Australia's rock lobster, and New Zealand's hoki fishery. Only products of a certified fishery that can be traced through a chain of rigorous standards can display the blue MSC logo when offered for sale; there is no need to join an organization or to donate to "a cause."

There is also the opportunity to either support or not support environmentally-responsible logging of forests. Louisiana-Pacific, once an environmental offender, has now accomplished a complete turnaround. They now also look at the "triple bottom line" of environmental sustainability (resource conservation), social responsibility (government standards compliance), and economic viability (improved profit margin). The Forest Stewardship Council is discussed in Diamond's *Collapse*. A similar certification process to that of the fisheries allows retailers to display the FSC logo. A list of big business participation includes Home Depot, Lowe's, Colombia Forest Products, Natural Resources Canada, *Canadian Boreal Initiative, Minnesota DNR*, Seven Islands Land Company, Collins Pine and Kane Hardwoods, Anderson Windows/Doors, Ikea, Penguin Group, Random House, Kinko's, Wells Fargo, Disney, Gibson Guitars, Kodak, and Dell computers.

The public should understand that spending money on the consumption of products or services can be an environmental problem in itself. That is, the consumption of goods and services is not environmentally-friendly. At the very least, an effort must be made to reduce excess consumption. Every time an item is purchased, resources are depleted. If there is to be a reduction in greenhouse gases produced, there must be a reduction in the amount of materials and energy required to support wasteful consumption. The affluent compulsive shopper that buys and buys items as a form of daily recreation is not often perceived as causing an environmental problem; however, this same shopper that buys and buys items that are simply wanted, but are not really needed, unknowingly doubly-depletes resources (materials and energy) and creates huge amounts of excess waste. And while the local economy depends upon the monetary exchange of goods and services, excess consumption of goods and services can be as environmentally-unfriendly as overpopulation. Think again about the tragedy of the commons; the environmental damage and loss of resources could be caused by grazing too many sheep, or by having half that number of sheep becoming pigs and eating twice as much. Excess consumption by the few can be as environmentally-damaging as conservative consumption by the many; the result is the same. To use a human example, the movie stars that overly-indulge themselves in material excess may be causing far more environmental damage than any particular protest issues championed by their favorite environmental extremist groups (e.g., PETA).

So, when it comes to public support of environmentally-friendly business, each individual should also control what they do to reduce excess consumption. This brings the familiar reduce, reuse and recycle (3 R's) of the real environmentalists into focus. Reducing consumption is, without a doubt, the 1st choice of environmental consideration for any individual. (Buy only what you need.) Reusing consumption is the 2nd best choice of environmental consideration. (Donate underutilized clothes and articles to the needy—like Goodwill.) From an environmental standpoint, it is far better to utilize reusable goods than it is support further consumption of new goods with cash to the needy. Recycling is only the 3rd best choice of environmental consideration, but it still is very important, as it saves on energy and materials that don't belong in the waste-stream. Recycling also encourages an individual awareness and a sense of responsibility for the refuse produced. It is cost-effective, even in residential areas. (Curbside pickup $35/ton vs. $80 landfill disposal.) While there is no shortage of sand to make glass products, the real savings comes from saving on energy required to produce the glass in the first place (40%). Plastic bottle recycling saves over 50% of the energy the energy required to produce a new plastic bottle. Recycling an aluminum can saves over 95% of the energy required to produce a new aluminum can. Saving on energy reduces greenhouse gas pollution. It also reduces other pollution. For the aquatic environment, burning coal for electric energy is the major source of mercury pollution. Mercury (Hg) pollution bio-accumulates and bio-magnifies, becoming a health threat to humanity. Mercury (Hg) pollution is also produced in the manufacture of paper from trees. Over 70% of the U.S. waste stream is paper; this is the pattern of most developed countries. Per day, 2M trees are lost to paper products in U.S; recycling paper saves over 60% of the energy required to produce new paper. Recycling 1 ton of paper saves 17 trees, 3 cubic yards of landfill space, 380 gallons of oil, 4000 KW of energy (will heat +/or cool average U.S. home 6 months), 7000 gallons of water, and 60 pounds of air pollution. A tree produces 200 lbs of O_2 and eliminates 154 lbs of CO_2 yearly. Recycling cuts waste drastically, reducing pressure on disposal systems (New York City goal is 50% reduction of waste stream—by recycling office paper, household and commercial waste; Japan has already reached 50% reduction of waste stream, mostly from paper recycling).

The free market can cause pollution to be relocated to another area. That is, production will always be relocated to the most cost-effective areas. The cost of labor may be just too high for production to offer a product that can compete in a free market. In addition, there are 3 other economic reasons for business outsourcing that do not even include the cost of labor to produce goods. They are excessive taxes, excessive regulation, and excessive liability, all of which drive up the cost of doing business, making it impossible to compete in a free market. If it is less expensive for a business to move overseas than it is to pay the taxes in the U.S., the business will relocate. If it is less expensive for a business to move overseas than it is to meet restrictive regulations in the U.S. (that do not exist elsewhere), the business will relocate. If it is less expensive for a business to move overseas than it is to risk incurring inappropriate litigation and excessive legal judgments in the U.S. (as with Dow Chemical silicone implants), the business will relocate. The U.S. is the only country on earth that does not have a "loser pays" assurance against frivolous lawsuits. The American legal system is the most expensive and least effective system in the world. The system works to benefit the overcrowded legal profession. The threat of liabilities destroying the air carriers involved prompted the U.S. government to pay millions in bribes to families of victims to avoid lawsuits from the September 11 World Trade Center disaster. In this case, the taxpayers really got the bill. Is the life of a U.S. soldier worth less (pun intended) than that of a civilian? Legislators, many of which are attorneys, tend to pass laws that benefit themselves and the overcrowded legal profession—at the expense of others (this is a blatant conflict of interest); tort reform is badly needed here. Legislators must do something to stop the hobbling of U.S. business. Why is this not campaign issue at election time? Liabilities for all business (including drug companies and health professionals) are passed on to the consumer, just as with shoplifting in retail. Having insurance companies become a "middle man" only further raises the cost. All of this results in excessively increased prices; this is the real reason the same medication in Canada is half the U.S. price. Even the cost of shipping goods halfway around the world can be far less expensive than producing goods in the U.S.; unfortunately, that wastes additional energy and further increases pollution from production. Yet, in a free trade market, if business does not relocate to another country, business cannot operate at a profit, and that will force any business to close. The public cannot "buy American" because it is no longer possible to do so. Everything has a price, and the price is often more than dollars.

BBC News rated China as the world's biggest polluter of greenhouse gases in 2007, and this role as the biggest polluter will only increase. Even today, as much as 40% of the air pollution over California comes from China. India is also becoming a major polluter. Much of this pollution is due to increased production (+ the increase in wealth of it's population from the increased production). In that more pollution receives little attention in an already badly polluted China, extensive utilization of China's abundant coal resource will significantly contribute to a short-term global dimming and a further long-term global warming, much like volcanism as discussed in the chapter on climate. It was no coincidence that China had such an unusually cold, snowy and harsh winter in 2007–2008. Jared Diamond, in *Collapse*, said it would be impossible for overpopulated second and third world countries to rise to the standard of living seen in the Western world, because there are simply not enough resources to do this. China and India have every intention of raising their standard of living through increased productivity, and they are well on their way.

What about greenhouse gas pollution in the U.S.?

The United States was previously ranked as the world's biggest polluter of greenhouse gases, but environmentalists have either failed to take the U.S. reduction of greenhouse gases through photosynthesis into account, or have erred in their calculations of U.S. greenhouse gas reduction; both biased views exaggerate U.S. greenhouse gas production. Actually, for most of the year, the United States removes more greenhouse gases than it creates, and the U.S. has never been properly credited for this. Unlike the extensive, brown, desert-like expanses of China and India, the United States is mostly green (crops and trees); this green removes more greenhouse gas CO_2 pollution than the U.S. creates during the entire year. Prevailing westerly winds (from the southwest) flow from the west to the east coast of the U.S. Compared to the east coast's offshore winds (from the land to the sea); the onshore west coast winds (from the sea to the land) average over 2 parts per million higher in CO_2. This means that for most of the year, along entire east coast, the same southwesterly winds blowing seaward have continuously lost greater than 1/2 billion tons (ton = 0.907+kg) of CO_2. And this means that the U.S. is taking up more CO_2 than it is producing. The United States is a major carbon sink! This is not a calculation—it is a repeatable fact and it has been so for at least 2 decades (NOAA). Why has everyone been criticizing the U.S. for being the main producer of CO_2 greenhouse emissions, if the U.S. is actually reducing more CO_2 than its producing? Perhaps countries criticizing the U.S. population for not paying a carbon tax to reduce emissions should pay the U.S. for its carbon reduction service with their carbon tax funds. And while this doesn't mean that greenhouse gas emissions are to be ignored, criticizing the U.S. population for its per-capita greenhouse footprint (individual population-member basis), is a biased statistic, unfairly favoring overpopulation. If the United States population must reduce its greenhouse gas emissions to 5% below 1990 levels by 2012, are overpopulated countries also reducing their excess population to 5% below 1990 levels by 2012?

Populations produce a great deal of CO_2 (respiration not even considered). Intentionally-started forest fires in Brazil and Indonesia can produce more atmospheric CO_2 than most industrialized nations. Recall that although not counted in fossil fuel emissions, the burning of the rain forest produces approximately 4 billion metric tons of carbon yearly, equivalent to over 14% of all yearly human fossil-fuel emissions (28 billion metric tons CO_2 in 2005). And of the 28B metric tons calculated emissions, 8.7B metric tons, or 30% of these CO_2 emissions, are released from burning to clear areas for subsistence agriculture or burning wood for cooking and/ or heat. Both sources total nearly 13B metric tons (comparable to half the fossil-fuel 28B metric tons produced worldwide). In addition, underground coal fires in China and elsewhere nearly match the CO_2 production of the forest fires. Therefore, **most greenhouse gas pollution actually comes from fires in overpopulated areas!** This was true even before China became the top pollution producer! Greenhouse gas reduction and overpopulation management are both important; both must have realistic goals and a continual positive improvement. At least China is limiting overpopulation.

In general, higher energy use correlates with a high standard of living. But Denmark and Switzerland use about half as much energy per capita as the U.S., and have a higher standard of living (but not much production). Oil-rich states, such as the United Arab Emirates, use about twice the energy per capita as the U.S., and are still relatively underdeveloped in most areas. Some claim the U.S. is "the Saudi Arabia of energy waste;" this actually offers a sizable and fairly painless opportunity to use less. California has recently become more energy-efficient (now ranked 47th of 50 states in energy waste per capita); it is a good start. This only happened because restrictive environmental regulations led to power shortages and its associated problems; Californians had no other option. Up to now, cheap energy has fueled the U.S. economy, the world's largest. But as the price of energy increases, the U.S. economy begins to suffer. Now, the balance of trade shows it is a consumer economy more than it is a productive economy. Yet, if the U.S. were self-reliant on energy, there would be no balance of trade problem. The U.S. has 5% of the global population and in 2005 produced 21% (6 billion metric tons) of the calculated CO_2 emissions (28B metric tons) from fossil fuel combustion. The U.S. is well aware that it needs to markedly reduce oil consumption; it must adapt to the change. In addition to reducing CO_2 emissions, the recently-mandated automobile and truck mileage increases will help save fuel and reduce dependence on foreign oil. Green energy is good energy, but green energy will never be able to meet the demand as the sole source of energy. In the U.S., non-polluting alternative energy resources have yet to even match the increased energy consumption from illegal immigrants! Worldwide, fossil fuels are still expected to provide 85–90% of the world's energy into the foreseeable future. The transition to alternative forms of energy will take some time; there is little time to waste. There is a long way to go; it is time to stop criticizing and get going. The bottom line is that the greenhouse gas pollution reduction should not to be done on a biased per-capita basis, but on a regional basis that accounts for the carrying capacity of the environment.

What is the "bottom line" for humanity on environmental biology?

The conditions of life will control the destiny of humanity. Overpopulation and environmental degradation create an environment that selects against humanity. Efforts for resource sustainability that do not address population sustainability cannot succeed. The environment ultimately determines the carrying capacity, the standard at which human overpopulation occurs. Environmental degradation causes society degradation (leads to war, disease and famine); it is the ultimate population check. Irresponsible overpopulation not only threatens the future of humanity, it also threatens the conditions of too many other lives; humans will suffer as a result of the ongoing loss of biodiversity. At issue is a choice between the quantity of human life and the quality of human life. A quality human life requires a quality environment. Survival tomorrow primarily depends upon increased consideration of the environment, today, as well as the education and empowerment of women, today- everywhere. If this cannot be done, the chances of survival in even the best environments will be fatally compromised.

References

ADF&G Division of Wildlife conservation. *Caribou, Management Report of Surv-Inv Act, study 3.0 Grants W-24-2 and W 24-3.* Alaska: Division of Game, Dec 1995.

Behe, Michael. *Darwin's Black Box—The Biochemical Challenge to Evolution.* The Free Press.

Benton, Michael. *When Life Nearly Died.* Thomas and Hudson Ltd.

Bray, R.S. *Armies of Pestilence.* Barns & Noble Books.

Carrol, Sean and others. *Regulating Evolution.* Scientific American. May 2008 60–67.

Carroll, Sean. *The Making of the Fittest: DNA and the Ultimate Forensic Record of Evolution.* W.W. Norton & Co.

Carroll, Robert. *Vertebrate Paleontology and Evolution.* W.H. Freeman and Co.

Darwin, Charles. (J Burrow Ed). *The Origin of Species.* Pelican Classics.

Darwin, Charles. *The Voyage of the Beagle.* Penguin Classics.

Dawkins, Richard. *Climbing Mount Improbable.* W.W. Norton & Co.

Dawkins, Richard. *The Ancestor's Tale.* Houghton Mifflin Co.

Diamond, Jared. *Collapse.* Viking (Penguin Group).

Diamond, Jared. *Guns, Germs & Steel.* W.W. Norton & Co.

Diamond, Jared. *The Third Chimpanzee.* Harper Perennial.

Enard, Wolfgang. *Molecular Evolution of FoxP2, a Gene involved in speech and language.* Nature August 2002 418: 869–872

Evans, Patrick. *Evidence that the Adaptive Allele of the Brain Size Gene Microcephalin introgressed into Homo sapiens from an Archaic Homo Lineage.* PNAS November Volume 103, 18178–18183.

Flannery, Tim. *The Weather Makers.* Atlantic Monthly Press.

Forsdyke, D.R. *Reciprocal Relationship between Stem-loop Potential and Substitution Density in Retroviral Quasispecies under Positive Darwinian Selection.* Journal of Molecular Evolution, 1995. Dec; 41(6):1022–37.

Fremling, Calvin. *Immortal River.* The University of Wisconsin Press.

Futuyma, Douglas J. *Evolutionary Biology.* Sinauer Associates, Inc.

Hardin, Garrett. *The Tragedy of the Commons.* Science 162:1243, (1968).

Hesse, Richard and others. *Ecological Animal Geography.* John Wiley & Sons.

Hilleman, Terry. *Environmental Lecture Notes* (unpublished—Copyright 2005).

Iowa State University Extension Office Publications: *Managing Iowa Habitats.*

Ippen, Arthur. *Estuary and Coastline Hydrodynamics.* McGraw-Hill.

Jannasch, H. W. *Interactions between the Carbon and Sulphur Cycles in the Marine Environment.* This paper was prepared under support from the National Science Foundation Grants OCE79–19178 and OCE81–24253 and is Contribution No. 4921 from the Woods Hole Oceanographic Institution.

Kingsley, David M. and others. *Widespread Evolution in Sticklebacks by Repeated Fixation of Ectodermal Alleles. Science* 25 March 2005 307: 1928–1933.

Kirschner, M. and Gerhart, John. *The Plausibility of Life.* Yale University Press.

Levinton, Jeffrey. *Genetics, Paleontology and Macroevolution*, 2/e. Cambridge University Press.

Mann, Charles. *1491.* Borzoi book (Alfred A Knopf).

Mekel-Bobrov, Nitzan. *Ongoing Evolution of ASPM, a Brain Size Determinant in Homo Sapiens.* Science 9 September 2005 309:1720–1722.

Mindell, David. *The Evolving World.* Harvard University Press.

Montgomery, Rex and others. *Biochemistry.* The C.V. Mosby Company.

Preston, Richard. *The Hot Zone.* Random House.

Ruttner, Franz. *Fundamentals of Limnology.* University of Toronto Press.

Shubin, Neil. *Your Inner Fish.* Pantheon Books.

Stedman, Hansell. *Myosin Gene Mutation Correlates with Anatomical changes in the Human Lineage.* Nature 2004 428: 414–418.

Stuppy, Wolfgang. *Seeds.* Firefly Books.

Sverdrup, H.U. and others. *The Oceans.* Prentice-Hall.

Turner, J. Scott. *The Tinkerer's Accomplice.* Harvard University Press.

Tyson, Neil DeGrasse and Donald Goldsmith. *Origins: Fourteen Billion Years of Cosmic Evolution.* W.W. Norton & Company.

Vallender, Eric. *Positive Selection on the Human Genome.* Human Molecular Genetics. 2004 PMID: 15358731 [PubMed—indexed for MEDLINE].

Vrba, Elizabeth and others. *Macroevolution: Diversity, Disparity, Contingency. Essays in Honor of Stephen J Gould.* The Palentological Society.

Ward, Peter. *Gorgon.* Viking (Penguin Group).

Ward, Peter and Donald Brownlee. *The Life & Death of Planet Earth.* Henry Holt & Company.

Waters, Ellen. *DNA is Not Destiny.* Discover. November 2006.

Watson, Lyall. *Dark Nature.* Harper Perennial.

West-Eberhard, Mary Jane. *Developmental Plasticity and Evolution.* Oxford University Press.

West-Eberhard, Mary Jane. *Developmental Plasticity and the Origin of Species Differences.* Proceedings of the National Academy of Science 2005 102: 6543–6549.

West-Eberhard, Mary Jane. *Phenotypic Accommodation: Adaptive Innovation Due to Developmental Plasticity.* Journal of Experimental Zoology (Mol. Dev. Evol.) 2005 304B: 610–618.

Young and Badyaev. Evolution of Ontogeny: Linking Epigenetic remodeling and Genetic Adaptation in Skeletal Structures. Integrative and Comparative Biology 2007; icm025v1–11.

Index